Introduction to Abstract Algebra

The Random House/Birkhäuser Mathematics Series:

INVITATION TO COMPLEX ANALYSIS
by R.P. Boas

BRIDGE TO ABSTRACT MATHEMATICS: MATHEMATICAL PROOF AND STRUCTURES
by Ronald P. Morash

ELEMENTARY NUMBER THEORY
by Charles Vanden Eynden

INTRODUCTION TO ABSTRACT ALGEBRA
by Elbert A. Walker

INTRODUCTION TO ABSTRACT ALGEBRA

Elbert A. Walker

New Mexico State University

The Random House/Birkhäuser Mathematics Series

Random House

New York

First Edition

9 8 7 6 5 4 3 2 1

 Published in the United States by Random House, Inc., and simultaneously in Canada by Random House of Canada Limited, Toronto.

Library of Congress Cataloging-in-Publication Data

Walker, E. (Elbert), 1930–
Introduction to abstract algebra.

(The Random House/Birkhäuser mathematics series)
(Random House College)
Bibliography: p.
Includes index.
1. Algebra, Abstract. I. Title. II. Series.
III. Series: Random House College.
QA162.W35 1987 512′.02 87–4454
ISBN 0–394–35611–X

Manufactured in the United States of America

Contents

Preface

In teaching a beginning course in abstract algebra, one must suppress the urge to cover a lot of material and to be as general as possible. The difficulties in teaching such a course are pedagogical, not mathematical. The subject matter is abstract, yet it must be kept meaningful for students meeting abstractness for perhaps the first time. It is better for a student to know what a theorem says than to be able to quote it and produce a proof of it detail by detail without having the faintest notion of what it's all about. Careful attention must be paid to rigor; sloppy thinking and incoherent writing cannot be tolerated. But rigor should be flavored with understanding. Understanding the content of a theorem is as important as being able to prove it. I have tried to keep these things in mind while writing this book.

The specific subject matter chosen here is standard, and the arrangement of topics is not particularly bizarre. In an algebra course, I believe one should get on with algebra as soon as possible. This is why I have kept Chapter 1 to a bare minimum. I didn't want to include it at all, but the material there is absolutely essential for Chapter 2, and the students' knowledge of it is apt to be a bit hazy. Other bits of *set theory* are expounded upon as the need arises. Zorn's Lemma and some of its applications are discussed in the Appendix.

Groups are chosen as the first algebraic system to study for several reasons. The notion is needed in subsequent topics. The axioms for a group are simpler than those for the other systems to be studied. Such basic notions as homomorphisms and quotient systems appear in their simplest and purest forms. The student can readily recognize many old friends as abstract groups in concrete disguise.

The first topic slated for thorough study is that of vector spaces. It is this material that is most universally useful, and it is important to present it as soon as is practical. In fact, the arrangement of Chapters 3 through 5 is a compromise between mathematical efficiency and getting the essentials of linear algebra done. Chapter 4 is where it is because its results are beautifully applicable in Chapter 5 and contain theorems one would do anyway. Besides, there are some nice applications of Chapter 3 in Chapter 4. I feel that Chapter 6, which presents the basics of Galois theory, should not be slighted in favor of Chapters 7 or 8. A feature of Chapter 7 is an algebraic proof of the fundamental theorem of algebra.

The text contains more than enough material for a one-year sequence, and a variety of one or two semester courses can be built around it. Further, there are a variety of choices of the order in which one may present the material. For example, Chapter 7, *Topics from Group Theory,* may be covered immediately after Chapter 2, and Chapter 8, *Topics in Ring Theory,* may be presented immediately after Chapter 4. Also, Section 2.7, on finite Abelian groups, may be omitted in favor of simply presenting the more general material in Section 4.7, on modules over principal ideal domains.

There are many exercises in this book. Mathematics is not a spectator sport, and is best learned by doing. The exercises are provided so that students can test their knowledge of the material in the body of the text, practice concocting proofs on their own, and pick up a few additional facts.

There is no need here to extol the virtues of the abstract approach and the importance of algebra to the various areas of mathematics. They are well known to the professional and will become fully appreciated by the student only through experience.

I would like to thank the reviewers for their thoughtful comments and suggestions: Joseph E. Adney, Michigan State University; Thomas Brahana, University of Georgia; Don E. Edmondson, University of Texas; Edith H. Luchins, Rensselaer Polytechnic Institute; Anne L. Ludington, Loyola College; Glen E. Mattingly, Sam Houston State University; Warren D. Nichols, Florida State University; and Carol L. Walker, New Mexico State University.

I also wish to express my deep appreciation to Valerie Reed for her many many hours of expert technical typing and retyping of the manuscript.

Finally, I am most grateful to Random House/Birkhauser and its Editorial Director for Mathematical Sciences, John R. Martindale, for their efforts in bringing this project to completion.

Elbert A. Walker
Las Cruces, New Mexico
December, 1986

Introduction to Abstract Algebra

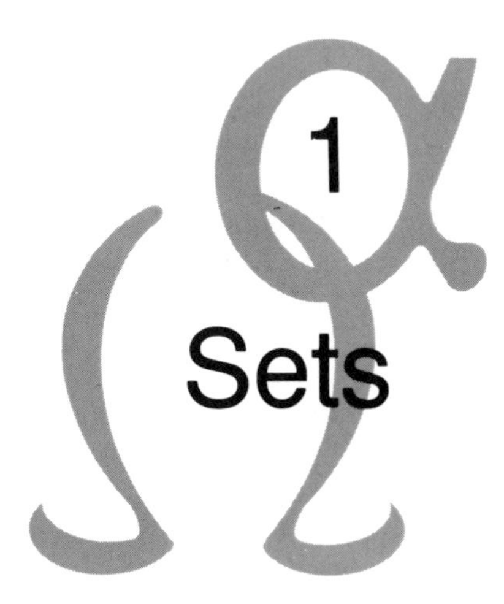

1 Sets

1.1 INTRODUCTION

The purpose of this book is to present some of the basic properties of groups, rings, fields, and vector spaces. However, some preliminaries are necessary. There are some facts about sets, mappings, equivalence relations, and the like that will be used throughout the text, and which are indispensable. This chapter presents those facts. The amount of material is kept to a minimum and includes only information that is necessary for beginning a study of algebra.

A few fundamental facts concerning the integers will be assumed. These are discussed in Section 1.6.

1.2 SETS AND SUBSETS

Our approach to sets is the naive or intuitive one. An axiomatic approach to sets here would lead us too far afield and would contribute nothing toward the understanding of the basic properties of a field, for example.

By the word ***set***, we mean any collection of objects. The objects in the collection are called the ***elements*** of the set and will usually be denoted by lowercase letters. Sets will usually be denoted by uppercase letters. If s is an element of a set S, we write $s \in S$, and if s is not an element of S, we write $s \notin S$. For any object s, either $s \in S$ or $s \notin S$. For the set $\mathbb{Z}$ of all integers, we have $-5 \in \mathbb{Z}$, but $\frac{1}{2} \notin \mathbb{Z}$. A set is determined by its elements. That is, two sets are the same if they have the same elements. Thus two sets S and T are equal, written $S = T$, if $s \in S$ implies $s \in T$, and $t \in T$ implies $t \in S$.

To specify a set we must tell what its elements are, and this may be done in various ways. One way is just to write out an ordinary sentence which does that. For example, we could say "S is the set whose elements are the integers 1, 2, and 3." A shorthand has been developed for this sort of thing, however. $S = \{1, 2, 3\}$ means that S is the set whose elements are 1, 2, and 3. Thus, one way to define a set is to list its elements and put braces around the list. If S is a large set, however, this procedure can be cumbersome. For example, it would be tiresome to describe the set of positive integers less than 1,000,000 in this way. The sentence we have just written would be more efficient, but there is a convention that is universally used and most convenient. Suppose S is a set that consists of all elements s which satisfy some property P. Then $S = \{s\text{: } s \text{ satisfies property } P\}$ means that S is the set of all elements s which satisfy property P. For example, $S = \{n\text{: } n \text{ is an integer, } 0 < n < 1{,}000{,}000\}$ is the set of all positive integers less than 1,000,000. Here, property P is the property of being a positive integer less than 1,000,000. If we already knew that $\mathbb{Z}$ was the set of all integers, we could write $S = \{n\text{: } n \in \mathbb{Z},\ 0 < n < 1{,}000{,}000\}$. This is sometimes written $S = \{n \in \mathbb{Z}\text{: } 0 < n < 1{,}000{,}000\}$. The set of even integers could be written in any of the following ways:

(a) $\{n \in \mathbb{Z}\text{: } n \text{ is even}\}$,
(b) $\{n\text{: } n \text{ is an even integer}\}$,
(c) $\{2n\text{: } n \in \mathbb{Z}\}$, or
(d) $\{n\text{: } n = 2m \text{ for some } m \in \mathbb{Z}\}$.

The letter $\mathbb{Z}$ will be used throughout this book to denote the set of all integers.

Suppose S and T are sets. If every element of S is an element of T, then S is a ***subset*** of T. This is denoted $S \subset T$, or $T \supset S$. Note that $S \subset T$ does not rule out the possibility that $S = T$. The symbol $S \subsetneqq T$ means $S \subset T$ and $S \neq T$, but we will not have much occasion to use this latter notation. If $S \subsetneqq T$, then S is a ***proper subset*** of T. Note that $S = T$ implies both $S \subset T$ and $T \subset S$. Also, if $S \subset T$ and $T \subset S$, then $S = T$. Thus, $S = T$ if and only if both $S \subset T$ and $T \subset S$.

It is convenient to allow the possibility that a set have no elements. Such a set is called the ***empty set***. There is only one such set, since, if S and T are both empty, then $S \subset T$ and $T \subset S$, so $S = T$. The empty set is denoted $\varnothing$. It has the property that $\varnothing \subset S$ for any set S.

There are various ways to construct new sets from old. For example, given sets S and T, there are various ways to associate with them a third set. The more common ones are listed below.

1.2.1 Definition

Let S and T be sets.

(a) $S \cup T = \{x\text{: } x \in S \text{ or } x \in T\}$. *This is the* **union** *of S and T.*
(b) $S \cap T = \{x\text{: } x \in S \text{ and } x \in T\}$. *This is the* **intersection** *of S and T.*

(c) $S \setminus T = \{x : x \in S \text{ and } x \notin T\}$. *This is the* **difference** *S minus T, or the* **complement** *of T in S.*

For example, if $\mathbb{R}$ denotes the set of real numbers,

$$S = \{x \in \mathbb{R} : 0 < x < 2\}$$

and

$$T = \{x \in \mathbb{R} : 1 \leq x \leq 3\}$$

then

$$S \cup T = \{x \in \mathbb{R} : 0 < x \leq 3\},$$

$$S \cap T = \{x \in \mathbb{R} : 1 \leq x < 2\},$$

and

$$S \setminus T = \{x \in \mathbb{R} : 0 < x < 1\}.$$

Let us note some properties of $\cup$, $\cap$, and $\setminus$. First, keep in mind what they mean. $S \cup T$ is that set consisting of all the elements of S along with all the elements of T. $S \cap T$ is that set consisting of the elements that are in both S and T. $S \setminus T$ is that set consisting of the elements that are in S but not in T.

Let A, B, and C be sets. The following are immediate from the definitions.

(a) $A \cup A = A$; $A \cap A = A$; $A \setminus A = \varnothing$.
(b) $A \cup B = B \cup A$; $A \cap B = B \cap A$.
(c) $(A \cup B) \cup C = A \cup (B \cup C)$;
$(A \cap B) \cap C = A \cap (B \cap C)$.
(d) $A \cup \varnothing = A$; $A \cap \varnothing = \varnothing$.

Less obvious are the following, which we will prove.

(e) $A \cap (B \cup C) = (A \cap B) \cup (A \cap C)$.
(f) $A \cup (B \cap C) = (A \cup B) \cap (A \cup C)$.
(g) $A \setminus (B \cap C) = (A \setminus B) \cup (A \setminus C)$.

To prove (e), we must show that each element of $A \cap (B \cup C)$ is an element of $(A \cap B) \cup (A \cap C)$ and that each element of $(A \cap B) \cup (A \cap C)$ is an element of $A \cap (B \cup C)$. Suppose that $x \in A \cap (B \cup C)$. Then $x \in A$ and $x \in B \cup C$. Therefore, $x \in A \cap B$ or $x \in A \cap C$. Thus, $x \in (A \cap B) \cup (A \cap C)$.

Now suppose that $x \in (A \cap B) \cup (A \cap C)$. Then $x \in A \cap B$ or $x \in A \cap C$. Thus, $x \in A$ and $x \in B \cup C$. Hence, $x \in A \cap (B \cup C)$. Therefore, (e) is true. One can prove (f) in a similar fashion, and you are asked to do so in Problem 9 at the end of this section.

We now prove (g). Suppose that $x \in A \setminus (B \cap C)$. Then $x \in A$ and $x \notin (B \cap C)$. Thus, $x \in A \setminus B$ or $x \in A \setminus C$. Therefore, $x \in (A \setminus B) \cup (A \setminus C)$.

Suppose that $x \in (A \backslash B) \cup (A \backslash C)$. Then $x \in A \backslash B$ or $x \in A \backslash C$. Thus, $x \in A$ and $x \notin B \cap C$. Therefore $x \in A \backslash (B \cap C)$, and this proves (g).

When B and C are subsets of A, (g) is one of the ***De Morgan Rules***. Problem 14 is another. For a subset S of A, letting S' denote $A \backslash S$, these two rules become

$$(B \cap C)' = B' \cup C'$$

and

$$(B \cup C)' = B' \cap C'.$$

Let us back up and consider (e). It asserts that

$$A \cap (B \cup C) = (A \cap B) \cup (A \cap C),$$

with nothing said about the sets A, B, and C. That is, this equality holds for all sets A, B, and C. That is what we proved. In our proof, nothing was used regarding A, B, and C except the fact that they were sets. No matter what sets we take for A, B, and C, the equality (e) holds. For example, let B, C, and D be any sets, and let $A = B \cup D$, so that $A \cap (B \cup C) = (B \cup D) \cap (B \cup C)$. From (e) we get the equality

$$(B \cup D) \cap (B \cup C) = ((B \cup D) \cap B) \cup ((B \cup D) \cap C).$$

Since $(B \cup D) \cap B = B$, the equality becomes

$$(B \cup D) \cap (B \cup C) = B \cup ((B \cup D) \cap C),$$

which is $B \cup (D \cap C)$. Hence, $(B \cup D) \cap (B \cup C) = B \cup (D \cap C)$ for any sets B, C, and D. This is the equality (f) above. The point to all this is, however, that once we have an equality such as (e) that holds for any sets A, B, and C, we can derive other equalities by taking A, B, and C to be any particular sets we wish.

PROBLEMS

1 List all the subsets of the set $\{a\}$.

2 List all the subsets of the set $\{a, b\}$.

3 List all the subsets of the set $\{a, b, c\}$.

4 List all the subsets of the set $\varnothing$.

5 List all the subsets of the set $\{\varnothing\}$.

6 List all the subsets of the set $\{a, b, \{a, b\}\}$.

7 List the elements in $\{n \in \mathbb{Z}: mn = 100 \text{ for some } m \in \mathbb{Z}\}$.

8 List the elements in $\{n \in \mathbb{Z}: n^2 - n < 211\}$.

9 Prove directly that $A \cup (B \cap C) = (A \cup B) \cap (A \cup C)$.

10 Prove that $A \cup ((A \cup B) \cap C) = A \cup (B \cap C)$.

11 Prove that $(A \setminus B) \cap (B \setminus A) = \varnothing$.

12 Prove that $A \setminus B = \varnothing$ if and only if $A \subset B$.

13 Prove that $(A \cap B) \setminus B = \varnothing$ and $(A \cup B) \setminus B = A \setminus B$.

14 Prove that $A \setminus (B \cup C) = (A \setminus B) \cap (A \setminus C)$.

15 Prove that $A \cap (B \setminus C) = (A \cap B) \setminus (A \cap C)$.

16 Prove that $A \cup B = (A \cap B) \cup (A \setminus B) \cup (B \setminus A)$.

17 Derive $(A \setminus B) \cup (B \setminus A) = (A \cup B) \setminus (A \cap B)$ from (g), page 3.

18 Prove that

$$A \cap ((B \setminus C) \cup (C \setminus B)) = ((A \cap B) \setminus (A \cap C)) \cup ((A \cap C) \setminus (A \cap B)).$$

19 Prove that

$$(A \cup ((B \cup C) \setminus (B \cap C))) \setminus (A \cap ((B \cup C) \setminus (B \cap C)))$$
$$= (((A \cup B) \setminus (A \cap B)) \cup C) \setminus (((A \cup B) \setminus (A \cap B)) \cap C).$$

20 Prove that if $(A \cup B) \setminus (A \cap B) = (A \cup C) \setminus (A \cap C)$, then $B = C$.

21 For two sets A and B, define

$$A + B = (A \cup B) \setminus (A \cap B), \quad \text{and} \quad A \cdot B = A \cap B.$$

(a) Prove that
  (i) $A + B = B + A$;
  (ii) $A + \varnothing = A$;
  (iii) $A + A = \varnothing$;
  (iv) $A \cdot A = A$.

(b) Express the equalities in Problems 18, 19, and 20 in terms of $+$ and $\cdot$.

1.3 MAPPINGS

One of the most important concepts in mathematics is that of a ***mapping***, or ***function***. We will use the two words interchangeably, with preference going to the former. Suppose that A and B are sets, and that with each element $a \in A$ is associated a unique element $f(a) \in B$. Then we say that f is a mapping, or function, from A into B. We wish to be a little more precise, but that is the general idea of a mapping.

Let A and B be sets. We can form the set of all ***ordered pairs*** (a, b), where $a \in A$ and $b \in B$. It is just the set of all pairs whose first member is an element of A and whose second member is an element of B. This is a fundamental construction.

1.3.1 Definition

Let A and B be sets. Then

$$A \times B = \{(a, b): a \in A, b \in B\}$$

is the **Cartesian product**, *or simply the* **product**, *of A and B.*

For example, if $A = \{1, 2, 3\}$ and $B = \{1, 4\}$, then

$$A \times B = \{(1, 1), (1, 4), (2, 1), (2, 4), (3, 1), (3, 4)\}.$$

Subsets of $A \times B$ are called ***relations in*** $A \times B$, and subsets of $A \times A$ are called ***relations on*** A. We will be particularly interested in two kinds of relations, one of which is a mapping, and the other of which is an equivalence relation.

1.3.2 Definition

A **mapping** *is a set A, a set B, and a subset f of $A \times B$ such that*
(a) *if $a \in A$, then there is an element $b \in B$ such that $(a, b) \in f$, and*
(b) *if $(a, b) \in f$ and $(a, c) \in f$, then $b = c$.*

This definition is just a careful way of saying that with each element of A is associated exactly one element of B. We say that "f is a mapping from A into B," that is, refer to the mapping as just f, although A and B are integral parts of the mapping.

Suppose f is a mapping from A into B. The set A is the ***domain*** of f, and B is the ***codomain***, or the ***range*** of f. The set f is the ***graph of the function***. Thus two mappings are equal if they have the same domain, the same codomain, and the same graph.

If $(a, b) \in f$, then we think of f as having associated the element $b \in B$ with the element $a \in A$, or as having taken the element a onto the element b. In fact, if $(a, b) \in f$, we say that "f takes a onto b." Instead of writing $(a, b) \in f$, we will write $f(a) = b$. If $f(a) = b$, then the element b is the ***image*** of the element a under the mapping f. The set $\{f(a): a \in A\}$ is the ***image*** of the mapping f, and it is denoted $\operatorname{Im} f$. If S is any subset of A, then $f(S) = \{f(a): a \in S\}$. Thus, $f(A) = \operatorname{Im} f$. If $\operatorname{Im} f = B$, then f is ***onto***, or ***surjective***. If $f(x) = f(y)$ implies $x = y$, then f is ***one-to-one***, or ***injective***. If f is ***both one-to-one and onto***, then it is ***bijective***, or a ***one-to-one correspondence between*** A and B, or an ***equivalence between*** A and B.

To define a function we must specify two sets A and B, and a subset f of $A \times B$ which satisfies conditions (a) and (b) of 1.3.2. Thus, having A and B, we must tell what $f(a)$ is for each $a \in A$ and make sure that this association of $f(a)$

with a satisfies (a) and (b) of 1.3.2. This means that with each $a \in A$ we must have exactly one $f(a)$.

A common situation is for f to be defined by a formula. To illustrate this, suppose that A and B both are the set of real numbers $\mathbb{R}$ and that we are only considering functions from $\mathbb{R}$ to $\mathbb{R}$. "The function $f(x) = x^2$" means that f is the function from $\mathbb{R}$ into $\mathbb{R}$ consisting of the set $\{(x, x^2): x \in \mathbb{R}\}$. That is, f associates with each real number x the real number x^2. This is clearly a function. In this vein, $g(x) = x - 1$, $h(x) = 69$, and $k(x) = -x/2$ determine functions g, h and k from $\mathbb{R}$ into $\mathbb{R}$, but we will freely use phrases such as "the function $g(x) = x - 1$."

The fact that f is a mapping from A into B will be denoted $f: A \to B$. A common notation for depicting a function is

$$f: A \to B: a \to f(a).$$

For example, the functions g, h, and k above could be denoted by

$$g: \mathbb{R} \to \mathbb{R}: x \to x - 1,$$

$$h: \mathbb{R} \to \mathbb{R}: x \to 69,$$

and

$$\to \mathbb{R}: x \to -x/2,$$

respectively. Thus, $f: A \to B: a \to f(a)$ means that f is the function from A to B that takes a onto the element $f(a)$. Let us consider some examples.

1.3.3 Examples

(a) Let A be any nonempty set. Then 1_A is the function $1_A: A \to A$: $a \to a$. That is, 1_A is the function from A into A that takes each $a \in A$ onto itself. 1_A is the ***identity function*** on A. It is one-to-one and onto, or bijective.

(b) Let A and B be any sets, and let $b \in B$. Then $f: A \to B: a \to b$ is the function from A into B that takes each $a \in A$ onto the (fixed) element $b \in B$. Such a function is said to be ***constant***.

(c) Let $\mathbb{Z}$ be the set of all integers. Then $\mathbb{Z} \to \mathbb{Z}: n \to n^2$ is a function. It is neither one-to-one nor onto.

(d) $+: \mathbb{Z} \times \mathbb{Z} \to \mathbb{Z}: (m, n) \to m + n$ is a function, called ***addition of integers***. $+(m, n)$ is usually written as $m + n$.

(e) Let $\mathbb{R}$ be the set of real numbers. Then $\mathbb{R} \to \mathbb{R}: x \to e^x$ is the well-known exponential function. It is one-to-one, and its image is the set $\mathbb{R}^+$ of positive real numbers.

(f) Let $\mathbb{R}$ and $\mathbb{R}^+$ be as in (e). Then $\mathbb{R}^+ \to \mathbb{R}: x \to \log_e(x)$ is the well-known natural logarithm function. It is one-to-one and onto.

(g) Let a and b be real numbers with $a < b$, and let I be the set of functions from the interval $[a, b]$ to $\mathbb{R}$ which are integrable on $[a, b]$. Then

$$\int : I \to \mathbb{R} : f \to \int_a^b f(x)\,dx$$

is a function. It is onto, but not one-to-one.

(h) Let D be the set of functions from $[a, b]$ to $\mathbb{R}$ which are differentiable at all points x in $[a, b]$, and let $\mathrm{Map}([a, b], \mathbb{R})$ be the set of all functions from $[a, b]$ to $\mathbb{R}$. Then

$$d: D \to \mathrm{Map}([a, b], \mathbb{R}): f \to f'$$

is a function. It is neither one-to-one nor onto.

(i) Let $\mathbb{C}$ be the set of complex numbers. The complex conjugate

$$\mathbb{C} \to \mathbb{C}: z = a + bi \to \bar{z} = a - bi$$

is a function that is both one-to-one and onto.

(j) Another well-known function is

$$\mathbb{C} \to \mathbb{R}: z = a + bi \to |z| = (a^2 + b^2)^{1/2},$$

the absolute value function.

Suppose $f: A \to B$, and $S \subset A$. Then $g: S \to B: s \to f(s)$ is obviously a mapping from S into B. The domain of f has just been cut down to S. The mapping g is denoted $f \mid S$, and read "f restricted to S."

If $\mathrm{Im} f \subset C$, then $g: A \to C: a \to f(c)$ is a mapping from A into C. The mappings $f: A \to B$ and $g: A \to C$ have the same domain and the same graph, but they are not equal unless $B = C$.

Composition of functions is an important notion. Suppose $f: A \to B$ and $g: B \to C$. From these two functions we can obtain a function from A into C in the following way. If $a \in A$, then $f(a) \in B$, and $g(f(a)) \in C$. Thus with $a \in A$, we can associate an element of C. This clearly defines a function.

1.3.4 Definition

If $f: A \to B$ and $g: B \to C$, then the function $g \circ f: A \to C: a \to g(f(a))$ is the **composition** *of f and g.*

Several useful properties of composition of functions are in the following theorem.

1.3.5 Theorem

Suppose that $f: A \to B$, $g: B \to C$, and $h: C \to D$. Then the following hold.
(a) $(h \circ g) \circ f = h \circ (g \circ f)$.
(b) *If f and g are onto, then $g \circ f$ is onto.*
(c) *If f and g are one-to-one, then $g \circ f$ is one-to-one.*
(d) *If f and g are equivalences, then $g \circ f$ is an equivalence.*
(e) $f \circ 1_A = 1_B \circ f = f$.

Proof Consider (a). The two functions clearly have the same domain and range, so it remains to show that for each $a \in A$, $((h \circ g) \circ f)(a) = (h \circ (g \circ f))(a)$. We have

$$((h \circ g) \circ f)(a) = (h \circ g)(f(a)) = h(g(f(a)))$$

and

$$(h \circ (g \circ f))(a) = h((g \circ f)(a)) = h(g(f(a))).$$

Thus (a) is proved. To prove (b), we need that if $c \in C$, then there is some $a \in A$ such that $(g \circ f)(a) = c$. Suppose $c \in C$. Then since g is onto, there is some $b \in B$ such that $g(b) = c$. Since F is onto, there is some $a \in A$ such that $f(a) = b$. Thus $g(f(a)) = (g \circ f)(a) = c$, and $g \circ f$ is onto.

To prove (c), we must show that if $(g \circ f)(x) = (g \circ f)(y)$ then $x = y$, where we know that both g and f are one-to-one. If $(g \circ f)(x) = (g \circ f)(y)$, then $g(f(x)) = g(f(y))$, and since g is one-to-one, $f(x) = f(y)$. Since f is one-to-one, $x = y$.

Part (d) follows from (b) and (c). Part (e) is easy.

Additional properties of composition of functions are in Problems 8–14.

Suppose that $f: A \to B$ is an equivalence. That is, f is one-to-one and onto. Since f is onto, for $b \in B$ there is an $a \in A$ such that $f(a) = b$. Since f is one-to-one, there is at most one such a. Hence, associating a with b gives a mapping $g: B \to A$. Notice that $(g \circ f)(a) = g(f(a)) = a$, so $g \circ f = 1_A$. Also, $(f \circ g)(b) = b$, so $f \circ g = 1_B$. The function g is denoted f^{-1}, and is called the ***inverse*** of f. Hence with every equivalence $f: A \to B$, we have associated another function $f^{-1}: B \to A$, and they are related by $f \circ f^{-1} = 1_B$ and $f^{-1} \circ f = 1_A$.

Conversely, suppose that $f: A \to B$ and $g: B \to A$, with $g \circ f = 1_A$ and $f \circ g = 1_B$. Then for $b \in B$,

$$f(g(b)) = (f \circ g)(b) = 1_B(b) = b,$$

so f is onto. If $f(x) = f(y)$, then

$$x = 1_A(x) = (g \circ f)(x) = g(f(x)) = g(f(y)) = (g \circ f)(y) = 1_A(y) = y.$$

Hence, f is one-to-one. Therefore, we have the function $f^{-1}: B \to A$. Suppose that $b \in B$. Then $f(f^{-1}(b)) = b = f(g(b))$, and since f is one-to-one, $f^{-1}(b) = g(b)$. Therefore, $f^{-1} = g$. To sum up, we have proved the following.

1.3.6 Theorem

The function $f: A \to B$ is one-to-one and onto (and hence has an inverse) if and only if there is a function $g: B \to A$ such that $g \circ f = 1_A$ and $f \circ g = 1_B$. In this case, g is one-to-one and onto, $g = f^{-1}$, and $g^{-1} = f$.

Of special interest are one-to-one functions of a set A onto itself. Such functions are called ***permutations*** of A, and will be studied in some detail in Section 2.5 for finite A.

An example of a one-to-one function from $\mathbb{R}$ to $\mathbb{R}$ is the function defined by $f(x) = bx + a$, with $b \neq 0$. The inverse f^{-1} is given by $f^{-1}(x) = (x - a)/b$. These particular one-to-one mappings of $\mathbb{R}$ onto $\mathbb{R}$ will appear in Chapter 2 (2.4, Problems 1–4, and 2.6, page 71).

We conclude this section with a short discussion of the number of elements in a set. A one-to-one mapping from a set A to a set B is called an ***equivalence***. In case there is such an equivalence, we say that A and B have the same ***cardinality***, and write $|A| = |B|$. Intuitively, A and B have the same ***number*** of elements. If there is a one-to-one mapping from A into B, we write $|A| \leq |B|$, and in addition if there is no one-to-one mapping from B into A, we write $|A| < |B|$. If $|A| < |B|$, we say that A has ***smaller cardinality*** than B. Intuitively, B is a larger set than A. It is true that for any sets A and B, exactly one of the conditions $|A| = |B|$, $|A| < |B|$, and $|B| < |A|$ holds, although we will not prove it.

Let $\mathbb{Z}^+(n) = \{1, 2, \ldots, n\}$. If $|A| = |\mathbb{Z}^+(n)|$, then we say that the cardinality of A is n, or that A has n elements, and we write $|A| = n$. If A is empty or has n elements for some positive integer n, then A is ***finite***, and A is ***infinite*** otherwise. A set A is ***denumerable***, or ***countably infinite***, if $|A| = |\mathbb{Z}^+|$, where $\mathbb{Z}^+$ is the set of positive integers. It turns out that $\mathbb{Z}$ and $\mathbb{Q}$ are denumerable, but $\mathbb{R}$ is not. A set is ***countable*** if it is finite or denumerable. Given any set, there is always a larger set (Problem 23). The reader is referred to Halmos, to Hungerford, page 15, and to Lang, page 688 for additional material on cardinality of sets.

PROBLEMS

1 Let $A = \{1, 2\}$ and $B = \{3, 4, 5\}$. Write down all the mappings from A to B. (Just write down the graphs.)

2 How many functions are there from a set with m elements into a set with n elements? Remember that sets can be empty.

3 Let $\mathbb{N}$ be the set $\{0, 1, 2, 3, \ldots\}$ of natural numbers. Show that the function $f: \mathbb{N} \to \mathbb{N}: n \to n+1$ is one-to-one but is not onto.

4 Let f be the function defined by $f: \mathbb{Z} \to \mathbb{Z}: n \to n+1$. Show that f is one-to-one and onto, and write down a formula for its inverse.

5 Let f be the function from $\mathbb{N}$ to $\mathbb{Z}$ defined by $f(n) = n/2$ if n is even and $f(n) = -(n+1)/2$ if n is odd. Show that f is a bijection, and find a formula for its inverse.

6 Let A be the open interval $(0, 1) = \{x \in \mathbb{R}: 0 < x < 1\}$, and let $\mathbb{R}^+$ be the set of all positive real numbers. Prove that the function given by $f: A \to \mathbb{R}^+: x \to (1-x)/x$ is an equivalence, and write down a formula for its inverse.

7 Let f be defined by $f: \mathbb{R} \to \mathbb{R}: x \to x^3$. Determine whether or not f is an equivalence, and if it is, find its inverse.

8 Let f be the function in Problem 7, and let g be the function defined by $g: \mathbb{R} \to \mathbb{R}: x \to x+1$.
(a) Find a formula for $f \circ g$.
(b) Find a formula for $g \circ f$.
(c) Prove that $f \circ g \neq g \circ f$.

9 Let $f: A \to B$ and $g: B \to C$. Show that
(a) if $g \circ f$ is injective, then f is injective;
(b) if $g \circ f$ is surjective, then g is surjective.

10 Let $f: A \to B$ and $g: B \to C$. Show that
(a) $g \circ f$ may be surjective without f being surjective;
(b) $g \circ f$ may be injective without g being injective.

11 Let $f: A \to B$ and $g: B \to C$. Show that
(a) $g \circ f$ may not be surjective even if g is surjective;
(b) $g \circ f$ may not be injective even if f is injective.

12 Let f be a function from A to B, and let g and h be functions from B to C. Show that if $g \circ f = h \circ f$ and f is surjective, then $g = h$.

13 Let g and h be functions from A to B, and let f be a function from B to C. Show that if $f \circ g = f \circ h$ and f is injective, then $g = h$.

14 Suppose that $f: A \to B$ and $g: B \to C$ are one-to-one and onto. Prove that $(g \circ f)^{-1} = f^{-1} \circ g^{-1}$.

15 Let A be a finite set, and let Perm(A) be the set of all permutations of A. For f in Perm(A) and n a positive integer, let f^n be f composed with itself n times.
(a) Prove that for f in Perm(A), there exist distinct positive integers m and n such that $f^m = f^n$.
(b) Prove that there exists a positive integer n such that $f^n = 1_A$.
(c) Prove that there exists a positive integer n such that $f^n = 1_A$ for all f in Perm(A).

16 Give an example to prove that $A \times B = C \times D$ does not imply that $A = C$ and $B = D$. What *is* the situation?

17 Prove that there is a one-to-one correspondence between $A \times B$ and $B \times A$.

18 Prove that there is a one-to-one correspondence between $A \times (B \times C)$ and $(A \times B) \times C$. Prove that there is a one-to-one correspondence between $A \times (B \times C)$ and

$$A \times B \times C = \{(a, b, c) : a \in A, b \in B, c \in C\}.$$

19 Prove that $|\mathbb{Z}| = |\mathbb{Z}^+|$ and that $|\mathbb{Z}| = |\mathbb{Z} \times \mathbb{Z}|$.

20 Let $A = \{q \in \mathbb{Q} : a < q < b\}$, where a and b are rational numbers with $a < b$. Let $\mathbb{Q}^+$ be the set of positive rational numbers. Prove that $|A| = |\mathbb{Q}^+|$.

21 Let $B = \{q \in \mathbb{Q} : q \leq b\}$, and let $C = \{q \in \mathbb{Q} : c \leq q\}$, where b and c are rational numbers. Prove that $|B| = |C|$. Prove that $|\mathbb{Q}| = |\mathbb{Q}^+|$.

22 Let A be a set, and let $P(A)$ denote the set of all subsets of A. For any sets A and B, let $\mathrm{Map}(A, B)$ denote the set of all mappings from A into B. Prove that there is a one-to-one correspondence between $P(A)$ and $\mathrm{Map}(A, \{1, 2\})$.

23 Prove that there is no mapping from A onto $P(A)$. *Hint*: If f were such a mapping, consider the set $\{a \in A : a \notin f(a)\}$. Prove that $|A| < |P(A)|$.

24 Let A and B be sets. Let $C = \{\{\{a\}, \{a, b\}\} : a \in A, b \in B\}$. That is, C is the set whose elements are the sets $\{\{a\}, \{a, b\}\}$ where $a \in A$ and $b \in B$. Prove that there is a one-to-one correspondence between $A \times B$ and C. (C is a rigorous way to define $A \times B$.)

25 Prove that there is a one-to-one correspondence between $A \times B$ and $\{f \in \mathrm{Map}(\{1, 2\}, A \cup B) : f(1) \in A, f(2) \in B\}$.

26 Let $f: B \to C$. Show that the mapping

$$f^*: \mathrm{Map}(A, B) \to \mathrm{Map}(A, C): g \to f \circ g$$

is (a) one-to-one if f is one-to-one, and (b) onto if f is onto.

27 Let $f: B \to C$. Show that the mapping

$$f_*: \mathrm{Map}(C, A) \to \mathrm{Map}(B, A): g \to g \circ f$$

is (a) one-to-one if f is onto, and (b) onto if f is one-to-one.

28 Let $f: A \to B$, and let $P(A)$ and $P(B)$ denote the set of all subsets of A and the set of all subsets of B, respectively. Prove that f induces a mapping $F: P(A) \to P(B)$ by $F(S) = \{f(a) : a \in S\}$. Prove that

(a) f is one-to-one if and only if F is one-to-one,

(b) f is onto if and only if F is onto, and

(c) $f = 1_A$ if and only if $F = 1_{P(A)}$.

1.4 FAMILIES

This section is devoted to some notation that will be useful throughout. Let A and B be sets. A ***family of elements of*** B ***indexed by A*** is a mapping $f: A \to B$. There is no new concept here. A family is a mapping. A notation for such a family is $\{b_a\}_{a \in A}$, where $f(a) = b_a$. Note that b_x can be b_y with $x \neq y$. Indeed, if f is not one-to-one, this is the case. We write such sentences as "Let $\{b_a\}_{a \in A}$ be a family of elements of B indexed by A." The set A is called the ***indexing*** set. The b_a's are called the ***members*** of the family. If the mapping is onto, then B is ***indexed*** by A. A familiar example of a family is a sequence of real numbers. It is just a family of real numbers indexed by the set $\mathbb{Z}^+$ of positive integers, that is, just a map $\mathbb{Z}^+ \to \mathbb{R}$. We do not think of sequences as mappings, but rather as real numbers with positive integers attached. Even though the concept of family is the same as that of mapping, families are thought of a little bit differently.

What does the sentence "Let $\{S_i\}_{i \in I}$ be a family of sets." mean? It means that for each $i \in I$ there is associated a set S_i. That is, $\{S_i\}_{i \in I}$ is a mapping f from I into a set of sets and $f(i) = S_i$. Remember that S_i may be S_j with $i \neq j$. A family of sets is not a set of sets.

With every family $\{b_a\}_{a \in A}$ there is associated a set, namely the image of the mapping. This set is $\{b_a : a \in A\}$. With every set B there is associated a family, namely the family $\{b_b\}_{b \in B}$ which is the identity mapping 1_B. For most purposes, there is no difference between the family $\{b_b\}_{b \in B}$ and the set B. From either we can recapture the other. Thus, in a sense, the concept of family is more general than that of set. (Family was defined in terms of set, however.)

The following situation arises fairly often. With each element $a \in A$, there is associated a subset S_a of a set B. That is, we have a map from A into the set $P(B)$ of subsets of B, or a family of subsets of B indexed by A. We define the ***intersection*** of such a family $\{S_a\}_{a \in A}$ to be

$$\bigcap_{a \in A} S_a = \{x : x \in S_a \text{ for all } a \in A\},$$

and the ***union*** of the family to be

$$\bigcup_{a \in A} S_a = \{x : x \in S_a \text{ for some } a \in A\}.$$

The members of the family are ***mutually disjoint*** if $S_x \cap S_y = \emptyset$ whenever $x \neq y$.

Let S be a set of sets. We can take the union and intersection of the sets in S. They are defined to be the sets

$$\bigcup_{A \in S} A = \{x : x \in A \text{ for some } A \in S\}$$

and

$$\bigcap_{A \in S} A = \{x : x \in A \text{ for all } A \in S\},$$

respectively. It should be clear that the union and intersection of a family $\{S_a\}_{a \in A}$ of sets is the same as the union and intersection, respectively, of the set of sets associated with that family. Furthermore, the union and intersection of a set of sets is the same as the union and intersection, respectively, of the family of sets associated with that set of sets.

If S is a set of sets, the sets in S are ***mutually disjoint*** if $A \cap B = \varnothing$ whenever A and B are in S and $A \neq B$. Note that the members of a family of sets may not be mutually disjoint even though the sets in the set of sets associated with the family are mutually disjoint.

PROBLEMS

1 Let $\{B_i\}_{i \in I}$ be a family of sets. Prove that

$$A \cap \left(\bigcup_{i \in I} B_i \right) = \bigcup_{i \in I} (A \cap B_i).$$

2 Let $\{B_i\}_{i \in I}$ be a family of sets. Prove that

$$A \cup \left(\bigcap_{i \in I} B_i \right) = \bigcap_{i \in I} (A \cup B_i).$$

3 Let $\{B_i\}_{i \in I}$ be a family of sets. Define

$$\sum_{i \in I} B_i = \left(\bigcup_{i \in I} B_i \right) \Big\backslash \left(\bigcap_{i \in I} B_i \right).$$

Prove that

$$A \cap \sum_{i \in I} B_i = \sum_{i \in I} (A \cap B_i).$$

4 For each $n \in \mathbb{Z}$, let

$$\mathbb{Z}_n = \{m \in \mathbb{Z} : m - n = 3k \text{ for some } k \in \mathbb{Z}\}.$$

Show that the members of the family $\{\mathbb{Z}_n\}_{n \in \mathbb{Z}}$ are not mutually disjoint, and that the sets in the set $\{\mathbb{Z}_n : n \in \mathbb{Z}\}$ are mutually disjoint. Determine how many elements there are in the set $\{\mathbb{Z}_n : n \in \mathbb{Z}\}$.

5 For each $x \in \mathbb{Q}$, the set of rational numbers, let $B_x = \{y \in \mathbb{Q} : x - y \in \mathbb{Z}\}$. Show that the members of the family $\{B_x\}_{x \in \mathbb{Q}}$ are not mutually disjoint and that the sets in the set $\{B_x : x \in \mathbb{Q}\}$ are mutually disjoint. Prove that

$$\{x \in \mathbb{Q} : 0 \leq x < 1\} \to \{B_x : x \in \mathbb{Q}\} : x \to B_x$$

is an equivalence.

1.5 EQUIVALENCE RELATIONS

The important concepts are those that arise often, and in many different places. Such a concept is that of *equivalence relation*. It is a simple concept, and it is the same as that of *partition*, as we shall see.

1.5.1 Definition

Let A be a set. An **equivalence relation** *on A is a subset $\approx$ of $A \times A$ satisfying*
(a) $(a, a) \in \approx$ *for all* $a \in A$,
(b) *if* $(a, b) \in \approx$, *then* $(b, a) \in \approx$, *and*
(c) *if* $(a, b) \in \approx$ *and* $(b, c) \in \approx$, *then* $(a, c) \in \approx$.

Condition (a) is expressed by saying that $\approx$ is ***reflexive***, (b) by saying that $\approx$ is ***symmetric***, and (c) by saying that $\approx$ is ***transitive***. Common practice is to write $a \approx b$ instead of $(a, b) \in \approx$. We will follow that practice.

In the examples below, the unverified statements are easy to verify.

1.5.2 Examples

(a) Let A be any set, and define $a \approx b$ if $a = b$. Then $\approx$ is an equivalence relation on A. That is, equality is an equivalence relation.
(b) Let A be any set, and define $a \approx b$ for all $a, b \in A$. Then $\approx$ is an equivalence relation on A.
(c) For $m, n \in \mathbb{Z}$, define $m \approx n$ if $m - n$ is even. Then $\approx$ is an equivalence relation on $\mathbb{Z}$.
(d) Let $k \in \mathbb{Z}$. For $m, n \in \mathbb{Z}$, define $m \approx n$ if $m - n$ is an integral multiple of k. Then $\approx$ is an equivalence relation on $\mathbb{Z}$.
(e) Let T be the set of all triangles in the plane. For $s, t \in T$, define $s \approx t$ if s is congruent to t. Then $\approx$ is an equivalence relation on T. Similarly, $\approx$ is an equivalence relation on T if we define $s \approx t$ if s is similar to t.
(f) Let S be a set of sets, and for $A, B \in S$ define $A \approx B$ if there is an equivalence $f: A \to B$. Then $\approx$ is an equivalence relation. In fact, 1_A is an equivalence, so $A \approx A$; if $f: A \to B$ is an equivalence, then $f^{-1}: B \to A$ is an equivalence, so $\approx$ is symmetric; and if $f: A \to B$ and $g: B \to C$ are equivalences, then $g \circ f: A \to C$ is an equivalence, so $\approx$ is transitive.
(g) Let $\mathbb{Q}$ be the set of rational numbers. For $x, y \in \mathbb{Q}$, define $x \approx y$ if $x - y \in \mathbb{Z}$. Then $\approx$ is an equivalence relation on $\mathbb{Q}$.

(h) Let A be a set and suppose that S is a set of mutually disjoint nonempty subsets of A such that $\bigcup_{B \in S} B = A$. For $a, b \in A$, define $a \approx b$ if a and b are in the same member B of S. Then $\approx$ is an equivalence relation on A. It will be seen that every equivalence relation can be obtained this way.

1.5.3 Definition

A **partition** *of a set A is a set P of mutually disjoint nonempty subsets of A such that $\bigcup_{S \in P} S = A$.*

A partition of a set is just what the name suggests—a dividing up of that set into nonoverlapping pieces. We now proceed to show that the concepts of partition and equivalence relation are the same.

1.5.4 Definition

Let $\approx$ be an equivalence relation on a set A. For $a \in A$, let $\mathrm{Cl}(a) = \{x \in A : a \approx x\}$. The set $\mathrm{Cl}(a)$ is the **equivalence class of $\approx$ determined by** *a. A subset S of A is an* **equivalence class of $\approx$** *if $S = \mathrm{Cl}(a)$ for some $a \in A$.*

The equivalence class $\mathrm{Cl}(a)$ is the set of all elements of A equivalent to a. In 1.5.2(a), the equivalence classes are the one-element subsets $\{a\}$ of A. In 1.5.2(b), there is only one equivalence class, namely A itself. In 1.5.2(c), there are two equivalence classes, the set of even integers and the set of odd integers. In 1.5.2(h), the equivalence classes are the sets $B \in S$.

We have seen in Example 1.5.2(h) that a partition P of a set A induces an equivalence relation on that set. Namely, two elements of A are defined to be equivalent if they are in the same member of P. That this is an equivalence relation is completely obvious. Now, given an equivalence relation on a set A, we can get from it a partition of A. These two processes are inverses of one another.

1.5.5 Theorem

Let $\approx$ be an equivalence relation on a set A. Then the set E of equivalence classes of $\approx$ is a partition of A.

Proof We must show that E is a set of mutually disjoint subsets of A whose union is A. Suppose $S, T \in E$. Then $S = \mathrm{Cl}(a)$, $T = \mathrm{Cl}(b)$ for some

$a, b \in A$. Since $a \in \mathrm{Cl}(a)$, no member of E is empty, and the union of the members of E is A. Suppose that $x \in \mathrm{Cl}(a) \cap \mathrm{Cl}(b)$. Then $a \approx x$ and $x \approx b$. Thus, $a \approx b$. If $y \in \mathrm{Cl}(b)$, then $b \approx y$, and hence $a \approx y$. Thus, $y \in \mathrm{Cl}(a)$, so that $\mathrm{Cl}(b) \subset \mathrm{Cl}(a)$. By the symmetry of the situation, $\mathrm{Cl}(a) \subset \mathrm{Cl}(b)$, so $\mathrm{Cl}(a) = \mathrm{Cl}(b)$. Therefore the sets in E are mutually disjoint, and E is a partition of A.

1.5.6 Corollary

Two equivalence classes of an equivalence relation are either equal or disjoint.

1.5.7 Theorem

Let A be a set. Let $E(\approx)$ be the set of equivalence classes of the equivalence relation $\approx$ on A. Then $\approx \to E(\approx)$ is a one-to-one correspondence between the set of equivalence relations on A and the set of partitions of A.

Proof By 1.5.5, $E(\approx)$ is indeed a partition of A if $\approx$ is an equivalence relation on A. Thus, $\approx \to E(\approx)$ gives a mapping from the set of equivalence relations on A into the set of partitions of A. We have seen that a partition gives an equivalence relation, so we have a mapping from the set of partitions of A into the set of equivalence relations on A. These two mappings are inverses of one another. By 1.3.6, the theorem follows.

The reader should devote enough thought to the content of 1.5.7 to understand it.

PROBLEMS

1 Determine which of the properties of "reflexive," "symmetric," and "transitive" the following relations on $\mathbb{Z}$ satisfy.
- (a) $m \approx n$ if $m \leq n$.
- (b) $m \approx n$ if $m - n$ is odd or is 0.
- (c) $m \approx n$ if m divides n.
- (d) $m \approx n$ if $|m - n| \leq 10$.

2 Prove that $\approx$ defined on $\mathbb{Z}$ by $m \approx n$ if $m = n$ or $m = -n + 5$ is an equivalence relation. Determine its equivalence classes.

3 Prove that $\approx$ defined on $\mathbb{Z}$ by $m \approx n$ if $m^2 + m = n^2 + n$ is an equivalence relation. Determine its equivalence classes.

4 Let two complex numbers be equivalent if their real parts are equal. Prove that this is an equivalence relation and that the equivalence classes are in one-to-one correspondence with the real numbers.

5 Let two real numbers be equivalent if their difference is rational. Prove that this is an equivalence relation.

6 Let two real numbers be equivalent if their difference is an integral multiple of 360. Prove that this is an equivalence relation.

7 Let $\approx$ be a relation which is reflexive and for which $a \approx b$ and $b \approx c$ imply that $c \approx a$. Prove that $\approx$ is an equivalence relation.

8 Let $\mathbb{Z}^+$ be the set of positive integers, and define $\approx$ on $\mathbb{Z}^+ \times \mathbb{Z}^+$ by $(a, b) \approx (c, d)$ if $a + d = b + c$. Prove that $\approx$ is an equivalence relation and that there is a natural one-to-one correspondence between the equivalence classes of $\approx$ and the set $\mathbb{Z}$ of all integers.

9 Do Problem 8 with the integers replaced by
(a) the rational numbers,
(b) the real numbers.

10 Let $\mathbb{Z}^*$ be the set of nonzero integers. Define $\approx$ on $\mathbb{Z} \times \mathbb{Z}^*$ by $(a, b) \approx (c, d)$ if $ad = bc$. Prove that $\approx$ is an equivalence relation and that the equivalence classes of $\approx$ are in one-to-one correspondence with the set $\mathbb{Q}$ of rational numbers.

11 Let $\mathbb{R}^*$ be the set of nonzero real numbers, and let $\mathbb{R}^+$ be the set of positive real numbers. Define $\approx$ on $\mathbb{R}^*$ by $a \approx b$ if $a/b \in \mathbb{R}^+$. Prove that $\approx$ is an equivalence relation, and determine the equivalence classes of $\approx$.

12 Do Problem 11 with the real numbers $\mathbb{R}$ replaced by the rational numbers $\mathbb{Q}$.

13 Determine all partitions and all equivalence relations on the empty set.

14 Let A be a set, and let $S \subset A$. For subsets B and C of A, define $B \approx C$ if $(B \cup C) \setminus (B \cap C) \subset S$. Prove that $\approx$ is an equivalence relation on the set $P(A)$ of all subsets of A. Prove that $P(S)$ is an equivalence class of $\approx$.

15 Prove that the set of equivalence classes of $\approx$ in Problem 14 is in one-to-one correspondence with $P(A \setminus S)$.

16 Let S be a nonempty set, and let Perm(S) be the set of all one-to-one mappings of S onto S. Let G be a nonempty subset of Perm(S) such that if f, g are in G then $g \circ f$ and f^{-1} are in G. Define $\approx$ on S by $a \approx b$ if there is an element $g \in G$ such that $g(a) = b$. Prove that $\approx$ is an equivalence relation on S. If $G = \text{Perm}(S)$, what are the equivalence classes of $\approx$? If $G = \{1_S\}$, what are the equivalence classes of $\approx$?

17 Let G and Perm(S) be as in Problem 16. Define $\approx$ on Perm(S) by $f \approx g$ if $g^{-1} \circ f \in G$. Prove that $\approx$ is an equivalence relation on Perm(S). Prove that G is an equivalence class of $\approx$. Determine the equivalence classes of $\approx$ in the cases $G = \mathrm{Perm}(S)$ and $G = \{1_S\}$.

18 Let $f: A \to B$. Define $\approx$ on A by $x \approx y$ if $f(x) = f(y)$. Prove that $\approx$ is an equivalence relation on A.

19 Let $f: A \to B$, and let $\cong$ be an equivalence relation on B. Define $\approx$ on A by $x \approx y$ if $f(x) \cong f(y)$. Prove that $\approx$ is an equivalence relation on A.

20 Let $f: A \to B$, and define $\approx$ on A by $x \approx y$ if $f(x) = f(y)$, as in Problem 18. Let $A/\approx$ denote the set of equivalence classes of $\approx$. Prove that

$$f^*: A/\approx \;\to \mathrm{Im}\, f: \mathrm{Cl}(a) \to f(a)$$

is an equivalence.

21 Let $\cong$ be an equivalence relation on a set A. Let $A/\cong$ be the set of equivalence classes of $\cong$. Let $f: A \to A/\cong : a \to \mathrm{Cl}(a)$. What are $\approx$ and f^* in Problem 20?

22 Suppose $\approx$ and $\cong$ are equivalence relations on a set A. Call $\cong$ ***weaker*** than $\approx$ if $a \approx b$ implies $a \cong b$. Suppose $\mathcal{S}$ and $\mathcal{T}$ are two partitions of A. Call $\mathcal{T}$ ***finer*** than $\mathcal{S}$ if each $C \in \mathcal{T}$ is a subset of some $D \in \mathcal{S}$. Prove that $\cong$ is weaker than $\approx$ if and only if the partition that $\approx$ determines is finer than the partition that $\cong$ determines.

23 Let $\approx$ be an equivalence relation on A, and let $S \subset A$. Let $\mathcal{P}$ be the partition of A induced by $\approx$. Let $\cong$ be the equivalence relation on S induced by $\approx$. That is, $\cong$ is the "restriction" of $\approx$ to S. Prove that the partition of S induced by $\cong$ is the set $\{T \cap S: T \in \mathcal{P}, T \cap S \neq \varnothing\}$.

24 Let $\mathcal{S}$ be a partition of a set A, and let $\mathcal{T} = \{A \setminus B: B \in \mathcal{S}\}$. Prove that
(a) $\bigcap_{C \in \mathcal{T}} C = \varnothing$,
(b) if $C, D \in \mathcal{T}$ and $C \neq D$, then $C \cup D = A$, and
(c) if $C \in \mathcal{T}$, then $C \neq A$.

25 Prove that the reflexive, symmetric, and transitive properties of an equivalence relation are independent; that is, prove that no two of the properties imply the third.

1.6 THE INTEGERS

Throughout this book, we will have occasion to use various properties of integers. This short section is devoted to putting down the more fundamental of these properties. This discussion is not meant to be complete or rigorous, but rather just a casual discussion of some of the properties of the system $\mathbb{Z}$ of integers.

A basic tool, used in proofs and in definitions, is ***mathematical induction***. Suppose that for each positive integer n, P_n is a statement. Now suppose that the statement P_1 is true, and suppose further that for each positive integer m, P_m implies P_{m+1}. Then the ***principle of mathematical induction*** asserts that all the statements P_n are true. Intuitively, the principle is obvious, because P_1 is given to be true; P_1 implies P_2, so P_2 is true; P_2 implies P_3, so P_3 is true; P_3 implies P_4, so P_4 is true; and so on.

The principle of mathematical induction has nothing to do with what the statements P_n are about, but rather it involves a basic property of the positive integers. The positive integers have the property that any nonempty subset of them has a smallest element. This is expressed by saying that the set of positive integers is ***well ordered***. The principle of mathematical induction follows from this property. Suppose, in fact, that not all of the P_n's above were true. Then the set $\{n: P_n \text{ is false}\}$ is nonempty, and so it has a smallest element t. Now $t > 1$ since P_1 is true. Thus, $t = s + 1$ with s a positive integer. Since $s < t$, P_s implies P_{s+1}, and so P_t is true. This is a contradiction. Hence, the assumption that not all the P_n's is true is false. Therefore, all the P_n's are true.

We now give an example of a proof by induction. Consider the statement $\sum_{i=1}^{n} i = n(n+1)/2$. Thus we have a statement $P(n)$ for each positive integer n. The principle of mathematical induction says that $P(n)$ is true for all positive integers n if $P(1)$ is true, and if for each positive integer m, $P(m)$ implies $P(m+1)$. The statement $P(1)$ is true because it just asserts that $1 = 1(1+1)/2$, which is indeed the case. Assume that $P(m)$ is true. That is, assume that $\sum_{i=1}^{m} i = m(m+1)/2$. We need to show that this implies that $\sum_{i=1}^{m+1} i = (m+1)(m+2)/2$. Now,

$$\begin{aligned}\sum_{i=1}^{m+1} i &= \sum_{i=1}^{m} i + (m+1) = m(m+1)/2 + (m+1) \\ &= (m(m+1) + 2(m+1))/2 = (m+1)(m+2)/2.\end{aligned}$$

Thus $P(m)$ implies $P(m+1)$, and so $P(n)$ is true for all positive integers n. Similar proofs are asked for in the Problems at the end of this section.

The principle of mathematical induction can be stated in a slightly different manner. Suppose we are given that P_n is true whenever P_m is true for all $m < n$. Then again, P_n is true for all n. In fact, suppose some P_n is false. Then $\{n: P_n \text{ is false}\}$ is nonempty, and so it has a smallest element t. But P_s is true for all $s < t$, and hence P_t is true. Again, we must conclude that P_n is true for all n. You may wonder why P_1 is true. It is true simply because for all positive integers $m < 1$, P_m is true.

Induction is useful in formulating definitions. We illustrate this with a couple of examples. Suppose we want to define x^n for each real number x and each positive integer n. Intuitively, we want x^n to be x times itself n times. This can be accomplished inductively by setting $x^1 = x$, and $x^{n+1} = x^n \cdot x$. Thus, x^1 is defined to be x, and assuming x^n is defined, we define x^{n+1} to be $x^n \cdot x$. It should be intuitively clear that this defines x^n for all n. To put this on a more rigorous basis is a bit troublesome. Problems 11 and 12 at the end of this section may be enlightening.

As a second example, we define the *Fibonacci sequence* 1, 1, 2, 3, 5, 8, 13, 21, . . . , where after the first two terms, each number is the sum of the two preceding it. It is defined inductively by

$$a_1 = 1,$$

$$a_2 = 1,$$

and

$$a_n = a_{n-1} + a_{n-2} \quad \text{for} \quad n > 2.$$

Suppose $m, n \in \mathbb{Z}$ with $m \neq 0$. Then there are unique integers q and r with $0 \leq r < |m|$, the absolute value of m, such that $n = mq + r$. This is the ***division algorithm*** for integers. We will simply assume it. If $m, n \in \mathbb{Z}$ and not both are zero, then the ***greatest common divisor*** (m, n) of m and n is the largest integer that divides them both. There clearly is such, because 1 divides them both, and those integers that divide both are no larger than the maximum of $|m|$ and $|n|$.

An important fact about greatest common divisors is this: (m, n) is the smallest positive integer of the form $am + bn$, where $a, b \in \mathbb{Z}$. In particular, the greatest common divisor of two integers m and n, not both zero, can be written in the form $am + bn$, with $a, b \in \mathbb{Z}$. To see this, let $m, n \in \mathbb{Z}$ with not both zero. The set of all positive integers of the form $am + bn$ with a and b in $\mathbb{Z}$ is not empty since it contains $m^2 + n^2$, which is positive. Now let d be the smallest positive integer of the form $d = am + bn$ with $a, b \in \mathbb{Z}$. Any positive integer that divides both m and n clearly divides d, so $d \geq (m, n)$. Write $m = dq + r$, with $0 \leq r < d$. Then

$$0 \leq r = m - dq = m - qam - qbn = (1 - qa)m - qbn < d,$$

whence $r = 0$. Therefore, d divides m. Similarly, d divides n. Hence, $d = (m, n)$.

Since any integer that divides both m and n divides d, it follows that the greatest common divisor d of m and n is characterized by the facts that it is positive, divides m and n, and is divisible by every integer that divides m and n.

If $(m, n) = 1$, then m and n are said to be ***relatively prime***. In this case, $1 = am + bn$ for suitable $a, b \in \mathbb{Z}$, a most handy fact.

There is an algorithm for computing the greatest common divisor of two integers m and n, and that algorithm actually gives the greatest common divisor as a linear combination of m and n. It is based on the division algorithm, and it is called the ***Euclidean Algorithm***. Here is how it works. Suppose that we want to find the greatest common divisor of n_1 and n_2. We may as well assume that n_2 is positive. Write $n_1 = n_2 q_1 + n_3$, with $0 \leq n_3 < n_2$. If n_3 is not zero, then write $n_2 = n_3 q_2 + n_4$ with $0 \leq n_4 < n_3$. Proceeding in this manner, we eventually get $n_{k+1}=0$ since for $i>1$, the n_i are nonnegative and strictly decreasing. Thus we have the following equations.

$$n_1=n_2q_1+n_3, \quad \text{with} \quad 0<n_3<n_2;$$

$$n_2=n_3q_2+n_4, \quad \text{with} \quad 0<n_4<n_3;$$

$$n_3=n_4q_3+n_5, \quad \text{with} \quad 0<n_5<n_4;$$

$$\vdots$$

$$n_{k-3}=n_{k-2}q_{k-3}+n_{k-1}, \quad \text{with} \quad 0<n_{k-1}<n_{k-2};$$

$$n_{k-2}=n_{k-1}q_{k-2}+n_k, \quad \text{with} \quad 0<n_k<n_{k-1};$$

$$n_{k-1}=n_kq_{k-1}.$$

Now n_k is the greatest common divisor of n_1 and n_2. From the last equation above, we see that n_k divides n_{k-1}, from the next to last, that n_k divides n_{k-2}, and so on, yielding the fact that n_k divides both n_1 and n_2. Starting with the first equation, if an integer n divides both n_1 and n_2, then n divides n_3, and the second equation shows that n divides n_4. Continuing, we see that n divides n_k. Thus, n_k is the greatest common divisor of n_1 and n_2.

Starting with the equation $n_i=n_{i+1}q_i+n_{i+2}$, we see that n_k is the greatest common divisor of n_i and n_{i+1} for any $i<k$.

To get n_k as a linear combination of n_1 and n_2, we show the more general fact that any of the n_i can be expressed as a linear combination of n_1 and n_2. Induct on i. Now n_1 and n_2 are certainly linear combinations of n_1 and n_2. If n_j is a linear combination of n_1 and n_2 for all $j<i$, with $i>2$, then using the equation $n_{i-2}=n_{i-1}q_{i-2}+n_i$, it follows readily that n_i is a linear combination of n_1 and n_2. For example,

$$n_3=n_1-q_1n_2;$$

$$\begin{aligned} n_4&=n_2-n_3q_2\\ &=n_2-(n_1-q_1n_2)q_2\\ &=-q_2n_1+(1+q_1q_2)n_2; \end{aligned}$$

$$\begin{aligned} n_5&=n_3-n_4q_3\\ &=n_1-q_1n_2-(-q_2n_1+(1+q_1q_2)n_2)q_3\\ &=(1+q_2q_3)n_1+(-q_3-q_1-q_1q_2q_3)n_2. \end{aligned}$$

We illustrate with $n_1=391$ and $n_2=102$. We get

$$391=102\cdot 3+85,$$

$$102=85\cdot 1+17,$$
$$85=17\cdot 5.$$

Thus, the greatest common divisor of 391 and 102 is 17. Further,

$$\begin{aligned}17 &= 102 - 85\\ &= 102 - (391 - 102\cdot 3)\\ &= 4\cdot 102 - 1\cdot 391.\end{aligned}$$

Later we will need the notion of greatest common divisor for any finite set of integers, not all zero. Let $m_1, m_2, \ldots, m_k \in \mathbb{Z}$, not all zero. The ***greatest common divisor*** of this set of integers is the largest positive integer d dividing them all. Such an integer d exists since 1 divides all the m_i, and the positive integers that divide them all are no larger than the maximum of the absolute values of the m_i. Furthermore, there exist integers a_i such that $d = \sum a_i m_i$. To see this, first note that the set of positive integers of the form $\sum a_i m_i$ with the a_i in $\mathbb{Z}$ is not empty since it contains $\sum m_i^2$, which is positive. Let d be the smallest positive integer of the form $\sum a_i m_i$ with the a_i in $\mathbb{Z}$. Then any positive integer that divides all the m_i divides d, so that d is at least as big as the greatest common divisor of the m_i. For any j, write $m_j = dq_j + r_j$, with $0 \le r_j < d$. Then $r_j = m_j - dq_j$, which is of the form $\sum a_i m_i$ since m_j and d are of that form. Thus $r_j = 0$, and so d divides each m_i. Therefore, d is the greatest common divisor of the m_i.

If the greatest common divisor of the integers m_i is 1, then the m_i are said to be ***relatively prime***. In this case, there exist integers a_i such that $1 = \sum a_i m_i$.

A positive integer $p \neq 1$ is a ***prime*** if p is divisible only by the integers ± 1 and $\pm p$. Equivalently, $p > 1$ is prime if whenever p divides a product mn of integers, it divides m or it divides n.

Every positive integer $n \neq 1$ is a product of primes. In fact, let S be the set of all positive integers $\neq 1$ which divide n. Since $n \in S$, S is not empty, so S has a smallest element p. But p must be a prime because otherwise p would be divisible by a positive integer $\neq 1$ and less than p, contradicting the fact that p is the smallest element of S. Thus, $n = pm$ with p prime. Since $m < n$, proceeding by induction, we get that $m = 1$ or m is a product of primes. Hence n is a product of primes, say $n = p_1 p_2 \cdots p_j$.

Suppose that $n = p_1 p_2 \cdots p_j = q_1 q_2 \cdots q_k$ with $q_1, q_2, \ldots, q_k$ also primes. Then p_1 divides $q_1(q_2 q_3 \cdots q_k)$, so p_1 divides q_1 or $q_2 q_3 \cdots q_k$. If p_1 divides q_1, then $p_1 = q_1$. If not, p_1 divides either q_2 or $q_3 q_4 \cdots q_k$, so that in any case, p_1 divides, and hence equals, some q_i. We may as well suppose $i = 1$. Thus $p_1 = q_1$, so $p_2 p_3 \cdots p_j = q_2 q_3 \cdots q_k$, and by induction on j we can conclude that $j = k$, and after rearrangement,

$$p_1 = q_1, p_2 = q_2, \ldots, p_j = q_j.$$

That is, "up to rearrangement," n can be written uniquely as a product of primes. We conclude that if $n \in \mathbb{Z}$ and $n \neq 0$ or 1, then

$$n = (\pm 1) p_1^{n_1} p_2^{n_2} \cdots p_k^{n_k}$$

with the p_i's distinct primes, and that any such representation of n is unique up to the order of the p_i's. In particular, the representation is unique if we insist that $p_1 < p_2 < \cdots < p_k$.

In Chapter 4 we will prove such "unique factorization theorems" for systems more general than that of the integers $\mathbb{Z}$.

Let $p_1, p_2, \ldots, p_n$ be primes. Then $p_1 p_2 \cdots p_n + 1$ is divisible by a prime, but is not divisible by any of the p_i. **Thus, there are an infinite number of primes.**

PROBLEMS

1 Prove that for all positive integers n,

$$1^3 + 2^3 + \cdots + n^3 = (1 + 2 + \cdots + n)^2.$$

2 Prove that for all positive integers n, $\sum_{i=1}^{n} 1/2^i < 1$.

3 Prove that $a_n < (7/4)^n$ for all terms a_n of the Fibonacci sequence.

4 Suppose x is a real number. Prove that $x^{m+n} = x^m \cdot x^n$ for all positive integers m and n.

5 Use the Euclidean Algorithm to find the greatest common divisors of the following pairs of integers. Express the greatest common divisors as linear combinations of the two given integers.
(a) 102 and 855,
(b) 101 and 57,
(c) 101 and -57.

6 Prove that for any integer n, $(a, b) = (a, b + na)$.

7 Prove that if m and n are both nonzero, then there are infinitely many pairs a and b for which $(m, n) = am + bn$.

8 Prove that the following two definitions of *prime* are equivalent.
(a) An integer $p > 1$ is prime if p is divisible only by the integers ± 1 and $\pm p$.
(b) An integer $p > 1$ is prime if whenever p divides a product mn, then p divides m or p divides n.

9 Let a, b, and c be nonzero integers. Prove that $(a, b, c) = ((a, b), c)$. How can you use the Euclidean Algorithm to write (a, b, c) as a linear combination of a, b, and c?

10 Let m and n be nonzero integers. The ***least common multiple*** of m and n is the smallest positive integer that both m and n divide. It is denoted $[m, n]$. Let $m = a(m, n)$, and let $n = b(m, n)$. Prove that
(a) $[m, n] = ab(m, n)$, and
(b) $mn = [m, n](m, n)$.

11 Let $\mathbb{Z}^+$ be the set of all positive integers, let $\mathbb{Z}_n = \{m \in \mathbb{Z}^+ : m \leq n\}$, and let $\mathbb{R}$ be the set of all real numbers. Let P be the set of all positive integers n such that there exists a unique function $f_n: \mathbb{R} \times \mathbb{Z}_n \to \mathbb{R}$ such that

(a) $f_n(x, 1) = x$, and

(b) $f_n(x, m+1) = f_n(x, m)f_n(x, 1)$ for all $1 \leq m < n$.

Prove that $P = \mathbb{Z}^+$.

12 Let $\mathbb{Z}^+$ be the set of all positive integers, and let $\mathbb{R}$ be the set of all real numbers. Prove that there is a unique function $f: \mathbb{R} \times \mathbb{Z}^+ \to \mathbb{R}$ such that

(a) $f(x, 1) = x$, and

(b) $f(x, n+1) = f(x, n)f(x, 1)$.

2 Groups

2.1 DEFINITIONS AND EXAMPLES

We choose to begin our study of algebra with that of groups because a group is one of the most basic algebraic systems. The various concepts that appear in algebra generally appear in their simplest form in the study of groups. The theory of groups is one of the oldest branches of algebra and has applications in many areas of science, particularly in physics and in other parts of mathematics. It arose in the theory of equations, specifically in trying to find roots of polynomials in terms of their coefficients, and we will see in Chapter 6 some of the uses of group theory in such endeavors. Suffice it to say that the richness, breadth, and usefulness of the subject has made group theory a central topic in mathematics.

A fundamental algebraic concept is that of ***binary operation*** on a set. It is just a way of putting two elements of a set together to get a third element of that set. That is, a binary operation on a set S is a mapping $S \times S \to S$. A group is a set with a particular kind of binary operation on it. Before we present the formal definition of a group, let us consider a couple of examples. Addition of integers is a binary operation on the set $\mathbb{Z}$ of integers. That is, $+: \mathbb{Z} \times \mathbb{Z} \to \mathbb{Z}$. Let $\operatorname{Perm}(S)$ be the set of all ***permutations*** of S, that is, the set of all one-to-one mappings of a set S onto itself. Now if f and g are in $\operatorname{Perm}(S)$, then by 1.3.5, so is $g \circ f$. Thus, $\circ: \operatorname{Perm}(S) \times \operatorname{Perm}(S) \to \operatorname{Perm}(S)$, and so $\circ$ is a binary operation on $\operatorname{Perm}(S)$. These two situations, that is, $\mathbb{Z}$ with its binary operation $+$, and $\operatorname{Perm}(S)$ with its binary operation $\circ$, may appear to have little in common. However, they are remarkably similar. Both binary operations are ***associative***, for example. If $a, b, c \in \mathbb{Z}$, then $(a+b)+c = a+(b+c)$. If $f, g, h \in \operatorname{Perm}(S)$, then $(f \circ g) \circ h = f \circ (g \circ h)$. Both have an ***identity***. For any $a \in \mathbb{Z}$, $a+0=0+a=a$, and for any $f \in \operatorname{Perm}(S)$, $1_S \circ f = f \circ 1_S = f$. In both cases, every element in the set has an ***inverse***. For $a \in \mathbb{Z}$, there is an element $b \in \mathbb{Z}$ such that $a+b=b+a=0$, namely $b=-a$. For $f \in \operatorname{Perm}(S)$, there is an element $g \in \operatorname{Perm}(S)$ such that $f \circ g = g \circ f = 1_S$, namely $g = f^{-1}$. Sets with binary operations on them which have these three properties arise often and in many different places in mathematics (and elsewhere). They are called ***groups***. They have been, and are still being, studied extensively. See Rotman and Scott—books devoted exclusively to group theory.

2.1.1 Definition

A **group** *is a set G and a binary operation* $\cdot\colon G \times G \to G$ *on G such that*

(a) $(g \cdot h) \cdot k = g \cdot (h \cdot k)$ *for all* $g, h, k \in G$;

(b) *there is an element* $e \in G$ *such that* $e \cdot g = g \cdot e = g$ *for all* $g \in G$; *and*

(c) *for* $g \in G$, *there is an* $h \in G$ *such that* $g \cdot h = h \cdot g = e$.

As is the custom, the image of the pair (g, h) under the mapping $\cdot$ is denoted $g \cdot h$ instead of $\cdot(g, h)$. Condition (a) is expressed by saying that $\cdot$ is ***associative***, condition (b) by saying that $\cdot$ has an ***identity*** element, and condition (c) by saying that each element has an ***inverse***.

There is only one identity element $e \in G$. Indeed, if e' were another such element, then $e = e \cdot e' = e' \cdot e = e'$. This unique element is called the ***identity element*** of G. It will typically be denoted by e, or e_G if we need to call attention to G. Note also that each $g \in G$ has only one inverse h. If h' were another, then

$$h = e \cdot h = (h' \cdot g) \cdot h = h' \cdot (g \cdot h) = h' \cdot e = h'.$$

This unique element associated with g is denoted g^{-1} and is called the ***inverse*** of g.

It is important to realize that $g \cdot h$ need not be $h \cdot g$. However, if $g \cdot h = h \cdot g$, for all $g, h \in G$, then the group is called ***commutative***, or ***Abelian***. A group that is not commutative is called ***noncommutative***, or ***non-Abelian***. Suppose that S is a subset of G. Consider the mapping $\cdot$ restricted to $S \times S$. If $\cdot$ maps $S \times S$ into S, then it is a binary operation on S, and S together with this binary operation might be a group. If it is, we say that S is a ***subgroup*** of G. Note that for any group G, G itself is a subgroup of G, and $\{e\}$ is a subgroup of G. A subgroup of G that is not $\{e\}$ or G is called a ***proper subgroup*** of G.

In order to give an example of a group, we must specify two things, a set G and a binary operation $\cdot$ on that set. Then we must make sure that $\cdot$ satisfies (a), (b), and (c) of 2.1.1.

2.1.2 Examples of groups

(a) We have seen two examples already—the set $\mathbb{Z}$ of integers with ordinary addition as the operation, and the set Perm(S) of all permutations of a set S with composition of mappings as the operation. If $S = \{1, 2, 3, \ldots, n\}$, then Perm($S$) is denoted S_n and is called the ***symmetric group of degree n***.

(b) The set $\mathbb{Q}^*$ of nonzero rational numbers with ordinary multiplication of numbers is an Abelian group. The identity is 1 and the inverse of an element $x \in \mathbb{Q}^*$ is $1/x$.

(c) The set $\mathbb{Q}^+$ of positive rational numbers with multiplication of numbers is an Abelian group. $\mathbb{Q}^+$ is a subgroup of $\mathbb{Q}^*$.

(d) The sets $\mathbb{R}^*$ and $\mathbb{R}^+$ of nonzero and positive real numbers, respectively, with ordinary multiplication of numbers, are groups. Note that $\mathbb{Q}^*$ and $\mathbb{R}^+$ are subgroups of $\mathbb{R}^*$, and that $\mathbb{Q}^+$ is a subgroup of $\mathbb{R}^+$. However, $\mathbb{Z}^+$ is not a subgroup of $\mathbb{Q}^+$.

(e) For $n > 0$, let $\mathbb{Z}(n) = \{0, 1, 2, \ldots, n-1\}$. Define the binary operations $+_n$ and $\cdot_n$ on $\mathbb{Z}(n)$ as follows. For x and y in $\mathbb{Z}(n)$, write $x + y = nq + r$ with $0 \leq r < n$, and let $x +_n y = r$. Write $xy = nq_1 + s$ with $0 \leq s < n$, and let $x \cdot_n y = s$. Then $+_n$ and $\cdot_n$ are associative operations. (See Problems 5–9 below.)

$\mathbb{Z}(n)$ with the operation $+_n$ is an Abelian group called the ***group of integers modulo*** n. We will view this group a bit differently later. (See 2.3.12.)

For $n > 1$, let $U(n)$ be the elements of $\mathbb{Z}(n)$ that are relatively prime to n. Then $U(n)$ with the operation $\cdot_n$ is a group called the ***group of units of*** $\mathbb{Z}(n)$.

(f) Let G be the set of all 2×2 matrices

$$\begin{pmatrix} a_{11} & a_{12} \\ a_{21} & a_{22} \end{pmatrix}$$

with $a_{ij} \in \mathbb{R}$, and whose determinant $a_{11}a_{22} - a_{12}a_{21} \neq 0$. Define $\cdot$ as the usual matrix multiplication

$$\begin{pmatrix} a_{11} & a_{12} \\ a_{21} & a_{22} \end{pmatrix} \cdot \begin{pmatrix} b_{11} & b_{12} \\ b_{21} & b_{22} \end{pmatrix} = \begin{pmatrix} a_{11}b_{11} + a_{12}b_{21} & a_{11}b_{12} + a_{12}b_{22} \\ a_{21}b_{11} + a_{22}b_{21} & a_{21}b_{12} + a_{22}b_{22} \end{pmatrix}.$$

Then G with $\cdot$ is a group. The matrix

$$\begin{pmatrix} 1 & 0 \\ 0 & 1 \end{pmatrix}$$

is the identity, and since $a_{11}a_{22} - a_{12}a_{21} \neq 0$, each element of G has an inverse. This group is called the ***general linear group*** of 2×2 real matrices and is denoted $\mathrm{GL}_2(\mathbb{R})$. Those elements of $\mathrm{GL}_2(\mathbb{R})$

with determinant 1 form a subgroup that is called the ***special linear group*** and is denoted $SL_2(\mathbb{R})$.

(g) Let $G = \{e, a\}$, and let $\cdot$ be given by the "multiplication" table

	e	a
e	e	a
a	a	e

Then G with the binary operation $\cdot$ is a group. This may be verified by brute force. For example, to see if $\cdot$ is associative, try all possibilities. Clearly e is the identity element of G. Also, $e = e^{-1}$ and $a = a^{-1}$.

(h) The set $\{1, -1, i, -i\}$ of complex numbers with ordinary multiplication of complex numbers is a group. More generally, the set $\{e^{k\pi i/n}: k = 0, 1, \ldots, n-1\}$ of all nth roots of 1 with ordinary multiplication of complex numbers is a group with n elements. ($e^{k\pi i/n} = \cos k\pi/n + i \sin k\pi/n$.)

The multiplication table for the group of 4th roots of 1 is

	1	-1	i	$-i$
1	1	-1	i	$-i$
-1	-1	1	$-i$	i
i	i	$-i$	-1	1
$-i$	$-i$	i	1	-1

Multiplication tables for groups are also called ***Cayley tables***.

(i) Let S be any set, and let $P(S)$ be the set of all subsets of S. For $A, B \in P(S)$, let $A \cdot B = (A \cup B) \setminus (A \cap B)$. The associativity of $\cdot$ is Problem 19, Section 1.2. Clearly $\varnothing$ is the identity element, and each element is its own inverse. (See Problem 21, Section 1.2.)

(j) Let S be any set, and let G be the set of all mappings from S into $\mathbb{R}$. For $g, h \in G$, define $(g \cdot h)(s) = g(s) + h(s)$. Then G with $\cdot$ is a group. Again, $\cdot$ is usually denoted $+$. If $S = \mathbb{R}$, then $H = \{g \in G: g$ is continuous$\}$ is a subgroup of G.

(k) Let p and q be in $\mathbb{R} \times \mathbb{R}$. That is, p and q are each pairs of real numbers. Let $d(p, q)$ denote the distance between p and q. Let G be the set of all functions f on $\mathbb{R} \times \mathbb{R}$ that preserve d, that is, all f such that $d(p, q) = d(f(p), f(q))$ for all p and q in $\mathbb{R} \times \mathbb{R}$. Let $\cdot$ be composition of functions. Then G with $\cdot$ is a non-Abelian group. That G is a group is clear once one proves that if $f \in G$, then $f^{-1} \in G$. (See Problem 10 below.) It is non-Abelian since the functions f and g given by $f(x, y) = (y, x)$ and $g(x, y) = (x + 1, y)$ do not commute and are in G. This group is called the ***group of isometries***, or the ***Euclidean group*** of the plane.

(l) Let G be the set whose elements are the following "motions" of the square S.

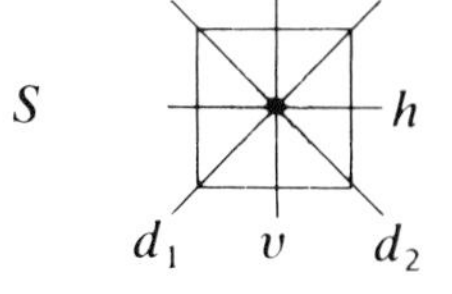

(1) The reflection H of S about h;
(2) the reflection V of S about v;
(3) the reflection D_1 of S about d_1;
(4) the reflection D_2 of S about d_2;
(5) the rotation R_1 of S counterclockwise through 90°;
(6) the rotation R_2 of S counterclockwise through 180°;
(7) the rotation R_3 of S counterclockwise through 270°; and
(8) the rotation E of S counterclockwise through 0°.

Let $\cdot$ be composition of motions. For example, $D_1 \cdot R_2$ is the motion of S resulting from the motion R_2 followed by D_1. It is easily seen that $D_1 \cdot R_2 = D_2$. With patience, it can be shown that G with $\cdot$ is a group. It is called the group of ***symmetries of a square***. It is non-Abelian.

Every regular polygon has an analogous group of symmetries. If a regular polygon has n sides, its group of symmetries has $2n$ elements, is called a ***dihedral group***, and is denoted D_n. Thus, the group of symmetries of the square is the dihedral group D_4 and indeed does have 8 elements. (See Problem 13 on page 34.)

The examples above should convince you that groups occur in many mathematical contexts. Since they do occur so frequently, it is worthwhile to study them "in the abstract." The knowledge gained in such a study can then be applied in many places, namely wherever a group occurs.

In spite of the fact that a group is a set G together with a binary operation on G satisfying certain conditions, most often just G itself is referred to as the group. Thus when we say "G is a group," we mean that G is a set on which there is a binary operation satisfying the requirements (a), (b), and (c) of 2.1.1.

There are many binary operations on a set which make that set into a group. For example, if G is a group with a binary operation $\cdot$, and f is a one-to-one mapping of G onto G, then $x * y = f^{-1}(f(x) \cdot f(y))$ defines a new binary operation $*$ on G, and G together with $*$ is a group. (See Problem 17 on page 34.)

For a, b in a group G, it is the usual custom to write ab instead of $a \cdot b$, and ab is read "a times b." We will follow this custom.

Suppose G is a group and $a, b, c, d \in G$. Then we can form the products $a(b(cd))$, $((ab)c)d)$, $(a(bc))d$, and $a((bc)d)$. Now using the associative law [(a) in 2.1.1], we can conclude that all these possibilities are equal. For example, $a(b(cd)) = a((bc)d) = (a(bc))d = ((ab)c)d$ by repeated application of the associative law. Thus, we may as well write $abcd$ for these products.

The "generalized" associative law holds. That is, suppose $a_1, a_2, \ldots, a_n \in G$. We may form the products

$$a_1(a_2(a_3 \cdots a_n)) \cdots),$$

$$(\cdots((a_1a_2) \cdots a_{n-1})a_n,$$

and

$$(a_1a_2)(a_3(a_4 \cdots a_n)) \cdots),$$

for example, and if n is very large then there are many, many such possibilities. We can get from any one of these products to another by repeated applications of 2.1.1(a). A formal proof may be effected by induction on n, and can be found in Jacobson, page 39, for example. Any product such as those above can be written simply $a_1a_2 \cdots a_n$ since there is no possible ambiguity.

Suppose $a \in G$. Then we have the products aa, aaa, $aaaa$, and so on. It is reasonable to denote these by a^2, a^3, a^4, and so on. Also a to negative powers may be defined, and certain useful laws of exponents hold. The following definition is an example of a simple definition by induction. Remember that a^{-1} is the inverse of the element a.

2.1.3 Definition

Let G be a group, and let $a \in G$. Then $a^0 = e$, and for $n \geq 1$, $a^n = a^{n-1}a$. For negative integers n, $a^n = (a^{-n})^{-1}$.

In 2.1.3, a^0 is defined, and for $n \geq 1$, a^n is defined whenever a^{n-1} is defined. Therefore, a^n is defined for all nonnegative integers n. We have already attached a meaning to a^{-1}. It is the inverse of a. Therefore, we can define $a^n = (a^{-n})^{-1}$ for negative integers n, and so we have defined a^n for all integers n. [Note that for $n = -1$, the definition $a^n = (a^{-n})^{-1}$ is compatible with our previous definition of a^{-1}.] The "laws of exponents" that can be expected to hold do hold.

2.1.4 Theorem

Let G be a group and let $a, b \in G$. Then

(a) $a^m a^n = a^{m+n}$ *for all* $m, n \in \mathbb{Z}$,

(b) $(a^m)^n = a^{mn}$ *for all* $m, n \in \mathbb{Z}$ *and*

(c) $(ab)^n = (b^{-1}a^{-1})^{-n}$ *for all* $n \in \mathbb{Z}$.

Proof To prove (a), we first prove that $aa^m = a^{m+1}$ for all $m \geq 0$. To do this, we induct on m. For $m = 0$, it is true. If $m \geq 0$ and $aa^m = a^{m+1}$, then

$$aa^{m+1} = (aa^m)a = a^{m+1}a = a^{m+1+1} = a^{m+2}.$$

Thus, $aa^m = a^{m+1}$ for all $m \geq 0$. Next, we show that

$$(a^{-1})^m = (a^m)^{-1} = a^{-m}$$

for all $m \in \mathbb{Z}$. First, we show by induction that this is so for $m \geq 0$. For $m = 0$, it is clearly true. If $m \geq 0$, and

$$(a^{-1})^m = (a^m)^{-1} = a^{-m},$$

then

$$\begin{aligned}(a^{-1})^{m+1} &= (a^{-1})^m a^{-1} = a^{-m}a^{-1} = (a^m)^{-1}a^{-1} \\ &= (aa^m)^{-1} = (a^{m+1})^{-1} = a^{-(m+1)}.\end{aligned}$$

Thus, $(a^{-1})^m = (a^m)^{-1} = a^{-m}$ for all $m \geq 0$. If $m < 0$, then $(a^{-1})^m = ((a^{-1})^{-m})^{-1} = ((a^{-m})^{-1})^{-1} = a^{-m} = (a^m)^{-1}$. Thus, $(a^{-1})^m = (a^m)^{-1} = a^{-m}$ for all $m \in \mathbb{Z}$. Now we show that

$$aa^m = a^{m+1}$$

for all $m < 0$. To show this we need to use the fact that

$$a^{-1}a^k = a^{k-1}$$

for all $k > 0$. Since $k - 1 \geq 0$, this follows from what we have already proved: $a^{k-1} = (a^{-1}a)a^{k-1} = a^{-1}(aa^{k-1}) = a^{-1}a^k$. Now if $m < 0$, then $aa^m = (a^{-1})^{-1}(a^{-1})^{-m} = (a^{-1})^{-m-1} = a^{m+1}$.

Finally, we show that

$$a^m a^n = a^{m+n}.$$

First we show that for all n, it is true for all $m \geq 0$. If $m = 0$, it is true. If $a^m a^n = a^{m+n}$ for all n, then

$$a^{m+1}a^n = a^m(aa^n) = a^m a^{n+1} = a^{m+n+1} = a^{(m+1)+n},$$

and so $a^m a^n = a^{m+n}$ for all $m \geq 0$ and for all n. Now suppose that $m < 0$. Then $a^m a^n = (a^{-1})^{-m}(a^{-1})^{-n} = (a^{-1})^{-m-n} = a^{m+n}$, and (a) is proved.

To prove (b), we show first that

$$(a^m)^n = a^{mn}$$

for all $n \geq 0$. For $n = 0$, it is true. If $n \geq 0$ and $(a^m)^n = a^{mn}$, then $(a^m)^{n+1} = (a^m)^n a^m = a^{mn}a^m = a^{mn+m} = a^{m(n+1)}$, whence $(a^m)^n = a^{mn}$ for all $m \in \mathbb{Z}$ and for all $n \geq 0$. If $n < 0$, then $(a^m)^n = ((a^m)^{-n})^{-1} = (a^{-mn})^{-1} = a^{mn}$.

To prove (c), note that $(ab)^{-1} = b^{-1}a^{-1}$. Then for any $n \in \mathbb{Z}$, we have $(ab)^n = ((ab)^{-1})^{-n} = (b^{-1}a^{-1})^{-n}$.

PROBLEMS

1 Prove that $\mathbb{Z}$ is not a group under subtraction.

2 Define the operation $*$ on $\mathbb{Z}$ by $a * b = a + b + 1$. Show that $\mathbb{Z}$ together wtih this operation is an Abelian group.

3 Let $G = \mathbb{Q}\backslash\{-1\}$. Define $*$ on G by $x * y = x + y + xy$. Prove that G together with $*$ is an Abelian group.

4 Let G be the rational numbers whose denominators are a power of the prime p, when reduced to lowest terms. Prove that G is a group under $+$.

5 Prove in detail that $\mathbb{Z}(n)$ is an Abelian group under the operation $+_n$.

6 Prove that $\mathbb{Z}(n)$ with the operation $\cdot_n$ is not a group unless $n = 1$.

7 Let $\mathbb{Z}(n)^* = \mathbb{Z}(n)\backslash\{0\}$. Prove that $\mathbb{Z}(n)^*$ under $\cdot_n$ is a group if and only if n is a prime.

8 Prove in detail that $U(n)$ is an Abelian group under the operation $\cdot_n$.

9 Write down a multiplication table for $\mathbb{Z}(6)$ and for $U(12)$.

10 Prove that in 2.1.2(k), G is a group.

11 Let G be a group with operation $*$, and let $g \in G$. Let $S = \{g^n : n \in \mathbb{Z}\}$. Prove that S is a group under $*$.

12 Prove that $\mathbb{R}^* \times \mathbb{R}$ is a group under the operation given by the rule $(a, b)(c, d) = (ac, bc + d)$.

13 A symmetry of a regular polygon can be viewed as a permutation of the vertices of that polygon that carries adjacent vertices to adjacent vertices. Prove that the set of symmetries of a regular polygon is a group under composition of permutations.

14 For elements a, b in a group G, prove that the equation $ax = b$ has exactly one solution for x in G. Prove a similar fact for the equation $xa = b$.

15 Prove that the *cancellation laws* hold for groups. That is, if G is a group, and $a, b, c \in G$, prove that $ab = ac$ implies that $b = c$, and that $ab = cb$ implies that $a = c$.

16 Let G be a group. Prove that left multiplication is a permutation of G. That is, for $a \in G$, prove that $G \to G: b \to ab$ is a one-to-one mapping from G onto G. Prove a similar fact for right multiplication.

17 Suppose G is a group and f is a one-to-one mapping from G onto the set S. Prove that defining a multiplication $*$ on S by $x * y = f(f^{-1}(x)f^{-1}(y))$ makes S into a group.

18 Let G be a group, $g \in G$, and define a new multiplication $*$ on G as follows. For $a, b \in G$, let $a * b = agb$. Prove that G with $*$ is a group.

19 Suppose G is a set and $\cdot$ is an associative binary operation on G such that

there is an element $e \in G$ with $e \cdot a = a$ for all $a \in G$, and such that for each $a \in G$ there is an element $b \in G$ with $b \cdot a = e$. Prove that G with such a binary operation is a group. That is, it is enough to assume associativity, a left identity, and left inverses, in order to have a group. Similarly, it is enough to assume associativity, a right identity, and right inverses.

20 Let G be the nonzero real numbers, and for x and y in G, define $x * y = |x|y$. Prove that this is an associative binary operation on G, there is a left identity, elements have right inverses, but that G is not a group under $*$.

21 Let G and H be groups. Prove that $G \times H$ with multiplication given by $(a_1, b_1)(a_2, b_2) = (a_1a_2, b_1b_2)$ is a group. ($G \times H$ is called the *direct product* of G and H. Direct products are studied in 2.6.)

22 Give an example of a group with 4 elements. Give an example of a group with 5 elements. Give an example of a group with 5^{1000} elements.

23 Prove that if $a^2 = e$ for all a in the group G, then G is Abelian.

24 Prove that if for all a, b in a group G, $(ab)^2 = a^2b^2$, then G is Abelian.

25 Prove that the group Perm(S) is commutative if and only if S has fewer than 3 elements.

26 For each positive integer $n \geq 3$, prove that there is a non-Abelian group with $n!$ elements.

27 Let a and b be elements of a group G, and suppose that $ab = ba$. Prove that for $n \in \mathbb{Z}$, $(ab)^n = a^nb^n$. Let $a_1, a_2, \ldots, a_k$ be elements of G with $a_ia_j = a_ja_i$ for all i and j. Prove that for $n \in \mathbb{Z}$, $(\prod_i a_i)^n = \prod_i a_i^n$.

28 Let G be a group, and let $n \in \mathbb{Z}$. Prove that the mapping $G \to G: a \to a^n$ is one-to-one and onto if $n = \pm 1$, and may or may not be otherwise.

29 Let a and b be elements of a group G. Derive from 2.1.4 that $(ab)^{-1} = b^{-1}a^{-1}$. Verify this for the elements H and V, and for the elements R_2 and D_2 of the dihedral group D_4 given in 2.1.2(l).

2.2 SUBGROUPS AND COSETS

Let G be a group. A subgroup of G has been defined to be a subset S of G such that when multiplication is restricted to elements of S, S is a group. Now suppose that S is a subgroup of G. If $s, t \in S$, then $st \in S$ since multiplication must be a binary operation on S. The fact that $st \in S$ whenever $s, t \in S$ is expressed by saying that S is ***closed*** under multiplication. Since S is a subgroup of G, S is a group, and so has an identity e_S. But then $e_Se_S = e_Se_G$, and cancelling e_S yields $e_S = e_G$. Furthermore, for $s \in S$, s has an inverse t as an element of the group S, and an inverse u as an element of the group G. But

$st = e_S = su$, and cancelling s gets $t = u$. Thus, for $s \in S$, s^{-1} is unambiguous. The following theorem is quite useful.

2.2.1 Theorem

Let G be a group, and let S be a nonempty subset of G. Then S is a subgroup of G if and only if for $s, t \in S$, $st \in S$ and $s^{-1} \in S$.

Proof We have already observed that if S is a subgroup of G, then st and s^{-1} are in S whenever $s, t \in S$. Conversely, suppose that st and s^{-1} are in S whenever $s, t \in S$. Since $st \in S$ if $s, t \in S$, multiplication is a binary operation on S. It is associative on S since it is associative on G. Since $S \neq \emptyset$, there is an element $s \in S$. Then $s^{-1} \in S$, and hence $ss^{-1} = e \in S$. Thus, S has an identity element e (the identity for G), and every element $s \in S$ has an inverse in S (its inverse as an element of G). Thus, S is a subgroup of G.

It is also true that a nonempty subset S of a group G is a subgroup if and only if for $s, t \in S$, $st^{-1} \in S$. Furthermore, a nonempty finite subset S of G is a subgroup if and only if S is closed under multiplication. These are left as exercises.

If S and T are subgroups of G, then $S \cap T$ is a subgroup of G. It is nonempty since $e \in S \cap T$. If $x, y \in S \cap T$, then $xy, x^{-1} \in S$ and $xy, x^{-1} \in T$ since $x, y \in S$ and $x, y \in T$, and S and T are subgroups. Therefore, we have $xy, x^{-1} \in S \cap T$, whence $S \cap T$ is a subgroup. Something much better holds.

2.2.2 Theorem

Let $\{S_i\}_{i \in I}$ be a family of subgroups of the group G, with $I \neq \emptyset$. Then $\bigcap_{i \in I} S_i$ is a subgroup of G.

Proof Suppose $x, y \in \bigcap_{i \in I} S_i = S$. Then $x, y \in S_i$ for all $i \in I$. Thus $xy, x^{-1} \in S_i$ for all $i \in I$, and so $xy, x^{-1} \in S$. Since $e \in S_i$ for all $i \in I$, $S \neq \emptyset$. Thus, S is a subgroup of G.

Of course 2.2.2 implies that the intersection of any nonempty set of subgroups of G is a subgroup of G.

As trivial as it is to prove, 2.2.2 has some worthwhile consequences. For example, let G be a group, and let S be a subset of G. Then there is a unique smallest subgroup T of G such that $S \subset T$. This means that there is a subgroup T of G such that $S \subset T$, and if V is any subgroup of G containing S, then $T \subset V$. We simply let T be the intersection of the set of all subgroups of G which contain S.

2.2.3 Definition

Let G be a group, and let S be a subset of G. Then the smallest subgroup of G containing S is called the **subgroup generated by** *S and is denoted $\langle S \rangle$. The group G is* **cyclic** *if there is an element $g \in G$ such that $G = \langle \{g\} \rangle$, and G is* **finitely generated** *if there is a finite subset S of G such that $G = \langle S \rangle$.*

If $g_1, g_2, \ldots, g_n \in G$, then the group

$$\langle \{g_1, g_2, \ldots, g_n\} \rangle$$

is written simply as $\langle g_1, g_2, \ldots, g_n \rangle$. In particular, $\langle \{g\} \rangle = \langle g \rangle$.

For example, the group $\mathbb{Z}$ of integers under $+$ is cyclic since $\mathbb{Z} = \langle 1 \rangle$. In that group, $\langle 6, 8 \rangle = \langle 2 \rangle$, which is the set of all even integers. By Problem 28 below, the additive group of rational numbers is not finitely generated.

It is useful to know the elements of $\langle S \rangle$ in terms of the elements of S.

2.2.4 Theorem

If S is a nonempty subset of a group G, then $\langle S \rangle$ is the set of all possible products of powers of elements of S.

Proof Since $\langle S \rangle$ is a group, then for $s \in S$ and $n \in \mathbb{Z}^+$, $s^n \in \langle S \rangle$, because $\langle S \rangle$ is closed under multiplication. But $(s^n)^{-1} = s^{-n} \in \langle S \rangle$, whence $s^m \in \langle S \rangle$ for all $m \in \mathbb{Z}$. Again, since $\langle S \rangle$ is closed under multiplication, $\langle S \rangle$ must contain all products of such s^m. It should be clear that the set of such products is a subgroup of G. For example, the product of two products of powers of elements of S is a product of powers of elements of S. The theorem follows.

2.2.5 Corollary

If $g \in G$, then $\langle g \rangle = \{g^n : n \in \mathbb{Z}\}$.

Thus the subgroup of G generated by g is just the set of all powers of g. In additive notation, which is often used if the group is Abelian, the subgroup of G generated by g is just the set of all multiples ng of g.

2.2.6 Theorem

A subgroup of a cyclic group is a cyclic group.

Proof Let S be a subgroup of the cyclic group $G = \langle g \rangle$. If $S = \{e\}$, then S is cyclic. If $S \neq \{e\}$, then $g^n \in S$ for some $n \neq 0$, whence $g^{-n} \in S$. Thus,

$g^n \in S$ for some $n > 0$. Let m be the smallest positive integer such that $g^m \in S$. Let $s \in S$. Then $s = g^n$ for some $n \in \mathbb{Z}$, and by the division algorithm, $n = mq + r$ with $0 \leq r < m$. Since g^n, $g^{mq} \in S$, then $g^n(g^{mq})^{-1} = g^n(g^{-mq}) = g^{n-mq} = g^r \in S$. Since $0 \leq r < m$ and m is the smallest positive integer such that $g^m \in S$, it follows that $r = 0$ and that $g^n = (g^m)^q$. We have shown that every element in S is a power of g^m. Since $\langle g^m \rangle \subset S$ is clear, $S = \langle g^m \rangle$, and so S is cyclic.

Note that 2.2.6 implies that every subgroup of the additive group $\mathbb{Z}$ is cyclic. (See Problem 5 below.)

There is a number associated with each element g in a group G, namely the number of elements in $\langle g \rangle$.

2.2.7 Definition

Let G be a group and $g \in G$. If $\langle g \rangle$ is finite, then the **order** *of g is the number of elements in $\langle g \rangle$. If $\langle g \rangle$ is infinite, then the order of g is infinite. The order of g is denoted $o(g)$, and if $o(g)$ is infinite, we write $o(g) = \infty$.*

2.2.8 Theorem

Let G be a group and $g \in G$. Then $o(g)$ is **finite** *if and only if there is a positive integer n such that $g^n = e$. In the case that $o(g)$ is finite, $o(g)$ is the smallest such n.*

Proof If $\langle g \rangle = \{g^m : m \in \mathbb{Z}\}$ is finite, then there are two distinct integers m, n such that $g^m = g^n$. Thus $g^{m-n} = e = g^{n-m}$, so there is a positive integer k such that $g^k = e$. Conversely, suppose that there is a positive integer k such that $g^k = e$. Let n be the smallest such positive integer. For $m \in \mathbb{Z}$, $m = nq + r$ with $0 \leq r < n$, and $g^m = (g^n)^q g^r = eg^r = g^r$. Therefore $\langle g \rangle = \{g^r : 0 \leq r < n\}$, and so $o(g)$ is finite. If $g^r = g^s$, with $0 \leq r < n$, $0 \leq s < n$, then $g^{r-s} = e = g^{s-r}$, and $0 \leq |r - s| < n$, whence $r = s$. Therefore, $o(g) = n$.

The number of elements in a finite group G is called the ***order*** of G and is denoted $o(G)$. If G is infinite, we write $o(G) = \infty$.

We pause here to discuss an example illustrating some of the concepts just introduced. We will see this example again from time to time.

Let D be the set of all 2×2 matrices with entries from the complex numbers $\mathbb{C}$, and of the form

$$\begin{pmatrix} a & b \\ -\bar{b} & \bar{a} \end{pmatrix},$$

where $\bar{x}$ denotes the complex conjugate of x. It is easily checked that the product of two such matrices has the same form. The determinant $a\bar{a} + b\bar{b}$ is not 0 unless both a and b are 0. Thus, a nonzero matrix of this form has an inverse, and that inverse turns out to be of the same form. It follows that the set D^* of all such nonzero matrices is a group under matrix multiplication. Note also that if a matrix

$$M = \begin{pmatrix} a & b \\ c & d \end{pmatrix}$$

is in D^*, then so is

$$-M = \begin{pmatrix} -a & -b \\ -c & -d \end{pmatrix}.$$

The matrices

$$I = \begin{pmatrix} 0 & 1 \\ -1 & 0 \end{pmatrix} \quad \text{and} \quad J = \begin{pmatrix} 0 & i \\ i & 0 \end{pmatrix}$$

are in D^*. What is the subgroup $\langle I, J \rangle$ of D^* generated by I and J? By 2.2.4, it is the set of all possible products of powers of I and J. Denoting

$$\begin{pmatrix} 1 & 0 \\ 0 & 1 \end{pmatrix} \quad \text{and} \quad \begin{pmatrix} i & 0 \\ 0 & -i \end{pmatrix}$$

simply by 1 and K, respectively, we have that $IJ = K$, $JK = I$, $KI = J$, $JI = -K$, $KJ = -I$, $IK = -J$, and $I^2 = J^2 = K^2 = -1$. Thus, any product of powers of I and J is one of the 8 elements ± 1, $\pm I$, $\pm J$, and $\pm K$. For example, $I^5J^6I^2 = IJ^2(-1) = -KJ = I$. The 8 elements are distinct, and this group of order 8 is the ***quaternion group***, which we will denote Q_8.

The elements $\pm I$, $\pm J$, and $\pm K$ are all of order 4, while 1 is of order 1 and -1 is of order 2. In the dihedral group D_4 [2.1.2(l)], which is also of order 8, there are only 2 elements of order 4, so that these two groups, Q_8 and D_4, both non-Abelian of order 8, are not alike.

Suppose that $\langle g \rangle$ is finite. If S is a subgroup of $\langle g \rangle$, then $S = \langle g^m \rangle$, where m is the smallest positive integer such that $g^m \in S$. (See 2.2.6.) If $o(g) = n$, then $n = mk + r$, with $0 \leq r < m$, and $g^{n-mk} = g^r \in S$. Thus $r = 0$, and hence m divides n. We then have the fact that if $\langle g \rangle$ is finite, then the number of elements in any subgroup of $\langle g \rangle$ divides the number of elements in $\langle g \rangle$. However, this is a very special case of the more general fact that the number of elements in a subgroup of a finite group divides the number of elements in that finite group. We now begin the development of this fact.

2.2.9 Definition

Let S be a subgroup of G, and let $g \in G$. Then $gS = \{gs : s \in S\}$ is the **left coset** *of S determined by g, and $Sg = \{sg : s \in S\}$ is the* **right coset** *of S determined by g.*

Of course, if G is Abelian, then the left coset gS is the same as the right coset Sg. Distinct elements of G may determine the same left coset of a subgroup S. For example, if $g \in S$ and $g \neq e$, then $gS = S = eS$.

For the group $\mathbb{Z}$ of integers under $+$, the subgroup of all multiples of 3 has just the three cosets

$$\{\ldots, -9, -6, -3, 0, 3, 6, 9, \ldots\},$$

$$\{\ldots, -8, -5, -2, 1, 4, 7, 10, \ldots\},$$

$$\{\ldots, -7, -4, -1, 2, 5, 8, 11, \ldots\}.$$

These cosets are disjoint from each other, and their union is all of $\mathbb{Z}$. This is a fundamental property of left (or right) cosets of any subgroup of any group.

2.2.10 Theorem

Let S be a subgroup of G. Then the set of left (or right) cosets of S is a partition of G.

Proof Since $g \in gS$, $\bigcup_{g \in G} gS = G$. We need only show that for $g, h \in G$, $gS \cap hS \neq \varnothing$ implies that $gS = hS$. If $gs = ht$ for $s, t \in S$, then $g = hts^{-1} \in hS$, and for any $x \in S$, $gx = hts^{-1}x \in hS$. Thus $gS \supset hS$. Similarly, $hS \subset gS$, and so $gS = hS$.

Note that the equivalence relation induced by this partition is given by $g \sim h$ if g and h are in the same left coset of S. Since $g \in gS$ and $h \in hS$, this means that $g \sim h$ if and only if $gS = hS$, and this is the case if and only if $g^{-1}h \in S$. (Keep in mind that two cosets are either equal or disjoint.) Thus, the equivalence relation is

$$g \sim h \quad \text{if and only if} \quad g^{-1}h \in S.$$

The mapping $S \to gS: s \to gs$ is one-to-one and onto. That it is onto is clear, and it is one-to-one since $gs = gt$ implies that $s = t$. Thus, there is a one-to-one correspondence between any two left cosets. If G is a finite group, and S a subgroup of G, the left cosets of S partition G into subsets all of the same size as S. Therefore, $o(G)$ equals $o(S)$ times the number of left cosets of S. Thus, we see that the following theorem is true.

2.2.11 Lagrange's Theorem

If S is a subgroup of the finite group G, then $o(S)$ divides $o(G)$.

2.2.12 Corollary

If G is a finite group and $g \in G$, then $g^{o(G)} = e$.

Proof By 2.2.7 and 2.2.11, $o(g)$ divides $o(G)$.

2.2.13 Corollary

If $o(G)$ is prime, then G is cyclic, and the only subgroups of G are $\{e\}$ and G.

For any subset S of a group G, let $S^{-1} = \{s^{-1}: s \in S\}$.

2.2.14 Theorem

Let L be the set of all left cosets of a subgroup S of a group G, and let R be the set of all right cosets of S. Then $L \to R: C \to C^{-1}$ is a one-to-one correspondence.

Proof The left coset C is gS for some $g \in G$. But $C^{-1} = \{(gs)^{-1}: s \in S\} = \{s^{-1}g^{-1}: s \in S\} = \{sg^{-1}: s \in S\} = Sg^{-1}$. Thus $L \to R: C \to C^{-1}$ is indeed a map from L into R, and it is clearly one-to-one and onto.

In particular, if S has a finite number of left cosets, it has the same number of right cosets. This number is denoted $G{:}S$, and called the ***index*** of S in G. Thus for finite groups G, we have

$$o(G) = o(S)(G{:}S).$$

Suppose S is a subgroup of G. An element in a coset of S is called a ***representative*** of that coset. A ***set of representatives*** of the left cosets of S in G is a set consisting of one representative of each left coset of S in G. Note that if g is a representative of a left coset C of S, then $C = gS$ since $g \in gS$.

Now suppose that H is of finite index in G and K is of finite index in H. We have the three indices $G{:}K$, $G{:}H$, and $H{:}K$. They are related by a simple formula.

2.2.15 Theorem

If H is a subgroup of finite index in G, and K is a subgroup of finite index in H, then

$$G{:}K = (G{:}H)(H{:}K).$$

This follows from the more explicit result below.

2.2.16 Theorem

Let H be a subgroup of G, and let $\{g_i\}_{i \in I}$ be a set of representatives of the left cosets of H in G. Let K be a subgroup of H, and let $\{h_j\}_{j \in J}$ be a set of representatives of the left cosets of K in H. Then $\{g_i h_j: (i, j) \in I \times J\}$ is a set of representatives of the left cosets of K in G.

Proof We need to show that if C is a left coset of K in G, then $C = g_i h_j K$ for some $(i, j) \in I \times J$, and that if $(i, j) \neq (m, n)$, then $g_i h_j K \neq g_m h_n K$. For any coset gK of K in G, $gH = g_i H$ for some $i \in I$. Thus, $g = g_i h$ for some $h \in H$. But $hK = h_j K$ for some $j \in J$. We have $gK = g_i h K = g_i h_j K$.

Now suppose that $g_i h_j K = g_m h_n K$. Then $g_i H = g_m H$ since $g_i h_j = g_m h_n k$ for some $k \in K$, and so $g_i h_j \in (g_i H) \cap (g_m H)$. Thus $g_i = g_m$, and hence $h_j K = h_n K$. Therefore, $h_j = h_n$ and the theorem is proved.

For the dihedral group D_4, the left cosets of its subgroup $\{E, H\}$ are

$$\{E, H\}, \{V, R_2\}, \{D_1, R_1\}, \quad \text{and} \quad \{D_2, R_3\},$$

and the right cosets of $\{E, H\}$ are

$$\{E, H\}, \{V, R_2\}, \{D_1, R_3\}, \quad \text{and} \quad \{D_2, R_1\}.$$

The left cosets of the subgroup $T = \{E, R_1, R_2, R_3\}$ are just T and its complement $\{V, H, D_1, D_2\}$, and its right cosets are also just T and its complement.

The subgroup $SL_2(\mathbb{R})$ of the general linear group $GL_2(\mathbb{R})$ has infinitely many left cosets. In fact, each nonzero diagonal matrix

$$\text{diag}(a) = \begin{pmatrix} a & 0 \\ 0 & a \end{pmatrix}$$

with positive a gives distinct left cosets of $SL_2(\mathbb{R})$. Simply note that every element in $\text{diag}(a)SL_2(\mathbb{R})$ has determinant a^2.

PROBLEMS

1 Write down all subgroups of the group $\{1, -1, i, -i\}$ of 4th roots of 1.

2 Write down all subgroups of the group of 6th roots of 1.

3 Write down all subgroups of the dihedral group D_4.

4 Write down all subgroups of the quaternion group Q_8.

5 Let $\mathbb{Z}n = \{nz : z \in \mathbb{Z}\}$.
(a) Prove that $\mathbb{Z}n$ is a subgroup of the additive group of integers $\mathbb{Z}$.
(b) Prove that every subgroup of $\mathbb{Z}$ is of this form.
(c) Prove that there is a natural one-to-one correspondence between the subgroups of $\mathbb{Z}$ and the natural numbers $\mathbb{N}$.
(d) Prove that $\mathbb{Z}n$ has exactly n cosets. Write them down.

6 In the dihedral group D_4, write down the left cosets and the right cosets of the subgroup $\{E, D_1\}$.

7 In the quaternion group Q_8, write down the left cosets and the right cosets of the subgroup $\{1, -1\}$.

8 Compute the order of each element in the group $U(15)$.

9 Find the order of each element in the multiplicative group $\mathbb{Q}^*$ of rational numbers.

10 Find a set of representatives of the cosets of $\mathbb{Q}^+$ in $\mathbb{Q}^*$.

11 Find a set of representatives of the left cosets of $\mathrm{SL}_2(\mathbb{R})$ in $\mathrm{GL}_2(\mathbb{R})$. Find such a set for the right cosets.

12 Prove that a nonempty subset S of a group G is a subgroup of G if and only if $st^{-1} \in S$ whenever $s, t \in S$.

13 Prove that in $\mathbb{Z}$, $\langle n_1, n_2, \ldots, n_k\rangle = \langle (n_1, n_2, \ldots, n_k)\rangle$.

14 Prove that a finite nonempty subset S of a group G is a subgroup of G if and only if $st \in S$ whenever $s, t \in S$.

15 What subgroup of a group does $\varnothing$ generate?

16 Let S and T be subsets of a group G. Prove that
(a) if $S \subset T$ then $\langle S\rangle \subset \langle T\rangle$,
(b) $\langle S \cup T\rangle = \langle \langle S\rangle \cup \langle T\rangle\rangle$,
(c) $\langle S \cap T\rangle \subset \langle S\rangle \cap \langle T\rangle$, and
(d) $\langle\langle S\rangle\rangle = \langle S\rangle$.

17 If S and T are subgroups of a group G, then $S \cup T$ is a subgroup if and only if $S \subset T$ or $T \subset S$.

18 Let $\{S_i\}_{i \in I}$ be a family of subgroups of a group G. Suppose that for every $i, j \in I$, there is a $k \in I$ such that $S_i, S_j \subset S_k$. Prove that if $I \neq \varnothing$, then $\bigcup_{i \in I} S_i$ is a subgroup of G.

19 Let $\{C_i\}_{i \in I}$ be a family of subgroups of $\mathbb{Z}$ such that for $i, j \in I$, either $C_i \subset C_j$ or $C_j \subset C_i$. Prove that if $I \neq \varnothing$, then there is a $k \in I$ such that $C_i \subset C_k$ for all $i \in I$. That is, there is a maximum member of $\{C_i\}_{i \in I}$.

20 Let $\{C_i\}_{i \in I}$ be a family of subgroups of $\mathbb{Z}$. Prove that if $I \neq \varnothing$, then there is a $k \in I$ such that for all $i \in I$, $C_k \subset C_i$ implies $C_k = C_i$. That is, there is a maximal member of $\{C_i\}_{i \in I}$.

21 Prove that there is a one-to-one correspondence between any left coset of a subgroup and any right coset of that subgroup.

22 Prove that if $o(g) = n$ is finite, then $o(g^m) = n/(m, n)$.

23 Let n be a positive integer. Prove that $o(g) = n$ if and only if
(a) $g^n = e$, and
(b) $g^m = e$ implies that n divides m.

24 Suppose that $g_1, g_2, \ldots, g_k$ are elements of finite order in a group and that any two of these elements commute. Prove that $o(\prod_i g_i)$ divides $\prod_i o(g_i)$, and that $o(\prod_i g_i) = \prod_i o(g_i)$ if the orders of the g_i are pairwise relatively prime.

25 Prove that a group G is finite if and only if it has only finitely many subgroups.

26 Prove that if a group $G \neq \{e\}$ has only the subgroups $\{e\}$ and G, then G is cyclic, and $o(G)$ is a prime.

27 Prove that the intersection of two subgroups of finite index is a subgroup of finite index. Prove that the intersection of finitely many subgroups of finite index is a subgroup of finite index.

28 Prove that a finitely generated subgroup of the additive group $\mathbb{Q}$ of rational numbers is cyclic. Prove that $\mathbb{Q}$ is not finitely generated.

29 Give an example of an infinite group G such that every subgroup $\neq G$ is of infinite index.

30 Give an example of an infinite group such that every subgroup $\neq \{e\}$ is of finite index.

31 Give an example of a group G which has subgroups $\neq \{e\}$ of infinite index and subgroups $\neq G$ of finite index.

32 Let $H_1 \subset H_2 \subset \cdots \subset H_n$ be groups with each $H_{i+1}\colon H_i$ finite. Prove that

$$H_n{:}H_1 = (H_n{:}H_{n-1})(H_{n-1}{:}H_{n-2}) \cdots (H_2{:}H_1).$$

33 Let S be a finite set, and let G be the group of all permutations of S. Let $s \in S$, and let $H = \{f \in G : f(s) = s\}$. What is $G{:}H$? If S is infinite, is $G{:}H$ infinite?

34 Let G be the group of all permutations of a finite set S, and let $T \subset S$. Let $H = \{f \in G : f(t) = t \text{ for all } t \in T\}$, and let $K = \{f \in G : f(t) \in T \text{ for all } t \in T\}$. What is $G{:}H$? What is $G{:}K$? What is $K{:}H$?

2.3 HOMOMORPHISMS AND NORMAL SUBGROUPS

A central idea in algebra is that of homomorphism. Here is a simple illustration of the concept. Consider the group $G = \{1, -1\}$, where the operation is ordinary multiplication, and consider the group S_2 of all permutations of the set $A = \{1, 2\}$. The group S_2 has 2 elements, 1_A and the mapping f that interchanges 1 and 2. Intuitively, these two groups are exactly alike, even though they have no elements in common. The element 1_A in S_2 behaves just like the element 1 does in G, and the element f behaves in S_2 just like the element -1 does in G. Here there is actually a one-to-one correspondence between the two groups, but homomorphisms allow more general circumstances. Consider the group $\mathbb{Z}$ of integers under addition and the mapping $f: \mathbb{Z} \to G$ given by

$$f(n) = 1 \quad \text{if } n \text{ is even,}$$

and

$$f(n) = -1 \quad \text{if } n \text{ is odd.}$$

Now f is certainly not one-to-one, but it does "preserve" the operations. That is, if $m, n \in \mathbb{Z}$, then the same result is obtained if m and n are combined in $\mathbb{Z}$ and then f is applied, or if f is applied to m and n and the results combined in G. In other words,

$$f(m+n) = f(m) \cdot f(n).$$

We will be interested in this type of mapping between groups: mappings from one group to another that pay attention to the operations of the groups.

2.3.1 Definition

Let G and H be groups. A **homomorphism** *from G into H is a mapping $f: G \to H$ such that $f(xy) = f(x)f(y)$ for all x, y in G.*

A homomorphism $f: G \to H$ is an ***epimorphism*** if it is onto, a ***monomorphism*** if it is one-to-one, and an ***isomorphism*** if it is both one-to-one and onto. If $f: G \to G$ is an isomorphism, we write $G \approx H$ and say that G is ***isomorphic*** to H. A homomorphism $f: G \to G$ is called an ***endomorphism***, and an isomorphism $f: G \to G$ is an ***automorphism***. Before looking at some examples, let us note some of the elementary facts about homomorphisms.

2.3.2 Theorem

Let $f: G \to H$ be a homomorphism. The following hold.

(a) $f(e_G) = e_H$.

(b) $f(x^{-1}) = (f(x))^{-1}$, *for all $x \in G$.*

(c) *If A is a subgroup of G, then $f(A)$ is a subgroup of H. In particular, $f(G) = \operatorname{Im} f$ is a subgroup of H.*

(d) *If B is a subgroup of H, then $f^{-1}(B)$ is a subgroup of G.*

(e) $f(x_1 x_2 \cdots x_n) = f(x_1)f(x_2) \cdots f(x_n)$ *for $x_i \in G$.*

(f) $f(x^n) = f(x)^n$ *for all $n \in \mathbb{Z}$ and all $x \in G$.*

(g) *If* $g: H \to K$ *is a homomorphism, then* $g \circ f: G \to K$ *is a homomorphism.*
(h) *If* $f: G \to H$ *is an isomorphism, then* $f^{-1}: H \to G$ *is an isomorphism.*

Proof
(a) $f(e_G) = f(e_G e_G) = f(e_G)f(e_G)$, and multiplying through by $f(e_G)^{-1}$ yields $e_H = f(e_G)$.
(b) $e_H = f(e_G) = f(xx^{-1}) = f(x)f(x^{-1})$, so $f(x^{-1}) = (f(x))^{-1}$.
(c) For $f(x)$, $f(y) \in f(A)$, $f(x)f(y) = f(xy) \in f(A)$, so $f(A)$ is closed under multiplication. Since $f(x)^{-1} = f(x^{-1})$, $f(A)$ is a subgroup of H.
(d) Let $x, y \in f^{-1}(B)$. Then $f(xy) = f(x)f(y) \in B$, and $f(x^{-1}) = f(x)^{-1} \in B$ since $f(x)$ and $f(y)$ are in B. Thus, $f^{-1}(B)$ is a subgroup of G.
(e) This follows readily by induction on n.
(f) This holds for $n > 0$ by (e). For $n < 0$, $f(x^n) = ((f(x^n))^{-1})^{-1} = (f((x^n)^{-1})^{-1} = (f(x^{-n}))^{-1} = ((f(x))^{-n})^{-1} = f(x)^n$.
(g) Let $x, y \in G$. Then $(g \circ f)(xy) = g(f(xy)) = g(f(x)f(y)) = g(f(x))g(f(y)) = (g \circ f)(x)(g \circ f)(y)$.
(h) Let $a, b \in H$. We need only that $f^{-1}(ab) = f^{-1}(a)f^{-1}(b)$. For suitable $x, y \in G$, $a = f(x)$ and $b = f(y)$. Thus, $f^{-1}(ab) = f^{-1}(f(x)f(y)) = f^{-1}(f(xy)) = xy = f^{-1}(a)f^{-1}(b)$.

Let S be a set of groups. Then $\approx$ is an equivalence relation on S. This follows from 2.3.2(g) and 2.3.2(h), and the fact that $1_G: G \to G$ is an isomorphism.

2.3.3 Examples of homomorphisms

(a) There are two trivial examples. For any group G, 1_G is a homomorphism, in fact, is an automorphism of G. For any groups G and H, the mapping $f: G \to H$ defined by $f(x) = e_H$ for all $x \in G$ is a homomorphism.
(b) Let G be any group, and let $g \in G$. The map given by $\mathbb{Z} \to G: n \to g^n$ is a homomorphism. This is a useful homomorphism to know about. It says that given any element g of any group, there is a homomorphism taking $\mathbb{Z}$ onto the cyclic subgroup generated by g. Note that the map is a homomorphism because by 2.1.4(a), $g^{m+n} = g^m g^n$.
(c) Let G be a group, and let $g \in G$. The mapping

$$g^*: G \to G: x \to gxg^{-1}$$

is an automorphism of G. Indeed, since

$$g^*(xy) = g(xy)g^{-1} = gxg^{-1}gyg^{-1} = g^*(x)g^*(y),$$

g^* is an endomorphism. It is one-to-one since $g^*(x) = g^*(y)$ implies that $gxg^{-1} = gyg^{-1}$, which implies in turn that $x = y$. It is onto since for any y, $y = g^*(g^{-1}yg)$. Such an automorphism is called an ***inner automorphism***.

(d) Let G be an Abelian group, and let $n \in \mathbb{Z}$. Then $G \to G: x \to x^n$ is an endomorphism of G. It is necessary that G be Abelian for multiplication to be preserved.

(e) Let $\mathbb{R}_2$ be the set of all 2×2 matrices

$$(a_{ij}) = \begin{pmatrix} a_{11} & a_{12} \\ a_{21} & a_{22} \end{pmatrix}$$

with real entries. Add two such matrices entry-wise, that is, let

$$(a_{ij}) + (b_{ij}) = (a_{ij} + b_{ij}).$$

This makes $\mathbb{R}_2$ into an Abelian group, and the mapping given by $\mathbb{R}_2 \to \mathbb{R}: (a_{ij}) \to a_{11} + a_{12}$ is a homomorphism from this group into the additive group of real numbers.

(f) For an element (a_{ij}) in the general linear group $GL_2(\mathbb{R})$, let $\det(a_{ij}) = a_{11}a_{22} - a_{21}a_{12}$. The mapping

$$GL_2(\mathbb{R}) \to \mathbb{R}^*: (a_{ij}) \to \det(a_{ij})$$

is a homomorphism from $GL_2(\mathbb{R})$ into the multiplicative group of nonzero real numbers. It is in fact an epimorphism. That it is a homomorphism is just the familiar fact that $\det((a_{ij})(b_{ij})) = \det(a_{ij}) \det(b_{ij})$.

(g) $\mathbb{R}^+ \to \mathbb{R}: x \to \ln x$ is an isomorphism between the group of positive reals under multiplication and the group of reals under addition. The relevant operations are preserved since $\ln(xy) = \ln(x) + \ln(y)$.

(h) $\mathbb{R} \to \mathbb{R}^+: x \to e^x$ is an isomorphism between the group of reals under addition and the group of positive reals under multiplication. This is the inverse of the isomorphism in (g).

Let $f: G \to H$ be a homomorphism. Defining $x \sim y$ if $f(x) = f(y)$, gives an equivalence relation on G. Furthermore, by 1.5, Problem 20, f induces a one-to-one correspondence f^* between the set $G/\sim$ of equivalence classes of $\sim$ and $\operatorname{Im} f$, namely

$$f^*: G/\sim \to \operatorname{Im} f: \operatorname{Cl}(x) \to f(x).$$

Since $\operatorname{Im} f$ is a group, this one-to-one correspondence induces a group structure on $G/\sim$. That is, multiply by

$$\operatorname{Cl}(x)\operatorname{Cl}(y) = (f^*)^{-1}(f^*(\operatorname{Cl}(x))f^*(\operatorname{Cl}(y))).$$

(See Problem 17 at the end of 2.1.) But

$$(f^*)^{-1}(f^*(\mathrm{Cl}(x))f^*(\mathrm{Cl}(y))) = (f^*)^{-1}(f(x)f(y)) = (f^*)^{-1}(f(xy))$$
$$= (f^*)^{-1}(f^*(\mathrm{Cl}(xy))) = \mathrm{Cl}(xy).$$

Thus, the multiplication in $G/\sim$ is $\mathrm{Cl}(x)\mathrm{Cl}(y) = \mathrm{Cl}(xy)$. In particular, this is well defined. It is conceivable that $\mathrm{Cl}(x) = \mathrm{Cl}(x')$, $\mathrm{Cl}(y) = \mathrm{Cl}(y')$ and $\mathrm{Cl}(xy) \neq \mathrm{Cl}(x'y')$, but this is not the case as we have seen.

We need to find out just what $\sim$ is. Its equivalence classes are the sets $f^{-1}(y)$ with $y \in \mathrm{Im}\, f$. The particular equivalence class $N = f^{-1}(e_H)$ is a subgroup by 2.3.2(d). We claim that the equivalence classes of $\sim$ are the left cosets of N. In fact, if $y \in \mathrm{Im}\, f$, then $f^{-1}(y) = xN$, where $f(x) = y$. To see this, let $a \in f^{-1}(y)$. Then $f(a) = y$. Since $f(x) = y$, we have that $f(a^{-1}x) = e_H$, and so $a^{-1}x \in N$. That is, $aN = xN$ and $a \in xN$. Thus $f^{-1}(y) \subset xN$, and similarly the other inclusion holds.

It is worth noticing at this point that for all $x \in G$, $xN = Nx$. If $a \in N$, then $xa = xax^{-1}x$, and $f(xax^{-1}) = f(x)f(a)f(x^{-1}) = f(x)e_H f(x^{-1}) = e_H$, and so $xa \in Nx$. Similarly, $Nx \subset xN$. Therefore, the equivalence classes of $\sim$ are also the right cosets of N. The multiplication in $G/\sim$ was given by $\mathrm{Cl}(x)\mathrm{Cl}(y) = \mathrm{Cl}(xy)$. But now we know that

$$\mathrm{Cl}(x) = xN.$$

Thus, the multiplication is $(xN)(yN) = (xy)N$.

All this can be summed up as follows. If $f\colon G \to H$ is a homomorphism and $N = f^{-1}(e_H)$, then

$$(xN)(yN) = (xy)N$$

makes the set of left cosets of the subgroup $N = f^{-1}(e_H)$ of G into a group, and f^* defined by $f^*(xN) = f(x)$ is an isomorphism from this group onto the subgroup $\mathrm{Im}\, f$ of H.

We do not need the homomorphism f in order to make the definition above, just the subgroup N. Having it, we can define the group whose elements are the left cosets of N and whose multiplication is given by $(xN)(yN) = (xy)N$. This procedure will not work for just any subgroup N of G. Recall that $N = f^{-1}(e_H)$ had the property that $xN = Nx$ for all $x \in G$. This is a special property that some subgroups have.

2.3.4 Definition

The subgroup N of the group G is **normal** *if $xN = Nx$ for all $x \in G$.*

Several conditions equivalent to being normal are given in Problem 21.

2.3.5 Definition

Let $f: G \to H$ be a homomorphism. Then $f^{-1}(e_H)$ is denoted Ker f *and called the* **kernel** *of f.*

Letting $B = \{e_H\}$ in 2.3.2(d), we get

2.3.6 Theorem

The kernel of a homomorphism is a normal subgroup.

Notice that for any group G, $\{e\}$ and G itself are normal subgroups of G. If G is Abelian, then any subgroup of G is normal. Keep in mind that one way to prove that a subgroup is normal is to get it to be the kernel of some homomorphism. For example, the kernel of the homomorphism in 2.3.3(f) is the special linear group $\mathrm{SL}_2(\mathbb{R})$. In particular then, $\mathrm{SL}_2(\mathbb{R})$ is a normal subgroup of $\mathrm{GL}_2(\mathbb{R})$.

A useful fact is that any subgroup of index 2 is normal. If $G{:}H = 2$, then the only left or right coset of H besides H itself is the complement of H in G. In the group Q_8 of quaternions, the subgroups $\langle i \rangle$, $\langle j \rangle$, and $\langle k \rangle$ are distinct subgroups of index 2, having order 4 in a group with 8 elements. Thus, these subgroups are all normal. It happens that all of the subgroups of Q_8 are normal (Problem 13).

In the dihedral group D_4, the subgroup $N = \{E, R_1, R_2, R_3\}$ is of index 2, hence normal, and so is the subgroup $M = \{E, V, H, R_2\}$. The subgroup $\{E, R_2\}$ of N is normal in N, and indeed normal in D_4, while the subgroup $\{E, H\}$ is normal in M, but it is not normal in D_4 since $D_1\{E, H\} = \{D_1, R_1\}$, while $\{E, H\}D_1 = \{D_1, R_3\}$.

Let N be a normal subgroup of a group G, and let G/N denote the set of all left cosets of N in G. Since $xN = Nx$, G/N is also the set of all right cosets of N. We will omit "left" or "right" when talking about cosets of normal subgroups. We wish to make G/N into a group. In the case where $N = \operatorname{Ker} f$ for some homomorphism f, we know that we can do it by defining

$$xNyN = xyN.$$

This will work for *any* normal subgroup N. Suppose $xN = x'N$, $yN = y'N$. Then $x = x'a$, $y = y'b$ for some $a, b \in N$. Then $xyN = x'ay'bN = x'ay'N = x'aNy' = x'Ny' = x'y'N$, using the facts that $bN = N$ for $b \in N$, and $y'N = Ny'$. Thus, $(xN)(yN) = (xy)N$ is unambiguous, that is, does define a binary operation on G/N, and G/N is a group under this operation. In fact,

$$(xN)(yNzN) = xN(yzN) = x(yz)N = (xy)zN = (xyN)(zN) = (xNyN)(zN),$$
$$(xN)(eN) = xeN = xN = exN = (eN)(xN),$$

and

$$(xN)(x^{-1}N) = xx^{-1}N = eN = x^{-1}xN = (x^{-1}N)(xN),$$

so that G/N is a group. Now notice that $G \to G/N: x \to xN$ is an epimorphism with kernel N.

2.3.7 Definition

If N is a normal subgroup of the group G, the group G/N is the **quotient group** *of G modulo N and is called "G modulo N," or "G over N." The homomorphism*

$$G \to G/N: x \to xN$$

is called the **natural homomorphism** *from G to G/N.*

Let $f: G \to H$ be a homomorphism. Then letting $N = \operatorname{Ker} f$ in 2.3.7, we have the natural homomorphism $G \to G/\operatorname{Ker} f$. It is, of course, an epimorphism. Analogous to this epimorphism, we have the monomorphism $\operatorname{Im} f \to H$ given by the inclusion $\operatorname{Im} f \subset H$. The following theorem says that the groups $G/\operatorname{Ker} f$ and $\operatorname{Im} f$ are essentially the same.

2.3.8 First isomorphism theorem

Let $f: G \to H$ be a homomorphism. Then

$$f^*: G/(\operatorname{Ker} f) \to \operatorname{Im} f: x(\operatorname{Ker} f) \to f(x)$$

is an isomorphism.

Proof If $x(\operatorname{Ker} f) = y(\operatorname{Ker} f)$, then $xa = y$ with $a \in \operatorname{Ker} f$, and $f(xa) = f(x)f(a) = f(x) = f(y)$, so that f^* is indeed a mapping. It is clearly onto. If $f^*(x(\operatorname{Ker} f)) = f^*(y(\operatorname{Ker} f))$, then $f(x) = f(y)$, so that $f(x^{-1}y) = e_H$. That is, $x^{-1}y \in \operatorname{Ker} f$, and so $x(\operatorname{Ker} f) = y(\operatorname{Ker} f)$. Thus, f^* is one-to-one. Since

$$\begin{aligned} f^*((x(\operatorname{Ker} f))(y(\operatorname{Ker} f))) &= f^*(xy(\operatorname{Ker} f)) = f(xy) = f(x)f(y) \\ &= f^*(x(\operatorname{Ker} f))f^*(y(\operatorname{Ker} f)), \end{aligned}$$

f^* is an isomorphism.

Thus given any homomorphism $f: G \to H$, it can be decomposed into a composition

$$G \to G/\operatorname{Ker} f \to \operatorname{Im} f \to H$$

of three homomorphisms, the first the natural homomorphism, the second the isomorphism in 2.3.8, and the third an inclusion.

Let N be a normal subgroup of G. We need to know what the subgroups of G/N are.

2.3.9 Theorem

Let N be a normal subgroup of the group G. Then $H \to H/N$ is a one-to-one correspondence between the set of subgroups of G containing N and the set of subgroups of G/N. Normal subgroups correspond to normal subgroups under this correspondence.

Proof Let H be a subgroup of G containing N. Then N is clearly a normal subgroup of H, so that H/N makes sense. Now the set of cosets of N in H does indeed form a subgroup of G/N since $aN, bN \in H/N$ imply $abN \in H/N$, and $a^{-1}N \in H/N$. If K is a subgroup of G containing N and $K \neq H$, then $H/N \neq K/N$ so that the association is one-to-one. Now let X be a subgroup of G/N. Then X is a set of cosets of N. Let $H = \{x \in G: xN \in X\}$. For $x, y \in H$, $xNyN = xyN \in X$, and $x^{-1}N \in X$, so that H is a subgroup of G. Now H clearly contains N.

Suppose that H is normal in G and that $N \subset H$. Then the map $G/N \to G/H: xN \to xH$ is a homomorphism, as can easily be checked. But its kernel is H/N. Thus H/N is normal in G/N. Now suppose that H/N is normal in G/N. The homomorphism given by the composition of the homomorphisms $G \to G/N$ and $G/N \to (G/N)/(H/N)$ has kernel H. Thus, H is normal in G.

In the process, we have proved the following theorem.

2.3.10 Second isomorphism theorem

Let G be a group, and let N and H be normal subgroups of G with $N \subset H$. Then

$$G/H \approx (G/N)/(H/N).$$

A simple illustration of 2.3.10 is the isomorphism

$$(\mathbb{Z}/\mathbb{Z}mn)/(\mathbb{Z}m/\mathbb{Z}mn) \approx \mathbb{Z}/\mathbb{Z}m.$$

Since $o(\mathbb{Z}/\mathbb{Z}mn) = mn$ and $o(\mathbb{Z}/\mathbb{Z}m) = m$, we have that $o(\mathbb{Z}m/\mathbb{Z}mn) = n$.

Suppose now that N is normal in G and that H is any subgroup of G. We have the natural homomorphism $G \to G/N$, and we can restrict it to H. The image of this restriction is $\{hN: h \in H\}$. Let $HN = \{hn: h \in H, n \in N\}$. Let $a, b \in H$ and let $c, d \in N$. Then $(ac)(bd) = ab(b^{-1}cb)d \in HN$, and $(ac)^{-1} = c^{-1}a^{-1} = a^{-1}(ac^{-1}a^{-1}) \in HN$, so that HN is a subgroup of G. Therefore, the image of $H \to G/N$ is $(HN)/N$. The kernel is $H \cap N$. Thus, from 2.3.8 we have the following theorem.

2.3.11 Third isomorphism theorem

Let H be a subgroup of the group G, and let N be a normal subgroup of G. Then HN is a subgroup of G, and

$$H/(H \cap N) \approx (HN)/N.$$

We now examine quotient groups of the additive group $\mathbb{Z}$ of integers. That group is cyclic. In fact, $\mathbb{Z} = \langle 1 \rangle$. Let S be a subgroup of $\mathbb{Z}$. Then we know from 2.2.6 that S is cyclic. The proof of 2.2.6 shows that if $S \neq \{0\}$, then the smallest positive integer in S generates S. If that integer were n, then this means that $S = \{mn: m \in \mathbb{Z}\}$, which we write as $\mathbb{Z}n$. Therefore, the subgroups of $\mathbb{Z}$ are precisely the subgroups $\mathbb{Z}n$ with $n \geq 0$. Distinct such n give distinct subgroups $\mathbb{Z}n$. (See Problem 5, Section 2.2.)

2.3.12 Definition

Let $n \geq 0$. The group $\mathbb{Z}/\mathbb{Z}n$ is called the **group of integers modulo** *n, and will also be denoted $\mathbb{Z}(n)$.*

We have already called another group the integers modulo n, denoted the same, namely the group 2.1.2(e). The two groups are isomorphic (Problem 2).

Let $G = \langle g \rangle$. That is, G is a cyclic group. The mapping $f: \mathbb{Z} \to G: n \to g^n$ is a homomorphism, in fact an epimorphism. Just note that $f(m+n) = g^{m+n} = g^m g^n = f(m)f(n)$. If $o(g) = m$, then $g^m = e$ and $\mathbb{Z}m = \operatorname{Ker} f$. If $o(g) = \infty$, then $\operatorname{Ker} f = 0$ and f is an isomorphism. In any case, $\mathbb{Z}/\operatorname{Ker} f \approx G$ and $\operatorname{Ker} f = \mathbb{Z}m$ for some $m \geq 0$. We thus have the following.

2.3.13 Theorem

If G is cyclic and $o(G) = n$, then $G \approx \mathbb{Z}/\mathbb{Z}n$. If $o(G) = \infty$, then $G \approx \mathbb{Z}$.

2.3.14 Corollary

Two cyclic groups are isomorphic if and only if they have the same order.

One way to express 2.3.13 is to say that the cyclic groups, up to isomorphism, are just the groups $\mathbb{Z}/\mathbb{Z}n$, $n \geq 0$. Notice that for $n > 0$, $\mathbb{Z}/\mathbb{Z}n$ has exactly n elements.

Let us see what 2.3.11 says in a familiar case. Consider two subgroups $\mathbb{Z}m$ and $\mathbb{Z}n$ of $\mathbb{Z}$ with m and n positive. We know that $\mathbb{Z}m \cap \mathbb{Z}n$ and $\mathbb{Z}m + \mathbb{Z}n$ are subgroups of $\mathbb{Z}$. The subgroup $\mathbb{Z}m \cap \mathbb{Z}n$ is the set of common multiples of m and n, and hence is generated by the least common multiple $[m, n]$ of m and n. (See 1.6, Problem 10.) Thus, $\mathbb{Z}m \cap \mathbb{Z}n = \mathbb{Z}[m, n]$. The subgroup $\mathbb{Z}m + \mathbb{Z}n = \{am + bn: a, b \in \mathbb{Z}\}$ is generated by the smallest positive integer contained in it. But the smallest positive integer of the form $am + bn$ is the greatest common divisor (m, n) of m and n. Thus, $\mathbb{Z}m + \mathbb{Z}n = \mathbb{Z}(m, n)$. Applying 2.3.11, we get

$$\mathbb{Z}n/\mathbb{Z}[m, n] \approx \mathbb{Z}(m, n)/\mathbb{Z}m.$$

Now suppose that $a, b \in \mathbb{Z}^+$ and a divides b. That is, suppose that $0 \neq \mathbb{Z}b \subset \mathbb{Z}a$. Then $o(\mathbb{Z}a/\mathbb{Z}b) = b/a$. Applying this to $\mathbb{Z}n/\mathbb{Z}[m, n]$ and $\mathbb{Z}(m, n)/\mathbb{Z}m$ yields $[m, n]/n = m/(m, n)$, or $mn = [m, n](m, n)$.

Let Aut(G) be the set of all automorphisms of a group G. Then Aut(G) is itself a group under composition of mappings. In fact, Aut(G) is a subset of Perm(G), the set of all permutations of G, [2.1.2(a)], and so 2.3.2(g) and 2.3.2(h) imply that Aut(G) is a subgroup of Perm(G).

Let $g \in G$. The mapping $\varphi(g): G \to G: x \to gxg^{-1}$ is an automorphism of G, as shown in 2.3.3(c). Such automorphisms are called ***inner automorphisms***, and $\varphi(g)$ is the inner automorphism induced by g. The set of inner automorphisms is denoted Inn(G). If $\varphi(g)$ is the identity automorphism of G, then $gxg^{-1} = g$ for all x in G, or equivalently, $gx = xg$ for all x in G. That is, g ***commutes*** with all the elements of G. The set of such g is called the ***center*** of G and is denoted $Z(G)$. If G is Abelian, then of course, $Z(G) = G$.

2.3.15 Theorem

Let G be a group, and let $\varphi(g)$ be the inner automorphism of G induced by the element g. Then

$$G \to \mathrm{Aut}(G): g \to \varphi(g)$$

is a homomorphism with image Inn(G) *and kernel* $Z(G)$. *In particular,* $G/Z(G) \approx \mathrm{Inn}(G)$.

Proof Since $\varphi(gh)(x) = ghx(gh)^{-1} = g(hxh^{-1})g^{-1} = \varphi(g)(\varphi(h)(x))$, $\varphi(gh) = \varphi(g)\varphi(h)$, and the mapping is a homomorphism. That the image is Inn(G) and the kernel is $Z(G)$ should be clear.

It is usually difficult to compute the automorphism group of a group.

However, there is one very special but important case where Aut(G) can be determined explicitly.

2.3.16 Theorem

$\text{Aut}(\mathbb{Z}(n)) \approx U(n)$.

Proof The mapping $\varphi_m: \mathbb{Z}(n) \to \mathbb{Z}(n): x \to x^m$ is an endomorphism of the cyclic, and hence Abelian, group $\mathbb{Z}(n)$ [2.3.3(d)]. Let g be a generator of $\mathbb{Z}(n)$. An automorphism φ of $\mathbb{Z}(n)$ is determined by the image of g, and the image of g is a power g^m of g, where $0 \leq m < n$. If $\varphi(g) = g^m$, then $\varphi = \varphi_m$, so that every automorphism is a φ_m. The image of an automorphism is generated by the image of g, so in order to be onto, which is equivalent to being one-to-one since $\mathbb{Z}(n)$ is finite, it must be that $(m, n) = 1$. Since $\varphi_k = \varphi_m$ if and only if $k - m$ is divisible by n, the automorphisms of $\mathbb{Z}(n)$ are in one-to-one correspondence with the elements of $U(n)$. Composition of the φ_m corresponds to multiplication in $U(n)$, and the theorem follows.

This theorem will be useful in the construction of examples of groups with certain properties (See 2.6, page 73.) Since $U(n)$ is Abelian, the automorphism group of a finite cyclic group is Abelian, although it may not be cyclic. For example, $U(8)$ is not cyclic. However, $U(p)$ is cyclic for p a prime, although this is not at all obvious. It follows from 6.2.12.

PROBLEMS

1 Let G be the group $\mathbb{R} \times \mathbb{R}^*$ under the operation $(a, b)(c, d) = (a + bc, bd)$. Prove that $f: G \to G: (a, b) \to (0, b)$ is an endomorphism with $f^2 = f$. Is $G \to G: (a, b) \to (a, 1)$ an endomorphism?

2 Prove that $\mathbb{Z}(n)$ as defined in 2.1.2(e) and in 2.3.12 are isomorphic.

3 How many homomorphisms are there from $\mathbb{Z}$ onto $\mathbb{Z}(8)$?

4 How many homomorphisms are there from $\mathbb{Z}(12)$ onto $\mathbb{Z}(5)$?

5 How many homomorphisms are there from $\mathbb{Z}$ onto $\mathbb{Z}$; from $\mathbb{Z}$ into $\mathbb{Z}$?

6 Let $|z|$ be the absolute value of a complex number z. Prove that $\mathbb{C}^* \to \mathbb{R}^+ : z \to |z|$ is an epimorphism.

7 Prove that multiplication by a nonzero rational number is an automorphism of the additive group of rational numbers.

8 Prove that if f is a homomorphism, and the order of x is finite, then $o(x)$ is a multiple of $o(f(x))$.

9 Prove that the dihedral group D_4 is isomorphic to a subgroup of the symmetric group S_4.

10 Let G be the additive group of all complex numbers of the form $m + ni$, where m and n are integers. Let H be the multiplicative group of all rational numbers of the form $2^m 3^n$, again where m and n are integers. Prove that $G \approx H$.

11 Let G be a non-Abelian group of order 6. Prove that $G \approx S_3$.

12 Prove that there are exactly two groups of order 4, up to isomorphism. Which one is $Q_8/\{1, -1\}$?

13 Prove that every subgroup of Q_8 is normal. Find all the normal subgroups of D_4.

14 Prove that the set of elements of $\mathrm{GL}_2(\mathbb{R})$ of the form

$$\begin{pmatrix} a & 0 \\ c & d \end{pmatrix}$$

is a subgroup, but is not normal.

15 Prove that $\mathrm{SL}_2(\mathbb{R})$ is a normal subgroup of the group $\mathrm{GL}_2(\mathbb{R})$, and that $\mathrm{GL}_2(\mathbb{R})/\mathrm{SL}_2(\mathbb{R})$ is isomorphic to the multiplicative group $\mathbb{R}^*$ of nonzero real numbers.

16 Let C be the group of all complex numbers of the form $e^{2\pi iq}$, where q is rational. Prove that $\mathbb{Q}/\mathbb{Z} \approx C$.

17 Let $H = \{f : f \in S_4, f(2) = 2\}$. Prove that H is a subgroup of S_4. Is H normal in S_4?

18 For groups A and B, define multiplication on the Cartesian product $A \times B$ by $(a, b)(c, d) = (ac, bd)$. Prove that
(a) $A \to A \times B : a \to (a, e)$ is a monomorphism, and
(b) $A \times B \to A : (a, b) \to a$ is an epimorphism whose kernel is isomorphic to B.

19 Let N and H be subgroups of G, and suppose that N is normal.
(a) Prove that $NH = \{nh : n \in N, h \in H\}$ is a subgroup of G, and that $NH = HN$.
(b) Prove that if H is also normal, then so is NH.
(c) Prove that if H is also normal, and if $N \cap H = \{e\}$, then $nh = hn$ for all $n \in N$ and $h \in H$.

20 Let G be a group of order p^2, where p is a prime.
(a) Prove that G has an element of order p.
(b) Prove that any subgroup of G of order p is normal.
(c) Prove that G is Abelian.

21 Let N be a subgroup of G, and for subsets S and T of G, let

$ST = \{st : s \in S, t \in T\}$. For g in G, let $gS = \{gs : s \in S\}$ and $Sg = \{sg : s \in S\}$. Prove that the following are equivalent.

(a) N is normal in G.
(b) $gNg^{-1} = N$ for all $g \in G$.
(c) $gNg^{-1} \subset N$ for all $g \in G$.
(d) $gNg^{-1} \supset N$ for all $g \in G$.
(e) $(xy)N = (xN)(yN)$.
(f) $Ng \subset gN$ for all $g \in G$.
(g) $gN \subset Ng$ for all $g \in G$.

22 Prove that if G has exactly one subgroup of order 50, then that subgroup is normal. Generalize.

23 Let g be an element of G, and let N be a normal subgroup of G. Prove that if $o(g)$ is finite, then it is a multiple of $o(gN)$.

24 Prove that if G/N and N are finitely generated, then so is G.

25 Prove that a homomorphic image of a cyclic group is cyclic.

26 Let G be an Abelian group. Let T be the set of elements of G which have finite order. Prove that T is a subgroup of G, and that in G/T, only the identity element has finite order.

27 Prove that the intersection of any family of normal subgroups of a group G is a normal subgroup of G.

28 Let $f: G \to H$ be an epimorphism, and let K be a normal subgroup of G. Prove that f induces an epimorphism $G/K \to H/f(K)$. Prove that this induced epimorphism is an isomorphism if and only if $\operatorname{Ker} f \subset K$.

29 Prove that if $G/Z(G)$ is cyclic, then $Z(G) = G$.

30 A subgroup H of G is called ***characteristic*** if for every automorphism f of G, $f(H) = H$. Prove that every subgroup of a cyclic group is characteristic.

31 Prove that a characteristic subgroup of a normal subgroup of G is a normal subgroup of G.

32 Let N be a cyclic normal subgroup of G. Prove that every subgroup of N is a normal subgroup of G.

33 A subgroup H of G is called ***fully invariant*** if for every endomorphism f of G, $f(H) \subset H$. Prove that the center $Z(G)$ of G is a characteristic but not necessarily a fully invariant subgroup of G.

34 Prove that if f is an automorphism of G and N is normal in G, then $f(N)$ is normal in G. Similarly for N characteristic, N fully invariant.

35 Find all endomorphisms of the additive group $\mathbb{Z}$. Which are automorphisms?

36 Let G' be the intersection of all the normal subgroups N of G such that G/N is Abelian. Prove that G' is the subgroup generated by all the elements of G of the form $x^{-1}y^{-1}xy$ with $x, y \in G$. Prove that every subgroup of G containing G' is normal in G. Prove that G' is fully invariant in G. (The subgroup G' is called the ***commutator subgroup*** of G.)

37 Prove that G/N and N Abelian does not imply that G is Abelian.

38 Find the center of the group of all 2×2 nonsingular real matrices.

39 Find the commutator subgroup of the group of all 2×2 nonsingular real matrices. *Hint*: Show that the commutator subgroup is generated by matrices of the form

$$\begin{pmatrix} 1 & a \\ 0 & 1 \end{pmatrix} \quad \text{and} \quad \begin{pmatrix} 1 & 0 \\ a & 1 \end{pmatrix}.$$

Then show that each of these matrices is a commutator.

2.4 PERMUTATION GROUPS

Let S be any nonempty set. We have noted that the set Perm(S) of all one-to-one mappings from S onto S is a group under composition of mappings. The elements of Perm(S) are called ***permutations***, and a ***permutation group*** is a subgroup of Perm(S) for some S. Out first objective is to show that every group is a permutation group—more precisely, that every group is isomorphic to a permutation group. This fact is known as ***Cayley's Theorem***. It has the virtue of representing an abstract group as something concrete. Every group can be represented as a group of one-to-one mappings of some set onto itself. In fact, groups arose in this way. Permutations groups were studied before the notion of an abstract group was formulated.

How is Cayley's Theorem proved? If G is a group, we must get a monomorphism from G into Perm(S) for some S. Where do we get a suitable set S? We use G. Then with each $g \in G$ we must associate a permutation of G. How can we do that? Multiplying each element of G on the left by g is a permutation of G.

2.4.1 Cayley's Theorem

Any group G is isomorphic to a subgroup of Perm(G).

Proof For each $g \in G$, let $\varphi(g)$ be the map from G into G defined by $\varphi(g)(x) = gx$. That is, $\varphi(g)$ is multiplication on the left by g. If $\varphi(g)(x) = \varphi(g)(y)$, then $gx = gy$, so that $x = y$. Hence, $\varphi(g)$ is one-to-one. If $y \in G$, then $\varphi(g)(g^{-1}y) = g(g^{-1}y) = y$, so that $\varphi(g)$ is onto. Thus, $\varphi(g) \in \text{Perm}(G)$. Therefore, we have a mapping $\varphi\colon G \to \text{Perm}(G)$. We will show that φ is a monomorphism. If $\varphi(g) = \varphi(h)$, then $\varphi(g)(x) = \varphi(h)(x)$ for all $x \in G$. In particular, $\varphi(g)(e) = ge = g = \varphi(h)(e) = he = h$. Thus φ is one-to-one. For $g, h \in G$,

$$\varphi(gh)(x) = ghx = g(\varphi(h)(x)) = \varphi(g)(\varphi(h)(x)) = (\varphi(g)\varphi(h))(x),$$

so that $\varphi(gh) = \varphi(g)\varphi(h)$. This proves Cayley's Theorem.

Cayley's Theorem can be generalized as follows. Each element $g \in G$ induces a permutation of the elements of G. The elements of G may be construed as left cosets of the subgroup $\{e\}$. That is, $x = y$ if and only if $x\{e\} = y\{e\}$. Thus, each element of G induces a permutation of the left cosets of the subgroup $\{e\}$. Now suppose we replace $\{e\}$ by any subgroup H of G. For any left coset xH of H and any $g \in G$, we can associate the left coset gxH. This association is a mapping from the set of left cosets of H into itself. In fact, if $xH = yH$, then $x = yh$ for some $h \in H$, and $gx = gyh$ is in both gxH and gyH. Hence, $gxH = gyH$. Since $g(g^{-1}xH) = xH$, the mapping is onto. If $gxH = gyH$, then $gx = gyh$ for some $h \in H$, and so $x = yh$. Thus $xH = yH$, and the mapping is therefore a permutation of the set of left cosets of H. Call this permutation $\varphi(g)$. Then $\varphi(gg')(xH) = gg'xH = (\varphi(g)\varphi(g'))(xH)$, so that $\varphi(gg') = \varphi(g)\varphi(g')$, and φ is a homomorphism from G into the group of permutations of the set of left cosets of H.

What is $\operatorname{Ker} \varphi$? It is clear that φ is not a monomorphism in the case $\{e\} \neq H$. If $g \in \operatorname{Ker} \varphi$, then $\varphi(g)eH = gH = eH$, so that $g \in H$. Thus, $\operatorname{Ker} \varphi \subset H$. Also, $\operatorname{Ker} \varphi$ is normal in G. Thus, $\operatorname{Ker} \varphi$ is some normal subgroup of G that is contained in H. Which one? The largest one, naturally. In fact, let N be any normal subgroup of G contained in H. For $h \in N$, $hxH = xx^{-1}hxH = xH$ since $x^{-1}hx \in N \subset H$. Thus, $h \in \operatorname{Ker} \varphi$. Therefore, $\operatorname{Ker} \varphi$ contains all the normal subgroups of G contained in H, and since $\operatorname{Ker} \varphi$ is a normal subgroup of G contained in H, it is the largest one. In particular, every subgroup H of a group G has a subgroup N which is normal in G and contains any other such subgroup. In fact, N is just the subgroup generated by all the subgroups of H which are normal in G. We have proved the following generalization of Cayley's Theorem.

2.4.2 Theorem

Let H be a subgroup of G, and let $L(H)$ be the set of left cosets of H. Then φ defined by $\varphi(g)(xH) = gxH$ is a homomorphism from G into $\operatorname{Perm}(L(H))$, *and* $\operatorname{Ker} \varphi$ *is the largest normal subgroup of G contained in H.*

Note that 2.4.1 follows from 2.4.2 by taking $H = \{e\}$. In 2.4.2 (and 2.4.1, of course), each element $g \in G$ induced a permutation of a set, and associating g with that permutation was a homomorphism.

The subgroup $S = \{E, H\}$ of D_4 [2.1.2(l)] is not normal, so that the largest normal subgroup of D_4 in S is $\{E\}$. The homomorphism $D_4 \to \text{Perm}(L(S))$ given by 2.4.2 then has kernel $\{E\}$, so is a monomorphism. Thus, we get a representation of D_4 as a group of permutations of the set $L(S)$ with 4 elements.

In Chapter 8 we will see how 2.4.2 is useful in proving that certain groups have nontrivial normal subgroups. Several of the following exercises also make use of 2.4.2.

PROBLEMS

1 Let $a, b \in \mathbb{R}$ with $b \neq 0$. Define $\Gamma_{a,b}$ by

$$\Gamma_{a,b}: \mathbb{R} \to \mathbb{R}: x \to bx + a.$$

Prove that the set of all such $\Gamma_{a,b}$ is a group of permutations of $\mathbb{R}$. Prove that this group is isomorphic to the group $G = \mathbb{R} \times \mathbb{R}^*$ with multiplication given by $(a, b)(c, d) = (a + bc, bd)$.

2 Let G be the group in Problem 1. Let H be the subgroup of elements of G of the form $(0, b)$. Use 2.4.2 to show that H has no nontrivial subgroups that are normal in G.

3 Let G be the group in Problem 1. Let N be the subgroup of elements of the form $(a, 1)$. Prove that N is normal in G.

4 Let G be the group in Problem 1. Prove that G is isomorphic to the multiplicative group of matrices of the form

$$\begin{pmatrix} b & a \\ 0 & 1 \end{pmatrix}$$

with a and b in $\mathbb{R}$, and $b \neq 0$. Which subgroups correspond to the subgroup H in Problem 2 and to the subgroup N in Problem 3?

5 Let G be a group. Prove that $\varphi: G \to \text{Perm}(G)$ given by $\varphi(g)(x) = xg^{-1}$ is a monomorphism. What happens if φ is defined by $\varphi(g)(x) = xg$?

6 Let G be a group, and let H be a subgroup of G. Let $R(H)$ be the set of all right cosets of H. Prove that

$$\varphi: G \to \text{Perm}(R(H)): \varphi(g)(Hx) \to Hxg^{-1}$$

is a homomorphism. What is its kernel?

7 Let H be a subgroup of G. Let $N(H) = \{g \in G: gHg^{-1} = H\}$, the *normalizer* of H in G. Prove that $N(H)$ is the largest subgroup of G in which H is normal.

8 Let H be a subgroup of G. For $g \in G$, prove that gHg^{-1} is a subgroup of G. Prove that gHg^{-1} is isomorphic to H.

9 Let f be a one-to-one mapping of a set A onto a set B. Prove that the mapping $\bar{f}\colon \text{Perm}(A) \to \text{Perm}(B)\colon \alpha \to f \circ \alpha \circ f^{-1}$ induced by f is an isomorphism.

10 Let H be a subgroup of G, and let $g \in G$. The subgroup gHg^{-1} is a ***conjugate*** of H. Let $\text{Cl}(H)$ denote the set of all conjugates of H, and let $L(N(H))$ denote the set of all left cosets of $N(H)$. Prove that

$$L(N(H)) \to \text{Cl}(H)\colon gN(H) \to gHg^{-1}$$

is an equivalence.

11 Let H be a subgroup of G. Prove that $\alpha\colon G \to \text{Perm}(\text{Cl}(H))$ given by $\alpha(g)(xHx^{-1}) = (gx)H(gx)^{-1}$ is a homomorphism. Prove that $\text{Ker}\,\alpha$ is the largest normal subgroup of G that is contained in $N(H)$.

12 Let H be a subgroup of G, and let φ be the mapping from G to $\text{Perm}(L(N(H)))$ given by $\varphi(g)(xN(H)) = (gx)N(H)$, as in 2.4.2. Let f be the mapping in Problem 10, let $\bar{f}$ be the isomorphism from $\text{Perm}(L(N(H)))$ to $\text{Perm}(\text{Cl}(H))$ induced by f, and let α be the mapping in Problem 11. Prove that $\bar{f} \circ \varphi = \alpha$. Here is a picture.

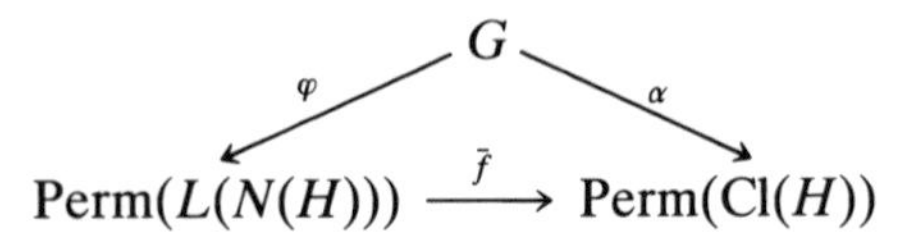

13 Prove that if H is a proper subgroup of the finite group G such that $o(G)$ does not divide $(G{:}H)!$, then G has a nontrivial normal subgroup contained in H.

14 Prove that if p is the smallest prime dividing $o(G)$, then any subgroup of G of index p is normal.

2.5 THE GROUPS S_n

Cayley's Theorem asserts that every group G is isomorphic to a subgroup of $\text{Perm}(G)$, the group of all permutations of the set G. If G is finite and $o(G) = n$, then $\text{Perm}(G)$ is isomorphic to $\text{Perm}(\{1, 2, \ldots, n\}) = S_n$, which is called the ***symmetric group of degree*** n. Thus, every group of order n is isomorphic to a subgroup of S_n. It is therefore of some interest to examine more closely the groups S_n.

An element α of S_n is a one-to-one mapping of $\{1, 2, \ldots, n\}$ onto itself. One way to represent such a mapping is

$$\begin{pmatrix} 1 & 2 & \cdots & n \\ \alpha(1) & \alpha(2) & \cdots & \alpha(n) \end{pmatrix}.$$

That is, under each integer k, $1 \le k \le n$, put its image. Thus, the elements of

S_3 are the six permutations

$$\begin{pmatrix}1 & 2 & 3\\1 & 2 & 3\end{pmatrix}, \quad \begin{pmatrix}1 & 2 & 3\\1 & 3 & 2\end{pmatrix}, \quad \begin{pmatrix}1 & 2 & 3\\2 & 1 & 3\end{pmatrix},$$

$$\begin{pmatrix}1 & 2 & 3\\2 & 3 & 1\end{pmatrix}, \quad \begin{pmatrix}1 & 2 & 3\\3 & 1 & 2\end{pmatrix}, \quad \begin{pmatrix}1 & 2 & 3\\3 & 2 & 1\end{pmatrix}.$$

For example,

$$\begin{pmatrix}1 & 2 & 3\\3 & 1 & 2\end{pmatrix}$$

is the mapping α such that $\alpha(1) = 3$, $\alpha(2) = 1$, and $\alpha(3) = 2$.

Multiplication in S_n is composition of mappings. We have defined composition $\alpha \circ \beta$ of two mappings α and β by $(\alpha \circ \beta)(x) = \alpha(\beta(x))$. That is, apply β and then apply α. *For the rest of this section we will adopt the convention that $\alpha \circ \beta$ means apply α and then apply β.* We will write $\alpha \circ \beta$ simply as $\alpha\beta$. Thus,

$$\begin{pmatrix}1 & 2 & 3\\2 & 1 & 3\end{pmatrix}\begin{pmatrix}1 & 2 & 3\\3 & 1 & 2\end{pmatrix} = \begin{pmatrix}1 & 2 & 3\\1 & 3 & 2\end{pmatrix}.$$

This convention is fairly standard for S_n.

There is another way to represent elements of S_n. We illustrate it with $n = 5$. The symbol $(1, 4, 3, 5, 2)$ means the permutation of $\{1, 2, 3, 4, 5\}$ that takes 1 to 4, 4 to 3, 3 to 5, 5 to 2, and 2 to 1. Thus in $(1, 4, 3, 5, 2)$, the image of an element is on its right, except that the image of the last element listed, in this case 2, is the first element listed. Thus,

$$(1, 4, 3, 5, 2) = \begin{pmatrix}1 & 2 & 3 & 4 & 5\\4 & 1 & 5 & 3 & 2\end{pmatrix}.$$

Notice that $(1, 4, 3, 5, 2) = (3, 5, 2, 1, 4)$.

What does $(4, 3, 2)$ mean? We are talking about elements of S_5, remember, and $(4, 3, 2)$ does not involve 1 or 5. This means that 1 and 5 are not moved by $(4, 3, 2)$. That is,

$$(4, 3, 2) = \begin{pmatrix}1 & 2 & 3 & 4 & 5\\1 & 4 & 2 & 3 & 5\end{pmatrix}.$$

Elements of S_n of the form $(i_1, i_2, \ldots, i_k)$ are called ***cycles***. Some cycles of S_7 are $(1, 2, 3, 4, 5, 6, 7)$, $(1, 2, 3, 4)$, $(7, 1, 4)$, $(1, 2)$ and $(3, 4, 1, 6)$. The cycle $(i_1, i_2, \ldots, i_k)$ is of ***length*** k and is a ***k-cycle***. For example, $(1, 7, 4, 2)$ is a 4-cycle. The cycle $(i_1, i_2, \ldots, i_k)$ of S_n may not involve every element of $1, 2, \ldots, n$. This means that those elements not involved are fixed. For example, $(7, 1, 4)$ is the element

$$\begin{pmatrix} 1 & 2 & 3 & 4 & 5 & 6 & 7 \\ 4 & 2 & 3 & 7 & 5 & 6 & 1 \end{pmatrix}$$

of S_7. Any cycle of length 1 represents the identity element.

Cycles are permutations, so two cycles of S_n may be multiplied. For example, if $n = 5$,

$$(3, 2, 4)(1, 3, 5) = \begin{pmatrix} 1 & 2 & 3 & 4 & 5 \\ 3 & 4 & 2 & 5 & 1 \end{pmatrix}.$$

Two cycles are ***disjoint*** if they have no element in common. Thus $(1, 2, 3)$ and $(4, 5, 6)$ are disjoint cycles, whereas $(7, 6, 1)$ and $(1, 2, 3)$ are not disjoint. Every element of S_n is a product of disjoint cycles. For example,

$$\begin{pmatrix} 1 & 2 & 3 & 4 & 5 & 6 & 7 \\ 5 & 3 & 2 & 7 & 1 & 4 & 6 \end{pmatrix}$$

is the product $(1, 5)(2, 3)(4, 7, 6)$. The way to write a permutation as a product of disjoint cycles is as follows. Let $\alpha \in S_n$, and $k \in \{1, 2, \ldots, n\}$. There is a smallest $m \geq 1$ such that $\alpha^m(k) = k$. This yields the cycle

$$(k, \alpha(k), \alpha^2(k), \ldots, \alpha^{m-1}(k)).$$

Two such cycles are either equal or disjoint. [Keep in mind that $(1, 2, 3) = (2, 3, 1) = (3, 1, 2)$]. The cycles so obtained are *the cycles of* α. It should be clear that α is the product of its cycles. Thus, we have the following theorem.

2.5.1 Theorem

Every element of S_n is a product of disjoint cycles.

Disjoint cycles commute. That is, if α and β are disjoint cycles, then $\alpha\beta = \beta\alpha$. Thus if α is a product $\alpha_1\alpha_2 \cdots \alpha_k$ of disjoint cycles, then $\alpha = \alpha_{\tau(1)}\alpha_{\tau(2)} \cdots \alpha_{\tau(k)}$, where τ is any permutation of $\{1, 2, \ldots, k\}$. If also $\alpha = \beta_1\beta_2 \cdots \beta_j$, with the β_i's disjoint cycles, then $j = k$, and (after rearrangement) $\alpha_1 = \beta_1$, $\alpha_2 = \beta_2, \ldots, \alpha_k = \beta_k$. This becomes clear by noting that each β_i is a cycle of α. Thus up to the order of the factors, each element α of S_n is uniquely a product of disjoint cycles, these disjoint cycles being simply the cycles of α.

Note that

$$(1, 2, 3, \ldots, n) = (1, 2)(1, 3)(1, 4) \cdots (1, n),$$

or more generally,

$$(a_1, a_2, \ldots, a_n) = (a_1, a_2)(a_1, a_3) \cdots (a_1, a_n).$$

That is, any cycle is a product of 2-cycles. The 2-cycles are called ***transpositions***. Therefore, by 2.5.1 we have the following fact.

2.5.2 Theorem

Every element of S_n is a product of transpositions.

There are many ways to write an element of S_n as a product of transpositions. For example, the identity of S_5 may be written as $(1,2)(1,2)$, or $(1,2)(1,2)(1,3)(1,3)$. Furthermore, two transpositions do not necessarily commute. For example, $(1,3)(1,2) \neq (1,2)(1,3)$. However, there is one very important thing that is unique about any representation of a permutation as a product of transpositions. That unique thing is the *parity* of the number of transpositions used. The parity of an integer is *even* or *odd* if the integer is even or odd, respectively. Thus we assert that if $\alpha_1\alpha_2 \cdots \alpha_j = \beta_1\beta_2 \cdots \beta_k$ with the α_i's and β_i's transpositions, then j and k are either both even or both odd, that is, i and j have the same parity. This fact is not obvious.

2.5.3 Theorem

If $\alpha_1\alpha_2 \cdots \alpha_j = \beta_1\beta_2 \cdots \beta_k$ with the α_i's and β_i's transpositions, then j and k have the same parity.

Proof If $\alpha_1\alpha_2 \cdots \alpha_j = \beta_1\beta_2 \cdots \beta_k$ with the α_i's and β_i's transpositions, then $\alpha_1\alpha_2 \cdots \alpha_j\beta_k\beta_{k-1} \cdots \beta_1 = e$. Now j and k have the same parity if and only if $j + k$ is even. Thus it suffices to show that if $\alpha_1\alpha_2 \cdots \alpha_m = e$ with the α_i transpositions, then m is even. Let $\alpha_1 = (a, b_1)$. Note that $(b_i, b_j)(a, b_j) = (a, b_j)(a, b_i)$ if a, b_i, and b_j are distinct. Thus, $e = \alpha_1\alpha_2 \cdots \alpha_m$ may be written in the form $(a, b_1)(a, b_2)(a, b_3) \cdots (a, b_r)\sigma_{r+1} \cdots \sigma_m$ with a in no σ_i for $i > r$. For example,

$$\begin{aligned}(a, b_1)(b_2, b_3)(b_4, b_5)(a, b_5) &= (a, b_1)(b_4, b_5)(a, b_5)(b_2, b_3)\\ &= (a, b_1)(a, b_5)(a, b_4)(b_2, b_3).\end{aligned}$$

For some $i < r$, $b_i = b_r$, else the identity carries b_r to a. The equation

$$(a, b_j)(a, b_r) = (a, b_r)(b_r, b_j)$$

allows us to move (a, b_r) over to the left next to (a, b_i), and since $i = r$, the product of these two is the identity. For example,

$$\begin{aligned}&(a, b_1)(a, b_2)(a, b_3)(a, b_2)\sigma_{r+1} \cdots \sigma_m\\ &= (a, b_1)(a, b_2)(a, b_2)(b_2, b_3)\sigma_{r+1} \cdots \sigma_m\\ &= (a, b_1)(b_2, b_3)\sigma_{r+1} \cdots \sigma_m.\end{aligned}$$

Thus, we have the identity e written as the product of $m-2$ transpositions, and by induction on n, we see that $m-2$, and hence m, is even. This concludes the proof.

2.5.4 Definition

A permutation $\alpha \in S_n$ is **even** *or* **odd**, *respectively, if α is a product of an even or an odd number of transpositions.*

For example,

$$\alpha = \begin{pmatrix} 1 & 2 & 3 & 4 & 5 & 6 & 7 \\ 5 & 3 & 2 & 7 & 1 & 4 & 6 \end{pmatrix}$$

is even since $\alpha = (1, 5)(2, 3)(4, 7, 6) = (1, 5)(2, 3)(4, 7)(4, 6)$, whereas $\alpha = (1, 2)$ is odd.

The subset of even permutations of S_n is a subgroup. (See 2.5.6 below.)

2.5.5 Definition

The subgroup of S_n consisting of all the even permutations of S_n is denoted A_n and is called the **alternating group** *of degree n.*

2.5.6 Theorem

A_n is a normal subgroup of S_n, and $S_n\colon A_n = 2$.

Proof The mapping from S_n into the multiplicative group $\{1, -1\}$ that takes $\alpha \in S_n$ onto 1 or -1 depending on whether α is even or odd is an epimorphism with kernel A_n.

The groups S_n and A_n have many interesting properties. One such property that we will need in Chapter 6 in our study of fields is that for $n \geq 5$, A_n is ***simple***. A group is simple if it has no nontrivial normal subgroups.

2.5.7 Theorem

If $n \geq 5$, then A_n is simple.

Proof Let N be a normal subgroup of A_n, $N \neq \{e\}$. We will show that $N = A_n$. Note that $(1, 2, 3) = (1, 2)(1, 3)$, so that a 3-cycle is even. That is, A_n contains all 3-cycles. The proof will proceed by showing in turn that N contains a 3-cycle, that N contains all 3-cycles, and then that A_n is generated by 3-cycles.

The proof uses the following observation, or technique, repeatedly: Suppose $\alpha \in N$ and $\beta \in A_n$. Then $\beta\alpha\beta^{-1} \in N$, since N is normal in A_n, and consequently $\beta\alpha\beta^{-1}\alpha^{-1} \in N$. Throughout the proof, the element $\beta \in A_n$ will be carefully selected in such a way that the element $\beta\alpha\beta^{-1}\alpha^{-1} \in N$ has a special form.

For example, suppose the 5-cycle $\alpha = (i_1, i_2, i_3, i_4, i_5)$ is an element of N. Let $\beta = (i_1, i_2, i_3)$. Then $\beta \in A_n$ so that the 3-cycle

$$\begin{aligned}
&\beta\alpha\beta^{-1}\alpha^{-1}\\
&\qquad = (i_1, i_2, i_3)(i_1, i_2, i_3, i_4, i_5)(i_1, i_2, i_3)^{-1}(i_1, i_2, i_3, i_4, i_5)^{-1}\\
&\qquad = (i_1, i_2, i_3)(i_1, i_2, i_3, i_4, i_5)(i_3, i_2, i_1)(i_5, i_4, i_3, i_2, i_1)\\
&\qquad = (i_2, i_3, i_5)
\end{aligned}$$

is also in N.

Let $\alpha \in N$ with $\alpha \neq e$, and write $\alpha = \alpha_1\alpha_2 \cdots \alpha_k$, with the α_i's disjoint cycles. Let $\alpha_1 = (i_1, i_2, \ldots, i_r)$. First suppose that $r > 3$, and let $\beta = (i_1, i_2, i_r)$. Then $\beta \in A_n$, and

$$\begin{aligned}
&\beta\alpha\beta^{-1}\alpha^{-1}\\
&\quad = (i_1, i_2, i_r)(i_1, i_2, \ldots, i_r)\alpha_2 \cdots \alpha_k(i_r, i_2, i_1)\alpha^{-1}\\
&\quad = (i_1, i_3, i_4, i_5, \ldots, i_{r-1}, i_2, i_r)\alpha_2 \cdots \alpha_k\alpha_k^{-1} \cdots \alpha_2^{-1}(i_r, , \ldots, i_2, i_1)\\
&\quad = (i_1, i_2, i_{r-1})
\end{aligned}$$

is in N.

Now suppose that $r = 3$. If $k = 1$, then N contains the 3-cycle (i_1, i_2, i_3). If $k > 1$, let $\alpha_2 = (j_1, j_2, \ldots, j_s)$, and let $\beta = (i_1, i_2, j_1)$. Then

$$\begin{aligned}
&\beta\alpha\beta^{-1}\alpha^{-1}\\
&\qquad = (i_1, i_2, j_1)(i_1, i_2, i_3)(j_1, j_2, \ldots, j_s)\alpha_3 \cdots \alpha_k(j_1, i_2, i_1)\alpha^{-1}\\
&\qquad = (i_1, i_3, j_1)(i_2, j_2, \ldots, j_s)\alpha_3 \cdots \alpha_k\alpha_k^{-1}\alpha_{k-1}^{-1} \cdots \alpha_2^{-1}\alpha_1^{-1}\\
&\qquad = (i_1, i_3, j_1)(i_2, j_2, \ldots, j_s)(j_s, j_{s-1}, \ldots, j_2, j_1)(i_3, i_2, i_1)\\
&\qquad = (i_1, i_2, j_1, i_3, j_s)
\end{aligned}$$

is in N. Thus by the argument above, there is a 3-cycle in N.

Finally, suppose that $r = 2$. Then $k > 1$ since α is even. Let

$\beta = (i_1, i_2, j_1)$. Then

$$\begin{aligned}
&\beta\alpha\beta^{-1}\alpha^{-1}\\
&\quad = (i_1, i_2, j_1)(i_1, i_2)(j_1, j_2, \ldots, j_s)\alpha_3\alpha_4 \cdots \alpha_k(j_1, i_2, i_1)\alpha^{-1}\\
&\quad = (i_1, j_1)(i_2, j_2, j_3, \ldots, j_s)\alpha_3\alpha_4 \cdots \alpha_k\alpha^{-1}\\
&\quad = (i_1, j_1)(i_2, j_2, j_3, \ldots, j_s)\alpha_2^{-1}\alpha_1^{-1}\\
&\quad = (i_1, j_1)(i_2, j_2, j_3, \ldots, j_s)(j_s, j_{s-1}, \ldots, j_2, j_1)(i_2, i_1)\\
&\quad = (i_1, j_s)(i_2, j_1)
\end{aligned}$$

is an element of N. Let $1 \le a \le n$ be different from i_1, i_2, j_1, and j_2. (This is where we use the hypothesis that $n \ge 5$.) Then

$$\begin{aligned}
&(i_1, i_2, a)(i_1, j_s)(i_2, j_1)(i_1, i_2, a)^{-1}((i_1, j_s)(i_2, j_1))^{-1}\\
&\quad = (i_1, i_2, a)(i_1, j_s)(i_2, j_1)(a, i_2, i_1)(i_1, j_s)(i_2, j_1)\\
&\quad = (i_1, i_2, j_1, j_s, a)
\end{aligned}$$

is an element of N. Again, by the argument above, there is a 3-cycle in N.

We have finally proved that N contains a 3-cycle. Now we will show that N contains every 3-cycle.

Let $(i_1, i_2, i_3) \in N$, $1 \le a \le n$, and $a \ne i_1, i_2, i_3$. Then

$$(i_1, a, i_2)(i_1, i_2, i_3)(i_1, a, i_2)^{-1} = (a, i_3, i_2)$$

is in N, whence

$$(a, i_3, i_2)^2 = (a, i_2, i_3)$$

is in N.

We have then that if $(i_1, i_2, i_3) \in N$, then for any $a \ne i_1, i_2, i_3$, the element $(a, i_2, i_3) \in N$. Suppose now that (a, b, c) is any 3-cycle, and that $(i_1, i_2, i_3) \in N$. If a is not one of i_1, i_2, or i_3, then $(a, i_2, i_3) \in N$, and

$$(a, i_2, i_3) = (i_2, i_3, a)$$

is in N. If a is one of i_1, i_2, or i_3, we may as well suppose that $a = i_1$. Thus in any case, $(a, i_2, i_3) = (i_2, i_3, a) \in N$. If $b \ne i_3$, then $(b, i_3, a) \in N$. If $b = i_3$, then $(i_2, b, a) \in N$. So either (i_3, a, b) or $(i_2, b, a) \in N$. Thus, either (c, a, b) or $(c, b, a) \in N$. But $(c, a, b) = (a, b, c)$, and $(c, b, a)^2 = (a, b, c)$, whence $(a, b, c) \in N$. We have then that any normal subgroup $N \ne \{e\}$ of A_n contains all 3-cycles. Noting that

$$(i_1, i_2)(i_1, i_3) = (i_1, i_2, i_3),$$

and that

$$(i_1, i_2)(i_3, i_4) = (i_1, i_3, i_4)(i_1, i_3, i_2),$$

and remembering that every element of A_n is a product of an even number of 2-cycles, we have that every element of A_n is a product of 3-cycles. Thus $A_n = N$, and our theorem is proved.

2.5.8 Corollary

If $n \geq 5$ and N is normal in S_n, then N is either $\{e\}$, or A_n, or S_n.

Proof Suppose $N \neq \{e\}$. Since $N \cap A_n$ is normal in A_n, $N \cap A_n = \{e\}$ or A_n. If $N \cap A_n = \{e\}$, then N has just 2 elements since A_n is of index 2 in S_n. Let $\alpha_1\alpha_2 \cdots \alpha_m$ be the element $\neq e$ in N with the α_i's disjoint cycles. This element is of order 2, so each α_i is a 2-cycle. Let $\alpha_1 = (i_1, i_2)$, and $\alpha_2 = (i_3, i_4)$ if $m > 1$. Then if $m > 1$,

$$(i_1, i_2, i_3, i_4)\alpha_1\alpha_2 \cdots \alpha_m(i_1, i_2, i_3, i_4)^{-1} = (i_1, i_4)(i_2, i_3)\alpha_3 \cdots \alpha_m$$

is in N, and is neither e nor $\alpha_1\alpha_2 \cdots \alpha_m$. If $m = 1$, let $i_1 \neq i_2 \neq i_3 \neq i_1$. Then

$$(i_1, i_2, i_3)(i_1, i_2)(i_3, i_2, i_1) = (i_1, i_3) \in N, \quad (i_1, i_3) \neq e,$$

and $(i_1, i_3) \neq (i_1, i_2)$. Thus $N \cap A_n \neq \{e\}$, and hence $N \supset A_n$. Since $S_n{:}A_n = 2$, N is either A_n or S_n. Therefore, $N = \{e\}$, or A_n, or S_n.

PROBLEMS

1 Write the following permutations as a product of disjoint cycles and as a product of transpositions.

(a) $\begin{pmatrix} 1 & 2 & 3 & 4 & 5 \\ 5 & 4 & 3 & 2 & 1 \end{pmatrix}$

(b) $\begin{pmatrix} 1 & 2 & 3 & 4 & 5 & 6 & 7 \\ 7 & 5 & 6 & 3 & 2 & 4 & 1 \end{pmatrix}$

(c) $\begin{pmatrix} 1 & 2 & 3 & 4 & 5 & 6 & 7 \\ 1 & 2 & 5 & 4 & 3 & 7 & 6 \end{pmatrix}$

(d) $(1, 2, 5, 3)(1, 2, 4, 5)(6, 3, 4)$

(e) $(1, 2, 3)(1, 2, 3, 4)(1, 2, 3, 4, 5)$

(f) $(1, 2, 3, 4, 5, 6, 7, 8)^{100}$.

2 Prove that the order of a product of disjoint cycles is the least common multiple of the lengths of those cycles.

3 Prove that in S_n,

$$\sigma^{-1}(i_1, i_2, \ldots, i_r)\sigma = (\sigma(i_1), \sigma(i_2), \ldots, \sigma(i_r)).$$

4 Prove that the elements α and β of S_n are conjugates if and only if they have the same cycle structure, that is, if and only if α and β have the same number of disjoint cycles of the same length.

5 View S_{n-1} as a subgroup of S_n, that is, as the subgroup of S_n fixing n. Write down a set of representatives of the left cosets of S_{n-1}. Write down a set of representatives of the right cosets of S_{n-1}.

6 What are the conjugates of S_{n-1} in S_n? (See Problem 5.)

7 What is $S_{n-1} \cap A_n$?

8 A_4 has nontrivial normal subgroups. Find them all.

9 Prove that if f is any endomorphism of S_n, then $f(A_n) \subset A_n$.

10 Prove that the commutator subgroup of S_n is A_n.

11 Show that every automorphism of S_3 is inner.

12 Find the automorphism group of S_3.

13 Find all subgroups of S_n of index 2.

14 Prove that S_n is generated by the $n-1$ transpositions $(1, 2), (1, 3), \ldots, (1, \mathrm{n})$.

15 Prove that S_n is generated by the $n-1$ transpositions $(1, 2), (2, 3), \ldots, (n-1, n)$.

16 Prove that S_5 is generated by $\{(1, 2), (1, 2, 3, 4, 5)\}$.

17 Prove that S_n is generated by 2 elements.

2.6 DIRECT PRODUCTS AND SEMIDIRECT PRODUCTS

In this section, we describe two constructions of new groups from old. These constructions will furnish additional examples of groups and are useful in giving complete descriptions of groups of certain types. For example, in 2.7, we will use our results in giving a description of all finite Abelian groups in terms of cyclic groups.

Let G and H be groups. In the set $G \times H$ define multiplication by

$$(a, b)(c, d) = (ac, bd).$$

This makes $G \times H$ into a group (2.1, Problem 21). The identity element is (e_G, e_H), and $(g, h)^{-1} = (g^{-1}, h^{-1})$. [From now on, we will simply write e for the identity of any group under consideration. Thus, (e_G, e_H) may be written (e, e) or even more simply just e.] The associative law holds since it holds in both G and H. This group is denoted $G \times H$ and is called the ***external direct product*** of G and H. In $G \times H$, let

$$G^* = \{(g, e) : g \in G\},$$

and

$$H^* = \{(e, h) : h \in H\}.$$

The mappings

$$G \times H \to H: (g, h) \to h$$

and

$$G \times H \to G: (g, h) \to g$$

are epimorphisms with kernels G^* and H^*, respectively. Therefore, G^* and H^* are normal subgroups of $G \times H$. For $(g, h) \in G \times H$, $(g, h) = (g, e)(e, h)$, so that every element in $G \times H$ can be written as the product of an element from G^* with one from H^*. Further, this can be done in exactly one way since $(g, h) = (g', e)(e, h')$ implies that $(e, e) = (g^{-1}g', h^{-1}h')$, so that $g = g'$ and $h = h'$. Thus, $G \times H$ has normal subgroups G^* and H^* such that every element of $G \times H$ can be written uniquely in the form gh with $g \in G^*$ and $h \in H^*$, and so $G \times H = G^*H^* = \{gh: g \in G^*, h \in H^*\}$. We say that $G \times H$ is the ***direct product*** of the normal subgroups G^* and H^*.

It is worth observing here that $G^* \cap H^* = (e, e)$, and that $gh = hg$ for $g \in G^*$ and $h \in H^*$. In fact, if M and N are normal subgroups of a group which intersect in just the identity element, then the elements of M commute with those of N. (See Problem 19 in Section 2.3.)

This discussion carries over to any finite number of groups. That is, let $G_1, G_2, \ldots, G_n$ be groups, and let $G_1 \times G_2 \times \cdots \times G_n$ be the group obtained from the set

$$G_1 \times G_2 \times \cdots \times G_n = \{(g_1, g_2, \ldots, g_n): g_i \in G_i \text{ for all } i\}$$

by defining

$$(g_1, g_2, \ldots, g_n)(h_1, h_2, \ldots, h_n) = (g_1h_1, g_2h_2, \ldots, g_nh_n).$$

The group $G = G_1 \times G_2 \times \cdots \times G_n$ is called the ***external direct product*** of the groups $G_1, G_2, \ldots, G_n$. Let

$$G_i^* = \{(e, e, \ldots, e, g_i, e, \ldots, e): g_i \in G_i\}.$$

The G_i^* are normal subgroups of G, and every element of G can be written uniquely as a product $g_1g_2 \cdots g_n$ with g_i in G_i^*.

2.6.1 Definition

Let $G_1, G_2, \ldots, G_n$ be normal subgroups of a group G. Then G is the **direct product** *of the subgroups $G_1, G_2, \ldots, G_n$ if every element $g \in G$ can be written uniquely in the form $g = g_1g_2 \cdots g_n$ with $g_i \in G_i$.*

If G is the direct product of normal subgroups $N_1, N_2, \ldots, N_n$, then we also write $G = N_1 \times N_2 \times \cdots \times N_n$. In the discussion above, the external direct product $G = G_1 \times G_2 \times \cdots \times G_n$ is the direct product of the normal subgroups

$G_1^*, G_2^*, \ldots, G_n^*$, so that

$$G_1 \times G_2 \times \cdots \times G_n = G_1^* \times G_2^* \times \cdots \times G_n^*,$$

although strictly speaking, the $\times$'s on the left have a different meaning than the $\times$'s on the right. The appropriate interpretation of $\times$ can always be inferred from the context. In general, when we write $G = G_1 \times G_2 \times \cdots \times G_n$, we mean it in the sense of 2.6.1. The group $G_1 \times G_2 \times \cdots \times G_n$ is also written $G = \prod_{i=1}^n G_i$.

If $G = G_1 \times G_2 \times \cdots \times G_n$, then $G_i \cap G_j = \{e\}$ for distinct i and j. Since G_i and G_j are normal, we have that for $g_i \in G_i$ and $g_j \in G_j$, $g_i g_j = g_j g_i$. Therefore, if $g_i, h_i \in G_i$, then

$$(g_1 g_2 \cdots g_n)(h_1 h_2 \cdots h_n) = (g_1 h_1)(g_2 h_2) \cdots (g_n h_n).$$

Direct products of groups serve two purposes. Given groups $G_1, G_2, \ldots, G_n$, we can make the new group $G_1 \times G_2 \times \cdots \times G_n$. Thus, it is a way to construct new groups from old and increases our repertoire of examples. On the other hand, a group is understood better if it is a direct product of subgroups that are less complicated than the group itself. For example, Section 2.7 is devoted to showing that every finite Abelian group is a direct product of cyclic groups. Since cyclic groups are relatively simple entities, finite Abelian groups themselves become better understood. Theorem 7.4.9 is another example of the same sort.

If $G = G_1 \times G_2 \times \cdots \times G_n$, then the G_i's are called ***direct factors*** of G. In fact, if H is any subgroup of G such that $G = H \times K$ for some subgroup K, then H is called a ***direct factor*** of G. If $G = H \times K$ implies that $H = \{e\}$ or $K = \{e\}$, then G is called ***indecomposable***.

Note that if $G = G_1 \times G_2 \times G_3$, then

$$G = (G_1 \times G_2) \times G_3 = G_1 \times (G_2 \times G_3).$$

Similarly, one can parenthesize $G_1 \times G_2 \times \cdots \times G_n$ at will. Since every element $g \in G = H \times K$ can be written uniquely in the form $g = hk$ with $h \in H$ and $k \in K$, and since for $h_i \in H$ and $k_i \in K$, $(h_1 k_1)(h_2 k_2) = (h_1 h_2)(k_1 k_2)$, $G \to K: hk \to k$ is an epimorphism with kernel H, yielding

$$(H \times K)/H \approx K.$$

Consider again the map $G \to K: hk \to k$. Denote it by f, and think of it as an endomorphism of G. Then f is ***idempotent***, that is, $f \circ f = f$. Thus, direct factors of G give rise to idempotent endomorphisms of G.

2.6.2 Theorem

A normal subgroup of a group G is a direct factor of G if and only if it is the image of an idempotent endomorphism of G. If f is any idempotent endomorphism of G such that $\operatorname{Im} f$ *is normal, then* $G = \operatorname{Ker} f \times \operatorname{Im} f$.

Proof Only the last statement remains to be proved. Let f be an idempotent endomorphism of G with $\operatorname{Im} f$ normal. We need to show that every element in G can be wrtten uniquely in the form xy with $x \in \operatorname{Ker} f$ and $y \in \operatorname{Im} f$. Let $g \in G$. Then $g = (gf(g)^{-1})f(g))$. Clearly, $f(g) \in \operatorname{Im} f$. Since $f(gf(g)^{-1}) = f(g)f \circ f(g^{-1})$ and $f \circ f = f$, then $gf(g)^{-1} \in \operatorname{Ker} f$. Suppose $g = xy$ with $f(x) = e$ and $f(a) = y$. Then $f(g) = f(x)f(y) = f \circ f(a) = f(a) = y$, and therefore also $x = gf(g)^{-1}$.

We now turn to semidirect products, a more general product of two groups than their direct product. If N is a normal subgroup of a group G, and H is any subgroup of G, then $NH = \{nh: n \in N, h \in H\}$ is a subgroup of G. (See 2.3.12.)

2.6.3 Definition

Let N be a normal subgroup of a group G, and let H be a subgroup of G. If $G = NH$ and $N \cap H = \{e\}$, then G is a **semidirect product** *of N and H, and G* **splits** *over N, or is a* **split extension** *of N by H.*

If H were also normal, then G would be the direct product of the two normal subgroups N and H. Thus, semidirect product is a generalization of direct product.

The group $G = \mathbb{R} \times \mathbb{R}^*$ with multiplication given by

$$(a, b)(c, d) = (a + bc, bd)$$

is a split extension of the normal subgroup $N = \{(a, 1): a \in \mathbb{R}\}$ by the subgroup $H = \{(0, b): b \in \mathbb{R}^*\}$. (See 2.3, Problem 1, and 2.4, Problems 1–4.)

The proof of the following theorem is very similar to that of 2.6.2 and is left to the reader.

2.6.4 Theorem

Let N and H be subgroups of G. Then G is a split extension of N by H if and only if H is the image of an idempotent endomorphism of G with kernel N.

Here is an illustration of 2.6.4. Let $\mathrm{GL}_2(\mathbb{R})$ and $\mathrm{SL}_2(\mathbb{R})$ be the general linear group and the special linear group of 2×2 matrices, respectively. [See 2.1.2(f) and 2.3.4(f).] The mapping

$$f: \mathrm{GL}_2(\mathbb{R}) \to \mathrm{GL}_2(\mathbb{R}): x \to \begin{pmatrix} \det(x) & 0 \\ 0 & 1 \end{pmatrix}$$

is an idempotent endomorphism of $GL_2(\mathbb{R})$ with kernel $SL_2(\mathbb{R})$, and so $GL_2(\mathbb{R})$ is a split extension of $SL_2(\mathbb{R})$ by $\operatorname{Im} f$, and $\operatorname{Im}(f)$ is isomorphic to the multiplicative group $\mathbb{R}^*$.

We now show how to construct split extensions. If G is a semidirect product NH, then for $n, m \in N$ and $h, k \in H$, $(nh)(mk) = (nhmh^{-1})(hk)$. The element hmh^{-1} is in N since N is normal in G. The mapping $\varphi: H \to \operatorname{Aut}(N)$ defined by $\varphi(h)(g) = hgh^{-1}$ is a homomorphism (2.3.15), and the multiplication above is $(nh)(mk) = (n\varphi(h)(m))(hk)$. Now, if N and H are *any* groups, and φ is *any* homomorphism from H to $\operatorname{Aut}(H)$, then we may construct a group that is a split extension of an isomorphic copy of N by an isomorphic copy of H.

2.6.5 Theorem

Let N and H be groups, and let φ be a homomorphism from H into the automorphism group $\operatorname{Aut}(N)$ *of N. Then*

(a) *The Cartesian product $G = N \times H$ is a group under the operation $(n_1, h_1)(n_2, h_2) = (n_1\varphi(h_1)(n_2), h_1h_2)$;*

(b) *$N^* = \{(n, e): n \in N\}$ is a normal subgroup, and $H^* = \{(e, h): h \in H\}$ is a subgroup of G; and*

(c) *G is a split extension of its normal subgroup N^* by its subgroup H^*.*

Proof The identity is (e, e), and the inverse of (n, h) is $(\varphi(h^{-1})(n^{-1}), h^{-1})$. The equations

$$\begin{aligned}
&(n_1, h_1)((n_2, h_2)(n_3, h_3)) \\
&\quad = (n_1, h_1)(n_2\varphi(h_2)(n_3), h_2h_3) \\
&\quad = (n_1\varphi(h_1)(n_2\varphi(h_2)(n_3)), h_1h_2h_3) \\
&\quad = (n_1(\varphi(h_1)(n_2))\varphi(h_1)(\varphi(h_2)(n_3)), h_1h_2h_3) \\
&\quad = (n_1\varphi(h_1)(n_2))(\varphi(h_1)(\varphi(h_2))(n_3)), h_1h_2h_3) \\
&\quad = (n_1(\varphi(h_1)(n_2))\varphi(h_1h_2)(n_3)), h_1h_2h_3) \\
&\quad = (n_1\varphi(h_1)(n_2), h_1h_2)(n_3, h_3) \\
&\quad = ((n_1, h_1)(n_2, h_2))(n_3, h_3)
\end{aligned}$$

establish the associative law. Thus, G is a group.

The map $N \times H \to H: (n, h) \to h$ is an epimorphism with kernel N^*, and the map $H \to N \times H: h \to (e, h)$ is a monomorphism with image H^*. Thus N^* is a normal subgroup of G, and H^* is a subgroup of G.

It is clear that $N \times H = N^*H^*$ and that $N^* \cap H^* = \{e\}$. This completes the proof.

The group in (a) is called the ***semidirect product of N and H depending on φ***.

Here are some special cases of interest. If φ is the trivial homomorphism $H \to \mathrm{Aut}(N)$: $h \to e$, then $N \times H$ is simply the external direct product of N and H (Problem 14). if $H = \mathrm{Aut}(N)$ and φ is the identity map, then $(n_1, h_1)(n_2, h_2) = (n_1 h_1(n_2), h_1 h_2)$. Noticing that $(h_1(n_2), e) = (e, h_1)(n_2, e)(e, h_1)^{-1}$, we see that every automorphism of N^* is induced by an inner automorphism of the semidirect product $N \times H$. Further, for $x_i \in N^*$ and $y_i \in H^*$,

$$(x_1 y_1)(x_2 y_2) = x_1 (y_1 x_2 y_1^{-1}) y_1 y_2.$$

In this case, that is, when $H = \mathrm{Aut}(N)$, the semidirect product is called the ***holomorph*** of N and is denoted $\mathrm{Hol}(N)$. Elements of a group N produce permutations of the set N by multiplication on the left, say. Automorphisms of N are special kinds of permutations of N. The subgroup of the group of all permutations of N generated by the left multiplications and the automorphisms of N is isomorphic to the holomorph of N. In this connection, see Problem 23.

Let N be any group, let H be any subgroup of $\mathrm{Aut}(N)$, and let φ be the homomorphism $H \to \mathrm{Aut}(N)$: $h \to h$. The semidirect product of N and H over φ is then the Cartesian product $G = N \times H$ with multiplication given by

$$(n_1, h_1)(n_2, h_2) = (n_1 h_1(n_2), h_1 h_2).$$

We say that G is the ***extension of N by the group H of automorphisms of N***, and that we have ***adjoined the group H of automorphisms to N***. If N has a nontrivial automorphism, then G is not just the direct product of N and H, and G is not Abelian. For example, if h is not the identity automorphism of N, then there is an n in N such that $h(n) \neq n$, and $(n, e)(e, h) = (n, h)$, whereas $(e, h)(n, e) = (h(n), h)$.

For any Abelian group N, the mapping $\alpha: N \to N: h \to h^{-1}$ is an automorphism. Since only elements of order 2 are their own inverses, α is the identity automorphism if and only if every element of N has order 2. So suppose that N is Abelian and not every one of its elements has order 2. For example, N can be any cyclic group of order greater than 2. Then the element α of $\mathrm{Aut}(N)$ has order 2, and $\{e, \alpha\}$ is a subgroup of $\mathrm{Aut}(N)$ of order 2. Let $H = \{e, \alpha\}$. The extension G of the Abelian group N by the 2-element group of automorphisms H is a non-Abelian group and is called a ***generalized dihedral group***. If N is cyclic, then the extension G is a ***dihedral group***, as defined in 2.1.2. If N is finite, then $o(G) = 2o(N)$. In particular, letting N be the cyclic group $\mathbb{Z}(n)$ of order n, we see that for $n > 2$, there is a non-Abelian splitting extension of $\mathbb{Z}(n)$ by a group of order 2. Thus for every such $n > 2$, there is a non-Abelian group of order $2n$. The simplest example is for $n = 3$, in which case the resulting group is isomorphic to S_3.

We conclude this section with the following example. Let N be a cyclic group of order p^2, where p is an odd prime. Raising each element of N to the

$p+1$ power is an automorphism α of N, and α has order p in $\mathrm{Aut}(N)$. Let H be the subgroup of $\mathrm{Aut}(N)$ generated by α. Then the semidirect product of N with H determined by α is a non-Abelian group of order p^3. Every group of order p^2 is Abelian (2.3, Problem 20). Also, there are groups of order p^3 not isomorphic to the one just constructed (Problem 20).

PROBLEMS

1 Let $\mathbb{Q}^*$ be the multiplicative group of nonzero rational numbers, and let $\mathbb{Q}^+$ be the subgroup of positive rational numbers. Prove directly from the definition that $\mathbb{Q}^* = \{1, -1\} \times \mathbb{Q}^+$. Prove also using 2.6.2.

2 Prove that a cyclic group of order 6 is a direct product of a group of order 3 and one of order 2.

3 Let p and q be distinct primes. Prove that a cyclic group of order pq is a direct product of a subgroup of order p and one of order q.

4 Prove that a cyclic group of order a power of a prime is indecomposable.

5 Prove that S_3 is indecomposable.

6 Prove that S_n is indecomposable.

7 Prove that the infinite cyclic group is indecomposable.

8 Prove that G is the direct product of the two normal subgroups H and K if and only if every $g \in G$ can be written $g = hk$ with $h \in H$ and $k \in K$, and $H \cap K = \{e\}$.

9 Let $G_1, G_2, \ldots, G_n$ be normal subgroups of G, and let G_j^* be the subgroup of G generated by all the G_i except G_j. Prove that G is the direct product of the normal subgroups $G_1, G_2, \ldots, G_n$ if and only if every $g \in G$ can be written $g = g_1 g_2 \cdots g_n$ with $g_i \in G_i$ for all i, and $G_i \cap G_i^* = \{e\}$ for all i.

10 Let $G = \prod_{i=1}^n G_i$, and $g = g_1 g_2 \cdots g_n$ with $g_i \in G_i$. Prove that $o(g) = \mathrm{lcm}\{o(g_1), o(g_2), \ldots, o(g_n)\}$ if $o(g)$ is finite. Prove that if G is a p-group, then $o(g)$ is the maximum of the $o(g_i)$.

11 Let $\alpha: G \to H$ be an isomorphism. Prove that if A is a direct factor of G, then $\alpha(A)$ is a direct factor of H. Prove that it is not enough to assume that α is a monomorphism, nor enough to assume that α is an epimorphism.

12 Prove that if $G_i \approx H_i$ for $i = 1, 2, \ldots, n$, then

$$\prod_{i=1}^n G_i \approx \prod_{i=1}^n H_i.$$

13 Prove that $G \approx G_1 \times G_2$ if and only if there are homomorphisms $\alpha_i: G \to G_i$ such that if H is any group and $\beta_i: H \to G_i$ are any homomorphisms, then there is a unique homomorphism $\alpha: H \to G$ such that $\alpha_i \circ \alpha = \beta_i$.

14 Let φ be a homomorphism $N \to \text{Aut}(N)$. Prove that the semidirect product of N and H depending on φ is the external direct product of N and H if and only if φ is the trivial homomorphism $N \to \text{Aut}(N)$: $h \to e$.

15 Prove that the group of quaternions Q_8 is not a nontrivial split extension.

16 Prove that if G has order 8 and is non-Abelian, then G is either Q_8 or D_4.

17 Prove that S_3 is a nontrivial split extension.

18 Prove that S_n splits over A_n.

19 Prove that the holomorph of $\mathbb{Z}(3)$ is S_3.

20 Construct two non-Abelian groups of order p^3, where p is an odd prime.

21 Construct all possible groups of order 30 which are split extensions of a cyclic group of order 10 by a group of order 3.

22 Construct a non-Abelian group which is a split extension of a cyclic group of order 3 by a cyclic group of order 10.

23 Let N be any group, and let P be the group of all permutations of the set N. Identify N with the set of left multiplications in P. Let G be the subgroup of P generated by N and the automorphisms of N. Prove that G is a split extension of N by $\text{Aut}(N)$. Prove that G is isomorphic to $\text{Hol}(N)$.

24 Prove that a normal subgroup of a direct factor of a group G is a normal subgroup of G.

25 Let G be a split extension of N by H. Let f be a homomorphism from H into N, and let $K = \{f(h)h: h \in H\}$. Prove that G is a split extension of N by K.

26 Let $\{G_i\}_{i \in I}$ be a family of groups. Let $G = \prod_{i \in I} G_i$ be the set of all mappings f of I into $\bigcup_{i \in I} G_i$ such that $f(i) \in G_i$ for all $i \in I$. For $f, g \in G$, let $(fg)(i) = f(i)g(i)$. Prove that G with this multiplication is a group. Let e_i be the identity of G_i. Prove that $\sum_{i \in I} G_i = \{f \in G: f(i) = e_i$ for all but finitely many $i\}$ is a subgroup of G. (G is the *direct product of the family* $\{G_i\}_{i \in I}$, and the subgroup described is the *direct sum*, or the *coproduct* of the family.)

2.7 FINITE ABELIAN GROUPS

Let G and H be finite Abelian groups. Is there some easy way to determine whether or not G and H are isomorphic? To lend meaning to the question, suppose that G and H are cyclic. Then the answer is yes: G and H are isomorphic if and only if they have the same number of elements. Thus, with each finite cyclic group G there is naturally associated something much simpler than G, namely the positive integer $o(G)$, that completely determines G up to isomorphism. That is, if G and H are finite cyclic groups, then $G \approx H$ if and only if $o(G) = o(H)$. (See 2.3.14.) As trivial as it is, this is a good theorem. In fact, a large portion of Abelian group theory is devoted to generalizing it. The goal of this section is a suitable generalization to finite Abelian groups. What we want to do is to associate, in a natural way, with each finite Abelian group G something simple, which we temporarily denote by Inv(G), such that $G \approx H$ if and only if Inv(G) = Inv(H). That is, we know G up to isomorphism if we know the relatively simple entity Inv(G). Note that Inv(G) cannot be just $o(G)$, since there are two nonisomorphic Abelian groups of order 4, the cyclic group of order 4, and the direct product of two cyclic groups of order 2. (Inv is an abbreviation for "invariant.")

To realize our objective, we will show that every finite Abelian group is a direct product of cyclic groups, and in essentially only one way. Once this is accomplished, a completely satisfying Inv(G) will be at hand.

When dealing with Abelian groups, it is customary to use additive notation. That is, the group operation will be called "addition," and we will "add" two elements rather than "multiply" them. We write $g + h$ instead of $g \cdot h$ or gh, and ng replaces g^n. The identity element is denoted 0 instead of 1 or e. Direct products are called direct sums, and direct factors are called direct summands. If $A_1, A_2, \ldots, A_n$ are Abelian groups, then their direct sum is written $A_1 \oplus A_2 \oplus \cdots \oplus A_n$, or $\bigoplus_{i=1}^{n} A_i$. *These are merely notational changes.* For example, 2.6.1 and 2.6.2 become, in additive notation, the following.

2.7.1 Definition

Let $G_1, G_2, \ldots, G_n$ be subgroups of the Abelian group G. Then G is the **direct sum** *of the subgroups $G_1, G_2, \ldots, G_n$ if every element $g \in G$ can be written uniquely in the form $g = g_1 + g_2 + \cdots + g_n$ with $g_i \in G_i$.*

2.7.2 Theorem

A subgroup of an Abelian group G is a summand of G if and only if it is the image of an idempotent endomorphism of G. If f is any idempotent endomorphism of G, then $G = \operatorname{Ker} f \oplus \operatorname{Im} f$.

If $G_1, G_2, \ldots, G_n$ are finite cyclic groups, then $\bigoplus_{i=1}^{n} G_i$ is of course a finite Abelian group. It may no longer be cyclic. [For example, if $\mathbb{Z}(2)$ denotes the cyclic group of order 2, then $\mathbb{Z}(2) \oplus \mathbb{Z}(2)$ is not cyclic since it has 4 elements and each element is of order at most 2.] In fact, every finite Abelian group can be realized in this way. That is, every finite Abelian group is a direct sum of cyclic groups. This is a marvelous theorem, because it says that every finite Abelian group can be constructed in an easy way from much less complicated groups. The proof of this will be broken up into several steps.

2.7.3 Definition

Let G be a group, and let p be a prime. Then G is a ***p*-group** *if every element in G has order a power of p.*

For example, $\mathbb{Z}(2) \oplus \mathbb{Z}(2)$ and $\mathbb{Z}(4)$ are 2-groups, and $\mathbb{Z}(27)$ is a 3-group. However, $\mathbb{Z}(6)$ is not a p-group for any prime p.

2.7.4 Theorem

Every finite Abelian group is a direct sum of p-groups.

Proof Let G be a finite Abelian group. If $o(G) = 1$, there is nothing to do. Otherwise, let P be the set of primes dividing $o(G)$. Let $G_p = \{g \in G: o(g) \text{ is a power of } p\}$. We will show that $G = \bigoplus_{p \in P} G_p$. The G_p are indeed subgroups of G since G is Abelian. We need to write every element $g \in G$ in the form $g = \sum_{p \in P} g_p$ with $g_p \in G_p$, and to show that this can be done in only one way. Let $o(g) = \prod_{p \in P} n_p$, where each n_p is a power of p. This is fair since $o(g)$ divides $o(G)$. Let $m_p = o(g)/n_p$. Then the m_p are relatively prime, so there exist integers a_p such that $1 = \sum a_p m_p$. (See 1.6, page 23.) Therefore $g = \sum a_p m_p g$, and since $n_p(a_p m_p g) = a_p o(g) g = 0$, then $a_p m_p g \in G_p$. Letting $g_p = a_p m_p g$, we have that $g = \sum g_p$ wtih $g_p \in G_p$.

Now suppose that $g = \sum h_p$ with $h_p \in G_p$. Then $0 = \sum_p (g_p - h_p)$. The order of $g_p - h_p$ is a power of p, and so the order of $\sum_p (g_p - h_p)$ is the least common multiple of the orders of the $(g_p - h_p)$. (See 2.2, Problem 24.) This must be 1 since $o(0) = 1$. It follows that $o(g_p - h_p) = 1$ and that $g_p = h_p$ for all $p \in P$.

The subgroup G_p of G is called the ***p-component*** of G, or the ***p-primary part*** of G. Thus, 2.7.4 says that *a finite Abelian group is the direct sum of its p-primary parts for the various primes p dividing $o(G)$*. If $G = H_1 \oplus H_2 \oplus \cdots \oplus H_m$ with each H_i a p_i-group for distinct primes p_i, then clearly H_i is the p_i-primary part of G. Therefore, there is only one way to write G as a direct sum of p-groups for distinct primes p, and that is as the direct sum of its p-primary parts.

To show that every finite Abelian group is a direct sum of cyclic groups, it now suffices to show that this is the case for finite Abelian p-groups. There are many ways to prove this. (See Problems 14 and 15, for example.) It suffices to write $G = A \oplus B$ with $A \neq \{0\} \neq B$ if G is not already cyclic. Induction on $o(G)$ does the rest. But how does one get hold of appropriate A and B? One way is to take A to be a cyclic subgroup of G of largest order. Then $G = A \oplus B$ for an appropriate B. This is a key fact, which we now prove, appealing to 2.7.2.

2.7.5 Lemma

Let G be a finite p-group, and let g be an element of G of maximum order. Then $\langle g \rangle$ is a summand of G.

Proof Let g be an element of largest order in G. If $\mathbb{Z}g \neq G$, then there is a nonzero subgroup S of G such that $S \cap \mathbb{Z}g = \{0\}$. To see this, suppose that $a \notin \mathbb{Z}g$. Then $p^n a \in \mathbb{Z}g$ for some positive integer n, so there is an element $b \in G$ such that $b \notin \mathbb{Z}g$ and $pb \in \mathbb{Z}g$. Since $o(g) > o(pb)$, we get that $pb = pmg$. Thus, $p(b - mg) = 0$ and $s = b - mg \notin \mathbb{Z}g$. But $\mathbb{Z}s$ has order p, so $\mathbb{Z}s \cap \mathbb{Z}g = \{0\}$.

So let S be any nonzero subgroup such that $S \cap \mathbb{Z}g = \{0\}$. In G/S, $o(g + S) = o(g)$. Now, $g + S$ is an element of maximum order in G/S. By induction on $o(G)$, $\langle g + S \rangle$ is a summand of G/S. The composition

$$G \to G/S \to \langle g + S \rangle \to \langle g \rangle \to G,$$

where the first map is the natural homomorphism, the second is the projection of G/S onto its summand $\langle g + S \rangle$, the third takes $m(g + S)$ to mg, and the last is the inclusion map, is an idempotent endomorphism of G with image $\langle g \rangle$. By 2.7.2, $\langle g \rangle$ is a summand of G.

2.7.6 Theorem

Every finite Abelian p-group is a direct sum of cyclic groups.

Proof As indicated above, this follows readily from 2.7.5. Let g be an element of G of maximum order. Then $G = \mathbb{Z}g \oplus B$ for some subgroup B. Since $o(B) < o(G)$, B is a direct sum of cyclic groups by induction on $o(G)$. The theorem follows.

From 2.7.4 and 2.7.6 we have

2.7.7 Fundamental theorem of finite abelian groups

Every finite Abelian group is a direct sum of cyclic groups.

Actually we proved more. We proved that *every finite Abelian group is a direct sum of cyclic p-groups.*

Now a fundamental question arises. A finite Abelian group is a direct sum of cyclic groups all right, but in how many ways? Let us restrict attention to p-groups. Consider the cyclic group $\mathbb{Z}(2)$ of order 2. Denote its elements by 0 and 1. Let

$$G = \mathbb{Z}(2) \oplus \mathbb{Z}(2),$$

the external direct sum of $\mathbb{Z}(2)$ and $\mathbb{Z}(2)$. Then

$$\begin{aligned} G &= \{(0, 0), (1, 0)\} \oplus \{(0, 0), (0, 1)\} \\ &= \{(0, 0), (1, 1)\} \oplus \{(0, 0), (0, 1)\} \\ &= \{(0, 0), (1, 1)\} \oplus \{(0, 0), (1, 0)\}. \end{aligned}$$

Thus, G can be written in several ways as the direct sum of cyclic groups. Even for a given cyclic subgroup, say $H = \{(0, 0), (1, 1)\}$, there were two distinct subgroups K and L such that $G = H \oplus K = H \oplus L$. This simple example shows that there will be, in general, many ways to write a finite Abelian group as a direct sum of cyclic groups. Is there *anything* worthwhile that is unique about such a direct sum decomposition? The answer is a resounding yes. Let G be a finite Abelian p-group, and let $G = G_1 \oplus G_2 \oplus \cdots \oplus G_m = H_1 \oplus H_2 \oplus \cdots \oplus H_n$ with each G_i and H_i cyclic and nonzero. We will show that $m = n$, and after a possible renumbering, $G_i \approx H_i$. Another way to say this is that for each positive integer k, the number of G_i of order p^k is the same as the number of H_i of order p^k. That is, this number depends only on G, not on a particular direct sum decomposition of it. These numbers will be our magic invariants.

2.7.8 Theorem

Let G be a finite p-group. If

$$G = G_1 \oplus G_2 \oplus \cdots \oplus G_m = H_1 \oplus H_2 \oplus \cdots \oplus H_n$$

with each G_i and H_i nonzero and cyclic, then $m = n$, and after suitable renumbering, $G_i \approx H_i$ for all i.

Proof The proof is by induction on m. If $m = 1$, then G is a cyclic p-group. Thus, G has an element g with $o(g) = o(G)$. Write $g = h_1 + h_2 + \cdots + h_n$. Then $o(h_i) = o(g)$ for some i. Thus $o(H_i) = o(G)$, so $H_i = G$. Therefore, $n = 1$ and obviously $G_1 \approx H_1$.

Now assume that $m > 1$. Number the G_i so that for all i, $o(G_1) \geq o(G_i)$. Let g_1 be a generator of G_1, and write $g_1 = h_1 + h_2 + \cdots + h_n$ with $h_i \in H_i$. The order of g_1 is the maximum of the $o(h_i)$. Thus, $o(g_1) = o(h_j)$ for some j. Since $o(g_1) = o(G_1)$ and $o(h_j) \leq o(H_j)$, it follows that $o(G_1) \leq o(H_j)$. Similarly some $o(G_k) \geq o(H_j)$, whence $o(G_1) = o(H_j)$. We may as well suppose that $j = 1$. Note that $G_1 \cap (H_2 \oplus \cdots \oplus H_n) = \{0\}$.

Indeed,

$$kg_1 = k(h_1 + \cdots + h_n) = x_2 + \cdots + x_n$$

with $x_i \in H_i$ yields $kh_1 = 0$, and hence $kg_1 = 0$. Since

$$o(G_1 \oplus (H_2 \oplus \cdots \oplus H_n)) = o(H_1 \oplus (H_2 \oplus \cdots \oplus H_n)),$$
$$G = G_1 \oplus H_2 \oplus \cdots \oplus H_n.$$

So we have

$$G/G_1 \approx G_2 \oplus \cdots \oplus G_m \approx H_2 \oplus \cdots \oplus H_n.$$

Now let α be an isomorphism from $H_2 \oplus \cdots \oplus H_n$ to $G_2 \oplus \cdots \oplus G_m$. We get $G_2 \oplus \cdots \oplus G_m = \alpha(H_2) \oplus \cdots \alpha(H_n)$. By the induction hypothesis, after suitable renumbering, $G_i \approx \alpha(H_i)$ for all $i \geq 2$. But $H_i \approx \alpha(H_i)$ for all $i \geq 2$, and so $G_i \approx H_i$ for all i. The theorem follows.

We now have the desired invariants for finite Abelian groups. If G is a finite p-group, then $G = G_1 \oplus \cdots \oplus G_m$ with G_i cyclic. For convenience, let us arrange the G_i so that

$$o(G_1) \geq o(G_2) \geq \cdots \geq o(G_m).$$

Now G_i, being cyclic, is determined by $o(G_i)$. Thus, the family $\{o(G_1), o(G_2), \ldots, o(G_m)\}$ determines G. Furthermore, by 2.7.8, it does not depend on the particular decomposition $G_1 \oplus \cdots \oplus G_m$. That is, $\{o(G_1), \ldots, o(G_m)\}$ is an invariant of G. Denote it $I(G)$. What all this means is that if H is a finite Abelian p-group, $H = H_1 \oplus \cdots \oplus H_n$ with H_i cyclic and

$$o(H_1) \geq o(H_2) \geq \cdots \geq o(H_n),$$

then $G \approx H$ if and only if

$$\{o(G_1), \ldots, o(G_m)\} = \{o(H_1), \ldots, o(H_n)\}.$$

This means that $m = n$, and $o(G_i) = o(H_i)$ for all i.

Now, given such an invariant, that is, a family $\{i_1, i_2, \ldots, i_n\}$ with i_j a power of p and $i_1 \geq i_2 \geq \cdots \geq i_n$, we can construct the finite p-group $\mathbb{Z}(i_1) \oplus \cdots \oplus \mathbb{Z}(i_n)$, where $\mathbb{Z}(i_j)$ denotes the cyclic group of order i_j. The invariant associated with this group is clearly $\{i_1, \ldots, i_n\}$. In summary, two finite p-groups G and H are isomorphic if and only if $I(G) = I(H)$, and given any family $\{i_1, \ldots, i_n\}$ of powers of p with $i_1 \geq \cdots \geq i_n$, there is a finite p-group G such that $I(G)$ is that family.

Consider now the general case. That is, let G be any finite Abelian group. First, $G = \bigoplus_{p \in P} G_p$, where P is the set of primes dividing $o(G)$, and G_p is the p-component of G. If H is another such group, then $G \approx H$ if and only if their corresponding p-components are isomorphic. Thus, we could define $I_p(G) = I(G_p)$, and then the finite Abelian groups G and H would be isomorphic if and only if $I_p(G) = I_p(H)$ for all p. There is a little neater way to do this, however. There is a complete invariant for G more analogous to the one when G is a p-group.

First, notice that if A and B are finite cyclic with $o(A)$ and $o(B)$ relatively prime, then $A \oplus B$ is cyclic. In fact, if a generates A and b generates B, then $a + b$ generates $A \oplus B$. This generalizes. If $A_1, A_2, \ldots, A_n$ are cyclic and $(o(A_i), o(A_j)) = 1$ for $i \neq j$, then $\bigoplus_{i=1}^{n} A_i$ is cyclic with generator $\sum_{i=1}^{n} a_i$, where a_i generates A_i.

Now let us consider an example. Suppose that

$$G = [\mathbb{Z}(5^3) \oplus \mathbb{Z}(5^2) \oplus \mathbb{Z}(5^2) \oplus \mathbb{Z}(5)] \\ \oplus [\mathbb{Z}(3^3) \oplus \mathbb{Z}(3^3) \oplus \mathbb{Z}(3^2)] \oplus [\mathbb{Z}(2^4) \oplus \mathbb{Z}(2)].$$

The subgroups in square brackets are the p-components of G. Regroup these summands of G like this:

$$G = [\mathbb{Z}(5^3) \oplus \mathbb{Z}(3^3) \oplus \mathbb{Z}(2^4)] \\ \oplus [\mathbb{Z}(5^2) \oplus \mathbb{Z}(3^3) \oplus \mathbb{Z}(2)] \oplus [\mathbb{Z}(5^2) \oplus \mathbb{Z}(3^2)] \oplus [\mathbb{Z}(5)].$$

Now the subgroups in square brackets are cyclic and the order of the first is divisible by the order of the second, the order of the second is divisible by the order of the third, and so on. These orders are, in fact, $5^3 \cdot 3^3 \cdot 2^4$, $5^2 \cdot 3^3 \cdot 2$, $5^2 \cdot 3^2$, and 5, respectively. Therefore, we have associated the family $\{5^3 \cdot 3^3 \cdot 2^4, 5^2 \cdot 3^3 \cdot 2, 5^2 \cdot 3^2, 5\}$ with G. We can reconstruct G (up to isomorphism) from this family, namely

$$G \approx \mathbb{Z}(5^3 \cdot 3^3 \cdot 2^4) \oplus \mathbb{Z}(5^2 \cdot 3^3 \cdot 2) \oplus \mathbb{Z}(5^2 \cdot 3^2) \oplus \mathbb{Z}(5).$$

Thus, this family determines G. What is not so clear is that if $G = \mathbb{Z}(n_1) \oplus \mathbb{Z}(n_2) \oplus \cdots \oplus \mathbb{Z}(n_k)$ with n_1 divisible by n_2, n_2 divisible by n_3, and so on, then $n_1 = 5^3 \cdot 3^3 \cdot 2^4$, $n_2 = 5^2 \cdot 3^3 \cdot 2$, and so on. However, this is indeed the case. What we are asserting is that the family $\{n_1, n_2, \ldots, n_k\}$ is an invariant of G. This is the same as saying that if

$$G = G_1 \oplus \cdots \oplus G_m = H_1 \oplus \cdots \oplus H_n,$$

with the G_i and H_i nonzero cyclic, and with $o(G_i)$ divisible by $o(G_{i+1})$ and $o(H_i)$ divisible by $o(H_{i+1})$ for all appropriate i, then $m = n$ and $o(G_i) = o(H_i)$, whence $G_i \approx H_i$. To see that this is the case, consider a p-component K of G. Then $K = \bigoplus_{i=1}^{m} (G_i)_p$. That is, the p-component of G is the sum of the p-components of the G_i's. But $(G_i)_P$ is cyclic since G_i is cyclic. Similarly, $K = \bigoplus_{i=1}^{n} (H_i)_p$. But $o(G_i)_p \geq o((G_{i+1})_p)$, and similarly for the $(H_i)_p$. Therefore, from the discussion of the invariant $I(K)$ for p-groups K, $(G_i)_p \approx (H_i)_p$ for all i. Since the p-component of G_i is isomorphic to the p-component of H_i for all p, it follows that $G_i \approx H_i$.

The result of all this discussion is this. With each finite Abelian group G there is associated, in exactly one way, a family $\{n_1, n_2, \ldots, n_k\}$ of positive integers such that n_i is divisible by n_{i+1} for $i < k$. The association is made by writing $G = \bigoplus_{i=1}^{k} G_i$ with G_i cyclic of order n_i, and we have shown that the n_i's are unique. If we know $\{n_1, \ldots, n_k\}$, we can retrieve G, up to isomorphism: $G \approx \bigoplus_{i=1}^{k} \mathbb{Z}(n_i)$. So for a finite Abelian group G we have a complete invariant $I(G)$, the family $\{n_1, \ldots, n_k\}$ of positive integers. In case G is a p-group, it is the same invariant as discussed for p-groups. In case G is cyclic, $I(G)$ is the family whose only member is $o(G)$.

2.7.9 Definition

The family $I(G)$ is called the family of **invariant factors** *of the finite Abelian group G. The invariant factors of the p-components G_p of G are the* **elementary divisors** *of G.*

Thus, the invariant factors of

$$\mathbb{Z}(2^2 \cdot 3^2 \cdot 5^3) \oplus \mathbb{Z}(2 \cdot 3^2 \cdot 5^3) \oplus \mathbb{Z}(2 \cdot 3 \cdot 5) \oplus \mathbb{Z}(3 \cdot 5) \oplus \mathbb{Z}(3)$$

are $\{2^2 \cdot 3^2 \cdot 5^3, 2 \cdot 3^2 \cdot 5^3, 2 \cdot 3 \cdot 3, 3 \cdot 5, 3\}$ and the elementary divisors are $\{2^2, 2, 2\}$, $\{3^2, 3^2, 3, 3, 3\}$, and $\{5^3, 5^3, 5, 5\}$.

Since the finite Abelian groups G and H are isomorphic if and only if $G_p \approx H_p$ for all primes p, and $G \approx H$ if and only if $I(G) = I(H)$, we have the following theorem.

2.7.10 Theorem

Let G and H be finite Abelian groups. The following are equivalent.

(a) *$G \approx H$.*
(b) *G and H have the same invariant factors.*
(c) *G and H have the same elementary divisors.*

Now that we have the invariant $I(G)$ for finite Abelian groups G, we can calculate how many nonisomorphic Abelian groups there are of a given order. Let n be any positive integer, and write $n = n_1 \cdot n_2 \cdots n_k$, where the n_i are powers of distinct primes p_i. If a finite Abelian group G has order n, then its p_i-component must have order n_i. Now two finite Abelian groups are isomorphic if and only if their corresponding p-components are isomorphic. Therefore, if $N(n)$ denotes the number of nonisomorphic Abelian groups of order n, then $N(n) = \prod_{i=1}^{k} N(n_i)$. So we must calculate $N(p^n)$, where p is an arbitrary prime. If G is an Abelian p-group of order p^n, then

$$I(G) = \{i_1, i_2, \ldots, i_k\},$$

where each i_j is a power of p and $i_1 \geq i_2 \geq \cdots \geq i_k$. Let $i_j = p^{n_j}$. Then $n = \sum_{i=1}^{k} n_i$. Thus $N(p^n)$ is the same as the number of distinct ways we can write n in the form $n_1 + n_2 + \cdots + n_r$ with $n_1 \geq n_2 \geq \cdots \geq n_r \geq 1$. That is, $N(p^n)$ is the same as the number of ***partitions*** of n. We have proved the following theorem.

2.7.11 Theorem

The number of nonisomorphic Abelian groups of order $n = p_1^{n_1} p_2^{n_2} \cdots p_k^{n_k}$, where the p_i's are distinct primes, is $P(n_1)P(n_2) \cdots P(n_k)$, where $P(n_i)$ is the number of partitions of n_i.

Thus the number of Abelian groups of order 4 is 2, of order 6 is 1, of order 100 is 4, and of order $17^5 \cdot 97^4$ is 35.

PROBLEMS

All groups in these problems are Abelian.

1 Prove that G is the direct sum of the two subgroups H and K if and only if every $g \in G$ can be written $g = h + k$ with $h \in H$ and $k \in K$, and $H \cap K = \{0\}$.

2 Prove that G is the direct sum of the subgroups $G_1, G_2, \ldots, G_n$ if and only if every $g \in G$ can be written $g = g_1 + g_2 + \cdots + g_n$ with $g_i \in G_i$ for all i, and $G_i \cap (G_1 + G_2 + \cdots + G_{i-1} + G_{i+1} + \cdots + G_n) = \{0\}$ for all i.

3 Let $G = \bigoplus_{i=1}^{n} G_i$, and $g = \sum g_i$ with $g_i \in G_i$. Prove that $o(g) = \text{lcm}\{o(g_1), o(g_2), \ldots, o(g_n)\}$ if $o(g)$ is finite. Prove that if G is a p-group, then $o(g)$ is the maximum of the $o(g_i)$.

4 Prove that every finite group G is a direct sum of p-groups in the following way. Let p divide $o(G)$, where p is a prime. Let G_p be the p-component of G, and let $G_p^* = \{g \in G: ((o(g), p) = 1.\}$ Prove that $G = G_p \oplus G_p^*$. Complete by inducting on the number of primes that divide the order of G.

5 Prove that every finite group is a direct sum of p-groups as follows.

(a) If G is a p-group and $(n, p) = 1$, then

$$n: G \to G: g \to ng$$

is an automorphism.

(b) Let $o(G) = i_1 i_2 \cdots i_r$, with the i_j powers of distinct primes p_j. Let $q = o(G)/i_1$. Then q is an automorphism of the p_1-component of G, and hence has an inverse f. Now q is an endomorphism of G, and

$f \circ q$ is an idempotent endomorphism of G with image the p_1-component of G. Apply 2.7.2 and induct.

6 Let G be a group, not necessarily finite, all of whose elements are of finite order. Suppose that only finitely many primes are relevant for G. That is, there are finitely many primes $p_1, p_2, \ldots, p_r$ such that if $g \in G$, then $o(g)$ is a product of powers of these primes. Prove that G is the direct sum of its p-components.

7 Prove that the plane is the direct sum of any two distinct straight lines through the origin.

8 Let $\alpha: G \to H$ be an isomorphism. Prove that if A is a summand of G, then $\alpha(A)$ is a summand of H. Prove that it is not enough to assume that α is a monomorphism, nor enough to assume that α is an epimorphism.

9 Prove that if $G_i \approx H_i$ for $i = 1, 2, \ldots, n$, then

$$\bigoplus_{i=1}^{n} G_i \approx \bigoplus_{i=1}^{n} H_i.$$

10 Prove that the subgroups of a cyclic p-group form a chain. That is, if A and B are two subgroups of $\mathbb{Z}(p^n)$, then $A \supset B$ or $A \subset B$.

11 Prove that $\mathbb{Z}(p^n)$ is indecomposable.

12 Prove that for each $m \le n$, $\mathbb{Z}(p^n)$ has exactly one quotient group isomorphic to $\mathbb{Z}(p^m)$.

13 Prove that for each $m \le n$, $\mathbb{Z}(p^n)$ has exactly one subgroup isomorphic to $\mathbb{Z}(p^m)$.

14 Prove that a finite p-group G is a direct sum of cyclic groups as follows. Let $g \in G$ be an element of maximum order. If $G \neq \{0\}$, then $o(G/\mathbb{Z}g) < o(G)$, so that $G/\mathbb{Z}g = \bigoplus_{i=1}^{n} (A_i/\mathbb{Z}g)$, with $A_i/\mathbb{Z}g$ cyclic. Each $A_i/\mathbb{Z}g = \mathbb{Z}(a_i + \mathbb{Z}g)$ with $o(a_i) = o(a_i + \mathbb{Z}g)$. Then $A_i = \mathbb{Z}a_i \oplus \mathbb{Z}g$, and

$$G = \mathbb{Z}g \oplus \mathbb{Z}a_1 \oplus \mathbb{Z}a_2 \oplus \cdots \oplus \mathbb{Z}a_n.$$

15 Prove that a finite p-group G is a direct sum of cyclic groups as follows. Let $g \in G$ be an element of maximum order. Let H be a subgroup of G such that $H \cap \mathbb{Z}g = \{0\}$ and such that if K is any subgroup of G properly containing H, then $K \cap \mathbb{Z}g \neq \{0\}$. Then $G = \mathbb{Z}g \oplus H$. Induct on $o(G)$.

16 Find the number of nonisomorphic groups of order 71, 71^2, 49^2, 10011, and $p_1^5 \cdot p_2^5 \cdot p_3^4 \cdot p_4^3 \cdot p_5^2 \cdot p_6$, where the p_i's are distinct primes.

17 Write down all nonisomorphic groups of order $7^3 \cdot 11^2$.

18 Let $G = H \oplus K$, with G finite. How does one get the invariant factors of G directly from those of H and K? Given the invariant factors of G and H, calculate those of K.

19 Prove that if $G_1 \oplus G_2 = H_1 \oplus H_2$ is finite, and if $G_1 \approx H_1$, then $G_2 \approx H_2$.

20 Let G and H be finite. How does one tell from the invariant factors of G

and H whether or not

(a) H is isomorphic to a summand of G;
(b) H is isomorphic to a subgroup of G;
(c) H is isomorphic to a quotient group of G?

21 Let G be a finite group and let p^n divide $o(G)$. Prove that G has a subgroup of order p^n. (This is true even if G is not necessarily Abelian, but is considerably more difficult to prove in that case.)

22 Prove that every finite group of order greater than 2 has a nontrivial automorphism. Prove this also for non-Abelian groups.

23 Prove that $G \approx G_1 \oplus G_2$ if and only if there are homomorphisms $\alpha_i: G_i \to G$ such that if H is any group and $\beta_i: G_i \to H$ are any homomorphisms, then there is a unique homomorphism $\alpha: G \to H$ such that $\alpha \circ \alpha_i = \beta_i$.

24 Let Hom(G, H) be the group of all homomorphisms from G into H. Prove that

$$\text{Hom}(G_1 \oplus G_2, H) \approx \text{Hom}(G_1, H) \oplus \text{Hom}(G_2, H),$$

and

$$\text{Hom}(G, H_1 \oplus H_2) \approx \text{Hom}(G, H_1) \oplus \text{Hom}(G, H_2).$$

Generalize.

25 Suppose that there is an epimorphism $G \to \mathbb{Z}$. Prove that $G = A \oplus B$ with $A \approx \mathbb{Z}$. Generalize.

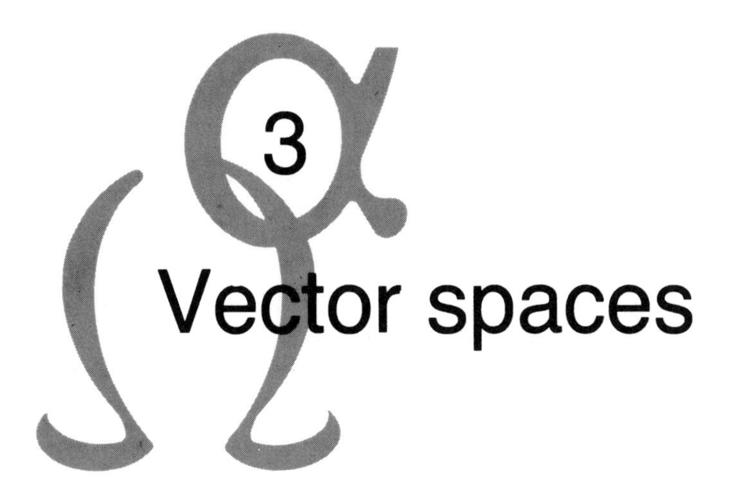

3 Vector spaces

3.1 DEFINITIONS AND EXAMPLES

In this chapter we begin the study of vector spaces, perhaps the most universally useful algebraic system. Let us start by looking at a familiar example. Consider the plane, that is, the set P of all pairs (a, b) of real numbers. Two such pairs may be added by the rule

$$(a, b) + (c, d) = (a + c, b + d).$$

There is another fundamental operation that is performed on the elements of P. An element $(a, b) \in P$ can be "multiplied" by a real number r by the rule

$$r(a, b) = (ra, rb).$$

Now real numbers themselves can be added and multiplied. We have four operations at hand—addition of elements of P, addition and multiplication of real numbers in $\mathbb{R}$, and multiplication of elements of P by elements of $\mathbb{R}$ to get elements of P. These four operations satisfy some basic rules. First, P is an Abelian group under addition of its elements. The real numbers $\mathbb{R}$, together with ordinary addition and multiplication of the elements of $\mathbb{R}$, constitute what is called a ***field***, an algebraic system we will formally define shortly. The multiplication of elements of P by elements of $\mathbb{R}$ satisfies several basic rules involving the other operations. For example, $r((a, b) + (c, d)) = r(a, b) + r(c, d)$. This type of a setup arises over and over again in mathematics (and elsewhere). It has been formalized and called a ***vector space***. We now proceed to its formal definition, but first we need to define a field.

3.1.1 Definition

A **field** *is a set* F *with two commutative binary operations, called addition and multiplication, and denoted* $+$ *and* $\cdot$ *respectively, such that*

(a) F *is a group under* $+$,
(b) $F^* = F\setminus\{0\}$ *is a group under* $\cdot$, *where* 0 *is the additive identity, and*
(c) $a\cdot(b+c) = a\cdot b + a\cdot c$ *for all* a, b, *and* $c \in F$.

As noted in (b), the additive identity is denoted by 0. The identity of the multiplicative group F^* is denoted by 1. Condition (c) is called the ***left distributive law***. Following the custom for groups, the negative of an element a in the additive group F is denoted $-a$, and the inverse of an element a in the multiplicative group F^* is denoted a^{-1}. Also, $a\cdot b$ is usually written simply as ab.

We now note some rather immediate but useful consequences of the axioms.

3.1.2 Theorem

Let F *be a field. Then for* $a, b, c \in F$, *the following hold.*

(a) $0\cdot a = a\cdot 0 = 0$.
(b) $(a+b)c = ac + bc$.
(c) $a(-b) = -(ab) = (-a)b$.
(d) $ab = (-a)(-b)$.
(e) $a(b-c) = ab - ac$.
(f) $(-1)a = -a$.
(g) $1 \neq 0$.

Proof

(a) $a\cdot 0 = a\cdot(0+0) = a\cdot 0 + a\cdot 0$, so $a\cdot 0 - a\cdot 0 = 0 = a\cdot 0 + a\cdot 0 - a\cdot 0 = a\cdot 0$. Also, $0\cdot a = a\cdot 0$ since multiplication is commutative.
(b) This follows from 3.1.1(c) and commutativity.
(c) $ab + a(-b) = a(b+(-b)) = a\cdot 0 = 0$. Thus, $-ab = a(-b)$. Now, $(-a)b = b(-a) = -(ba) = -(ab)$.
(d) $(-a)(-b) = -(a(-b)) = -(-(ab)) = ab$, using (c).
(e) $a(b-c) = a(b+(-c)) = ab + a(-c) = ab + (-(ac)) = ab - ac$.
(f) $0 = 0\cdot a = (1+(-1))a = 1\cdot a + (-1)a = a + (-1)a$. Thus, $-a = (-1)a$.
(g) $1 \in F^*$ and $0 \notin F^*$.

Let S be a subset of a field F. Restricting addition and multiplication to elements of S might make S into a field. If so, S is called a ***subfield*** of F.

3.1.3 Theorem

Let S be a subset of a field F. Then S is a subfield of F if and only if S is a subgroup of the additive group F, and S^* is a subgroup of the multiplicative group F^*.

Proof Suppose S is a subfield of F. Then addition and multiplication must be binary operations on S which make S into a field. Thus S must be a group under addition, hence an additive subgroup of F. Similarly, S^* is a subgroup of F^*.

Now suppose S is an additive subgroup of G and S^* is a subgroup of F^*. Since $0 \cdot a = 0$ for all $a \in F$, and $0 \in S$, multiplication is a binary operation on S. Addition is a binary operation on S, since S is an additive subgroup of F. So S is an Abelian group under $+$, S^* is an Abelian group under $\cdot$, and $a \cdot (b + c) = a \cdot b + a \cdot c$ for all $a, b, c \in S$. That is, S is a field. Thus, S is a subfield of F.

Suppose that S and T are subfields of a field F. Then $S \cap T$ is a subfield of F, since $S \cap T$ is an additive subgroup of F, and $(S \cap T)^* = S^* \cap T^*$ is a subgroup of the multiplicative group F^*. Similarly, the intersection of any set of subfields of F is a subfield of F. One important subfield of a field F is the intersection of ***all*** subfields of F. That subfield is called the ***prime subfield*** of F.

3.1.4 Examples of fields

(a) The set of rational numbers $\mathbb{Q}$ with ordinary addition and multiplication is a field.

(b) The set of real numbers $\mathbb{R}$ with ordinary addition and multiplication is a field.

(c) The set of complex numbers $\mathbb{C}$ with ordinary addition and multiplication is a field. Notice that $\mathbb{Q}$ is a subfield of $\mathbb{R}$ and $\mathbb{C}$, and $\mathbb{R}$ is a subfield of $\mathbb{R}$ and $\mathbb{C}$. The field $\mathbb{Q}$ is the prime subfield of all three. (See Problem 4 at the end of this section.)

(d) Let $\mathbb{Q}[\sqrt{2}] = \{a + b\sqrt{2}: a, b \in \mathbb{Q}\}$. Then $\mathbb{Q}[\sqrt{2}]$ is a subfield of $\mathbb{R}$. This is easy to check. For example, the inverse of a nonzero element $a + b\sqrt{2}$ is $a/(a^2 - 2b^2) - (b/(a^2 - 2b^2))\sqrt{2}$. Note that $a^2 - 2b^2 \neq 0$. (In this example, $\sqrt{3}$ works just as well as $\sqrt{2}$.)

(e) Let $\mathbb{Q}[i] = \{a + bi: a, b \in \mathbb{Q}\}$. Then $\mathbb{Q}[i]$ is a subfield of $\mathbb{C}$. ($i^2 = -1$.)

(f) Let F be any subfield of $\mathbb{C}$ and let c be any element of $\mathbb{C}$ such that $c^2 \in F$. Then $F[c] = \{a + bc: a, b \in F\}$ is a subfield of $\mathbb{C}$. This example can be generalized. What makes this work is that c satisfies a polynomial of degree 2 with coefficients in F. Suppose F is a subfield of $\mathbb{C}$ and c is an element of $\mathbb{C}$ that satisfies a polynomial of degree n with coefficients in F. Then

$$F[c] = \{a_0 + a_1c + \cdots + a_{n-1}c^{n-1}: a_i \in F\}$$

is a subfield of $\mathbb{C}$. However, delving into examples such as this takes us too far afield at the moment.

(g) Let p be a prime. Then the "integers modulo p" form a field. Specifically, in the Abelian group $\mathbb{Z}/\mathbb{Z}p$, multiply by the rule $(m + \mathbb{Z}p)(n + \mathbb{Z}p) = mn + \mathbb{Z}p$. This makes $\mathbb{Z}/\mathbb{Z}p$ into a field. Two things need checking—that multiplication is well defined, and that every nonzero element has an inverse. No tricks are needed to do either, but the fact that p is a prime is necessary to establish the latter.

(h) Let F be a field, and let K be a set. Suppose $f: K \to F$ is a one-to-one map from K onto F. For $a, b \in K$, define $a + b = f^{-1}(f(a) + f(b))$, and $a \cdot b = f^{-1}(f(a)f(b))$. This makes K into a field.

Now that we have the definition of a field, we can formally define a vector space. A vector space consists of several things. Recall the situation with the plane. We had the Abelian group P of pairs of real numbers, the field $\mathbb{R}$ of real numbers, and a way to multiply elements of P by elements of $\mathbb{R}$ to get elements of P. This last multiplication satisfies some basic rules. Here is the definition.

3.1.5 Definition

A **vector space** *is an Abelian group V, a field F, and a mapping $\cdot: F \times V \to V$ satisfying* (a)–(d) *below. (The image of $(a, v) \in F \times V$ under the mapping $\cdot$ is denoted $a \cdot v$.)*

(a) $a \cdot (v + w) = a \cdot v + a \cdot w$ *for $a \in F$, and $v, w \in V$.*
(b) $(a + b) \cdot v = a \cdot v + b \cdot v$ *for $a, b \in F$, and $v \in V$.*
(c) $(ab) \cdot v = a \cdot (b \cdot v)$ *for $a, b \in F$, and $v \in V$.*
(d) $1 \cdot v = v$ *for all $v \in V$.*

The elements of V are called ***vectors***, and the elements of F are called ***scalars***. The mapping $F \times V \to V$ is called ***scalar multiplication***. Vectors may be multiplied by elements of F (scalar multiplication). Scalars may be added and multiplied since F is a field. We denote both addition in V and addition in F by $+$. The context will make it clear which operation is intended. For example, in (a), both of the $+$ signs denote addition in V. In (b), however, the $+$ on the left is in F, and the $+$ on the right is in V. We will habitually drop the $\cdot$ denoting scalar multiplication. Thus, $a \cdot v$ will be written simply as

av. There are also two zeroes here, the additive identity of F and the identity of V. Both will be denoted by 0. Again, the context will make it clear which zero is intended.

A vector space may be viewed as a special kind of Abelian group, namely one where a multiplication of its elements by those of a field is given which satisfies (a)–(d) above. A vector space is sometimes denoted simply by one letter, such as V, but V actually is just the underlying set of vectors. It is understood that V is an Abelian group and that there is a field F of scalars around. Also, we say that V is a ***vector space over*** F.

The following theorem gives some elementary consequences of the vector space axioms.

3.1.6 Theorem

Let V be a vector space over F, let a, $b \in F$, and let v, $w \in V$. Then the following hold.

(a) $a \cdot 0 = 0$.
(b) $0 \cdot v = 0$.
(c) $(-a) \cdot v = -(a \cdot v) = a \cdot (-v)$.
(d) $(-a) \cdot (-v) = a \cdot v$.
(e) $(-1) \cdot v = -v$.
(f) $a(v - w) = av - aw$.
(g) $(a - b)v = av - bv$.
(h) $av = 0$ *if and only if either* $a = 0$ *or* $v = 0$.

Proof

(a) $a \cdot 0 = a(0 + 0) = a \cdot 0 + a \cdot 0$. Thus, $a \cdot 0 - a \cdot 0 = 0 = a \cdot 0 + a \cdot 0 - a \cdot 0 = a \cdot 0$.
(b) $0 \cdot v = (0 + 0)v = 0 \cdot v + 0 \cdot v$, whence $0 = 0 \cdot v$.
(c) $(a + (-a))v = 0 \cdot v = 0 = av + (-a)v$, so $(-a)v = -(av)$. Similarly, $a(-v) = -(av)$.
(d) $(-a)(-v) = -(a(-v)) = -(-(av)) = av$, using (c).
(e) $(-1)v = -(1 \cdot v) = -v$.
(f) $a(v - w) = a(v + (-w)) = av + a(-w) = av + (-(aw)) = av - aw$.
(g) The proof of (g) is similar to that of (f).
(h) If $av = 0$ and $a \neq 0$, then $a^{-1}(av) = (a^{-1}a)v = 1 \cdot v = v = 0$. If either $a = 0$ or $v = 0$, then $av = 0$ by (a) and (b).

Let V be a vector space over F, and let S be a subset of V. Restricting vector addition to elements of S, and restricting scalar multiplication to $F \times S$ might make S into a vector space over F. If so, S is called a ***subspace*** of V. A vector space V always has the trivial subspaces, V itself and $\{0\}$.

3.1.7 Theorem

Let V be a vector space over F, and let S be a nonempty subset of V. Then S is a subspace of V if and only if S is closed under vector addition and scalar multiplication, that is, if and only if $s + t \in S$ whenever s, $t \in S$ and $as \in S$ whenever $a \in F$ and $s \in S$.

Proof Suppose S is a subspace of V. Then S is a group under vector addition, so is closed under that addition. Scalar multiplication must map $F \times S$ into S. In other words, $as \in S$ for all $a \in F$ and $s \in S$.

Conversely, suppose that S is a nonempty subset of V closed under vector addition and scalar multiplication. Since $(-1)s = -s \in S$ if $s \in S$, S is a subgroup of V. Since S is closed under scalar multiplication, the restriction of scalar multiplication to $F \times S$ maps $F \times S$ into S. Now it is clear that S is a subspace of V.

3.1.8 Corollary

The intersection of any family of subspaces of a vector space V is a subspace of V.

Proof Just note that the intersection of any family of subspaces is closed under addition and scalar multiplication, so it is a subspace by 3.1.7.

Thus by 3.1.8, if S is any subset of a vector space V, then there is a unique smallest subspace of V containing S, namely the intersection of all the subspaces of V which contain S. As in the case for groups (2.2.3), we define this subspace to be the ***subspace generated by S***. A vector space is ***finitely generated*** if it is generated by a finite set. If $S = \{v\}$, where v is an element of V, then the subspace generated by S is $Fv = \{av: a \in F\}$. If $V_1, V_2, \ldots, V_n$ are subspaces of V, then the smallest subspace containing them all is

$$V_1 + V_2 + \cdots + V_n = \{v_1 + v_2 + \cdots + v_n: v_i \in V_i\}.$$

Therefore, if $S = \{v_1, v_2, \ldots, v_n\}$, then the subspace generated by S is $Fv_1 + Fv_2 + \cdots + Fv_n$. In vector spaces, the word ***span*** is often used in place of ***generate***. Thus, we also speak of the subspace ***spanned*** by a set. (See Problems 17 and 18.)

3.1.9 Examples

To give an example of a vector space, we must specify three things—an Abelian group V, a field F, and a mapping $F \times V \to V$, and we must ensure that this mapping satisfies conditions (a)–(d) in 3.1.5.

(a) Let V be the Abelian group of all pairs of real numbers with addition given by

$$(a, b) + (c, d) = (a + c, b + d).$$

Let $\mathbb{R}$ be the field of real numbers, and let $\mathbb{R} \times V \to V$ be given by

$$c(a, b) = (ca, cb).$$

Then V is a vector space over $\mathbb{R}$. It is finitely generated since $V = \mathbb{R}(1, 0) + \mathbb{R}(0, 1)$. We have a convenient geometric representation of it, namely the Cartesian plane. Its nontrivial subspaces are precisely the straight lines through the origin. (See Problem 10.)

(b) Let F be any field. Let V be the set of all pairs of elements of F, and add by the rule

$$(a, b) + (c, d) = (a + c, b + d).$$

Define $F \times V \to V$ by

$$c(a, b) = (ca, cb).$$

This makes V into a vector space over F. Every subspace of V that is not V itself is of the form Fv for some v in V. (See Problem 10.)

(c) Let F be any field. For any integer $n \geq 1$, let V be the set of all n-tuples $(a_1, a_2, \ldots, a_n)$ with $a_i \in F$. Add by the rule

$$(a_1, a_2, \ldots, a_n) + (b_1, b_2, \ldots, b_n) = (a_1 + b_1, a_2 + b_2, \ldots, a_n + b_n).$$

Define $F \times V \to V$ by

$$a(a_1, a_2, \ldots, a_n) = (aa_1, aa_2, \ldots, aa_n).$$

Then V is a vector space over F. This vector space will be denoted F^n. It is finitely generated since

$$F^n = F(1, 0, 0, \ldots, 0) + F(0, 1, \ldots, 0) + \cdots + F(0, 0, \ldots, 1).$$

We will show later that any (finite-dimensional) vector space is essentially one of these. Notice that when $n = 1$, V is just the additive group of F, and scalar multiplication is just multiplication in F. *Thus, any field may be regarded as a vector space over itself.*

(d) Let F be any field, and let $F[x]$ be the set of all polynomials in x with coefficients in F. Add polynomials with coefficients in F just as you would add polynomials with real coefficients. Scalar multiplication is given by

$$a(a_0 + a_1x + a_2x^2 + \cdots + a_nx^n) = aa_0 + aa_1x + aa_2x^2 + \cdots + aa_nx^n.$$

Then $F[x]$ is a vector space over F. It is not finitely generated. (See Problem 11.) For a given n, all polynomials of degree $\leq n$ form a subspace, as do all those polynomials which have given elements a and b of F as roots.

(e) Let V be all continuous functions from the real numbers into the real numbers. Add by the rule

$$(f+g)(x)=f(x)+g(x).$$

For $r \in \mathbb{R}$ and $f \in V$, define rf by

$$(rf)(x)=r(f(x)).$$

This makes V into a vector space over the field $\mathbb{R}$ of real numbers. (It is proved in calculus that $f+g$ and rf, as defined above, are again continuous functions.) Instead of all continuous functions from $\mathbb{R}$ into $\mathbb{R}$, we could just as well take all integrable functions, or all differentiable functions, or even all functions from $\mathbb{R}$ into $\mathbb{R}$. The set of differentiable functions is a subspace of the vector space of continuous functions, of course.

(f) Let V be the additive group of the complex numbers. Then multiplication of complex numbers makes V into a vector space over the field of real numbers. In fact, if V is any vector space over any field F, and if K is a subfield of F, then V is automatically a vector space over K. In particular, the additive group of any field is a vector space over any subfield of that field.

(g) Consider the system

$$\begin{aligned} a_{11}x_1+a_{12}x_2+\cdots+a_{1n}x_n &= 0 \\ a_{21}x_1+a_{22}x_2+\cdots+a_{2n}x_n &= 0 \\ &\vdots \\ a_{m1}x_1+a_{m2}x_2+\cdots+a_{mn}x_n &= 0 \end{aligned}$$

of linear homogeneous equations with $a_{ij} \in \mathbb{R}$, and unknowns x_i. Let S be the set of all solutions of this system. That is, S is the set of all n-tuples $(s_1, s_2, \ldots, s_n)$ with $s_i \in \mathbb{R}$, such that $\sum_j a_{ij}s_j=0$ for all i. Then S is a vector space over $\mathbb{R}$, in fact, a subspace of $\mathbb{R}^n$. Of course, we could replace $\mathbb{R}$ by any field.

PROBLEMS

1 Prove that there is no field with 6 elements.

2 Prove that there is a field with 4 elements.

3 Find all subfields of a field with 4 elements.

4 Prove that $\mathbb{Q}$ is the only subfield of $\mathbb{Q}$.

5 Find all subfields of $\mathbb{Q}[\sqrt{2}]$; of $\mathbb{Q}[i]$.

6 Let F be a field, and let S be a subset of F. Prove that there is a subfield K of F such that $S \subset K$ and such that every subfield of F that contains S contains K.

7 Let V be the set of all functions f from $\mathbb{R}$ into $\mathbb{R}$ such that $f(x) = f(x + 2\pi)$. Prove that V is a vector space with the usual definitions of addition and scalar multiplication.

8 Let V be the set of all functions f from the closed interval $[0, 1]$ into itself such that $f(x) = f(1 - x)$. Prove that V is a vector space under the usual definitions of addition and scalar multiplication.

9 Let V be the set of all functions f from $\mathbb{R}$ into $\mathbb{R}$ which have second derivatives f'', and such that $f'' = -f$. Prove that V is a vector space under the usual definitions of addition and scalar multiplication. (It is actually a subspace of the vector space in Problem 7.)

10 Prove that every subspace of F^2 not F^2 itself is Fv for some v in F^2. Prove that every nontrivial subspace of the plane is a straight line through the origin.

11 Prove that the vector space of all polynomials over any field is not finitely generated.

12 Let S be a set, and let F be a field. Let V be the set of all mappings from S into F. Add such mappings by the rule $(v + w)(s) = v(s) + w(s)$, and define scalar multiplication by the rule $(af)(s) = a \cdot f(s)$.
(a) Prove that V is a vector space over F.
(b) Prove that V is finitely generated if and only if S is finite.

13 Prove that the subspace of $\mathbb{R}^3$ generated by $\{(1, 2, 3), (4, 5, 0), (0, 6, 7)\}$ is $\mathbb{R}^3$ itself.

14 Find the intersection of the two subspaces $\mathbb{R}(1, 2, 3) + \mathbb{R}(4, 5, 6)$ and $\mathbb{R}(1, 2, 0) + \mathbb{R}(4, 5, 0)$ of $\mathbb{R}^3$.

15 Let $F = \mathbb{Z}/\mathbb{Z}2$. Write down all the subspaces of F^2. Write down all the subspaces of F^3.

16 Let $F = \mathbb{Z}/\mathbb{Z}3$. Write down all the subspaces of F^2.

17 Let S and T be subspaces of a vector space V. Prove that

$$S + T = \{s + t \colon s \in S, t \in T\}$$

is a subspace of V. This is called the ***sum of the subspaces*** S and T. Prove that $S + T$ is the intersection of all subspaces of V that contain both S and T.

18 Generalize Problem 17 to any finite set of subspaces of a vector space.

19 Let $\{S_i\}_{i \in I}$ be a family of subspaces of V such that for any $i, j \in I$, there is a $k \in I$ such that $S_i \subset S_k$ and $S_j \subset S_k$. Prove that $\bigcup_{i \in I} S_i$ is a subspace of V.

20 Let $\{S_i\}_{i \in I}$ be any family of subspaces of a vector space V. For any finite subset A of I, let S_A be the sum of the subspaces S_i for i in A. Let $\mathscr{F}$ be the set of all finite subsets of I. Prove that $\{S_A\}_{A \in \mathcal{F}}$ satisfies the hypothesis of Problem 19 and that the union of the S_A's is the subspace generated by the family $\{S_i\}_{i \in I}$.

21 Let A, B, and C be subspaces of a vector space V such that $B \subset A$. Prove that

$$A \cap (B + C) = B + (A \cap C).$$

22 Prove that in the Definition 3.1.5 of a vector space, $1 \cdot v = v$ cannot be proved from the other postulates.

3.2 HOMOMORPHISMS OF VECTOR SPACES

In the study of any algebraic system, the concept of a homomorphism from one of those systems to another is fundamental. A homomorphism from one group to another was a map that preserved the group operations. In the case of vector spaces, we not only have the Abelian group of vectors around, but also the field of scalars, and it may not be clear just what a homomorphism should be. It turns out that the proper concept is that of a suitable mapping between two vector spaces over the same field. The mapping is required to preserve addition and scalar multiplication.

3.2.1 Definition

Let V and W be vector spaces over a field F. A **homomorphism** *from V into W is a mapping $f: V \to W$ such that*

(a) $f(v_1 + v_2) = f(v_1) + f(v_2)$ *for all v_1, v_2 in V, and*
(b) $f(a \cdot v) = a \cdot f(v)$ *for all $a \in F$ and $v \in V$.*

Condition (a) just says that f is a homomorphism from the group V into the group W, and condition (b) makes the map pay some attention to scalar multiplication. Condition (b) is expressed by saying that f ***preserves*** scalar multiplication. As in the case of groups, a homomorphism is an ***epimorphism*** if it is onto, a ***monomorphism*** if it is one-to-one, and an ***isomorphism*** if it is one-to-one and onto. If there is an isomorphism $f: V \to W$, we say that V is ***isomorphic*** to W, and we write $V \approx W$. As in the case of groups, $V \approx V$, $V \approx W$ implies $W \approx V$ and $V \approx W$ and $W \approx U$ imply that $V \approx U$. An ***automorphism*** of V is an isomorphism of V with itself. A homomorphism from V into V is an ***endomorphism.***

Compositions of homomorphisms are homomorphisms. That is, if $f: U \to V$ and $g: V \to W$ are homomorphisms, then so is $g \circ f$. This is Problem 1.

The set $\mathrm{Hom}(V, W)$ of all homomorphisms from V into W is itself a vector space in a natural way, and it will be studied from that point of view in Section 3.4.

In the case of vector spaces, homomorphisms are usually called ***linear transformations***. An automorphism of a vector space is called a ***nonsingular*** linear transformation, and an endomorphism that is not an automorphism is called ***singular***. From now on, we will call homomorphisms of vector spaces linear transformations.

If f is a linear transformation from V into W, then the ***kernel*** *of* f is denoted $\operatorname{Ker} f$ and is defined by

$$\operatorname{Ker} f = \{v \in V : f(v) = 0\},$$

and the ***image*** *of* f is denoted $\operatorname{Im} f$ and is defined by

$$\operatorname{Im} f = \{f(v) : v \in V\}.$$

3.2.2 Theorem

Let $f: V \to W$ be a linear transformation. Then the following hold.

(a) $f(0) = 0$.
(b) $f(-v) = -f(v)$.
(c) *f is one-to-one if and only if* $\operatorname{Ker} f = \{0\}$.
(d) $\operatorname{Ker} f$ *is a subspace of V.*
(e) $\operatorname{Im} f$ *is a subspace of W.*

Proof Parts (a), (b), and (c) hold because f is a homomorphism from the group V into the group W. For the same reason, $\operatorname{Ker} f$ is a subgroup of V and $\operatorname{Im} f$ is a subgroup of W. Let $a \in F$ and let $v \in \operatorname{Ker} f$. Then $f(av) = af(v) = 0$. Thus $av \in \operatorname{Ker} f$, and so $\operatorname{Ker} f$ is a subspace of V. Let $a \in F$ and let $w \in \operatorname{Im} f$. Then there is some $v \in V$ such that $f(v) = w$. But $f(av) = af(v) = aw \in \operatorname{Im} f$. Thus, (d) and (e) hold.

3.2.3 Examples

In these examples, it is straightforward to verify that what are claimed to be linear transformations are indeed linear transformations.

(a) Let V and W be vector spaces over any field F. The maps $V \to W : v \to 0$ and $1_V : V \to V : v \to v$ are linear transformations.
(b) Let V be a vector space over F, and let $a \in F$. The map $f: V \to V$ defined by $f(v) = av$ is a linear transformation. Thus, each scalar induces a linear transformation $V \to V$. If $a \neq 0$, then this linear transformation is nonsingular.

(c) Let F be any field, let F^n be the vector space of n-tuples of elements of F, and let $(a_1, a_2, \ldots, a_n) \in F^n$. The map $f: F^n \to F^n$ defined by

$$f(b_1, b_2, \ldots, b_n) = (a_1 b_1, a_2 b_2, \ldots, a_n b_n)$$

is a linear transformation. It is nonsingular if and only if each a_i is nonzero.

A special case is the linear transformation

$$(a_1, a_2, \ldots, a_n) \to (0, a_2, a_3, \ldots, a_n)$$

(d) The map $F^n \to F^n$ defined by

$$(a_1, a_2, \ldots, a_n) \to (a_{\sigma(1)}, a_{\sigma(2)}, a_{\sigma(3)}, \ldots, a_{\sigma(n)}),$$

where σ is an element of S_n, is a linear transformation. It is nonsingular.

(e) Consider F as a vector space over itself. Let $(a_1, a_2, \ldots, a_n) \in F^n$. The map $f: F^n \to F$ given by

$$f(b_1, b_2, \ldots, b_n) = \sum_{i=1}^{n} a_i b_i$$

is a linear transformation. Its kernel consists of those elements of F^n that are "orthogonal" to $(a_1, a_2, \ldots, a_n)$.

(f) Let a_{ij}, $i, j = 1, 2, \ldots, n$, be elements of a field F. Then the map $F^n \to F^n$ given by

$$(a_1, a_2, \ldots, a_n) \to \left(\sum_{j=1}^{n} a_{1j} a_j, \ldots, \sum_{j=1}^{n} a_{nj} a_j\right)$$

is a linear transformation.

(g) Let a_{ij}, $i = 1, 2, \ldots, m$ and $j = 1, 2, \ldots, n$ be elements of a field F. The map $F^n \to F^m$ given by

$$(a_1, a_2, \ldots, a_n) \to \left(\sum_{j=1}^{n} a_{1j} a_j, \ldots, \sum_{j=1}^{n} a_{mj} a_j\right)$$

is a linear transformation. In a sense, every linear transformation between finite-dimensional vector spaces is like this one, and the relation between linear transformations and matrices will be studied in detail in Chapter 5.

(h) Rotation of the plane through any number of degrees about the origin is an automorphism of the vector space $\mathbb{R}^2$. The formula for rotation counterclockwise through $\theta°$ is given by $(x, y) \to (x \cdot \cos \theta° + y \cdot \sin \theta°, y \cdot \cos \theta° - x \cdot \sin \theta°)$. This example is a special case of (f).

(i) Let V be the vector space of all sequences $\{a_1, a_2, a_3, \ldots\}$ of real numbers. Define f and g by the formulas

$$f(\{a_1, a_2, a_3, \ldots\}) = \{0, a_1, a_2, a_3, \ldots\},$$
$$g(\{a_1, a_2, a_3, \ldots\}) = \{a_2, a_3, a_4, \ldots\}.$$

Then f and g are linear transformations, f is one-to-one but not onto, g is onto but not one-to-one, $g \circ f = 1_V$, and $f \circ g \neq 1_V$.

(j) Let W be the subspace of V in (i) consisting of all those sequences which converge. For $\{a_i\}$ in W, let

$$f(\{a_i\}) = \lim_{i \to \infty} \{a_i\}.$$

Then f is a linear transformation from W onto $\mathbb{R}$, and its kernel consists of those sequences which converge to zero.

Kernels of linear transformations are subspaces. It will turn out that every subspace of a vector space is the kernel of a linear transformation from that vector space to another one. In fact, given a subspace S of a vector space V over a field F, we will construct a vector space V/S over F, the quotient space of V with respect to S. There will be a natural epimorphism $V \to V/S$, with kernel S. This construction is similar to that of quotient groups, in fact, uses that of quotient groups, and is fundamental.

3.2.4 The construction of quotient spaces

Let V be a vector space over a field F, and let S be a subspace of V. Then S is, in particular, a subgroup of the Abelian group V, so we have the quotient group V/S. Its elements are cosets $v + S$, and one adds by the rule

$$(v + S) + (w + S) = (v + w) + S.$$

We will make V/S into a vector space over F by defining a scalar multiplication. Let $a \in F$, and let $v + S \in V/S$. The most natural thing to do is to let $a(v + S) = av + S$. However, we must make sure that this defines a map $F \times (V/S) \to V/S$. To do this, we must show that if $v + S = w + S$, then $av + S = aw + S$. So suppose that $v + S = w + S$. Then $v = w + s$ for some $s \in S$. Then

$$av = a(w + s) = aw + as,$$

and $as \in S$ since S is closed under scalar multiplication. Thus $av + S = aw + S$, and we at least have a map $F \times (V/S) \to V/S$. But it is easy to check that this makes V/S into a vector space over F. For example,

$$\begin{aligned} a((v + S) + (w + S)) &= a((v + w) + S) = (a(v + w)) + S = (av + aw) + S \\ &= (av + S) + (aw + S) = a(v + S) + a(w + S), \end{aligned}$$

and the other required properties follow just as readily. The vector space V/S is called the ***quotient space of V with respect to S***, or more simply, V ***modulo*** S, or V ***over*** S. The mapping $V \to V/S: v \to v + S$ is an epimorphism, called the ***natural linear transformation from*** V to V/S. Note that its kernel is S.

Consider the vector space $V = \mathbb{R}^2$ and the subspace $S = \mathbb{R}v$, where v is a nonzero vector in V. Then S is a straight line through the origin, and a coset $(a, b) + S$ is a line in the plane parallel to S. Thus, the cosets of S are just the lines parallel to S.

Recall that the subgroups of V/S are in one-to-one correspondence with the subgroups of V containing S. That is, every subgroup of V/S is of the form W/S where W is a subgroup of V containing S, and distinct such W give distinct W/S. But the subgroup W/S is a ***subspace*** of V/S if and only if W/S is closed under scalar multiplication. Thus W/S is a subspace of V/S if and only if for $a \in F$ and $w + S \in W/S$, it is true that $a(w + S) = aw + S \in W/S$. This is true if and only if $aw \in W$. Thus, the subspaces of V/S are in one-to-one correspondence with the subspaces W of V such that $W \supset S$. The correspondence is $W \to W/S$.

Let $V = \mathbb{R}^3$, which we view as ordinary 3-space. The nontrivial subspaces of V are the straight lines through the origin and the planes through the origin. Let S be a straight line through the origin. That is, $S = \mathbb{R}(a, b, c)$, where not all of a, b, and c are 0. Then the subspaces of V/S are the T/S, where T is a plane through the origin containing the line S. (See Problem 7.)

The various isomorphism theorems for groups have analogues for vector spaces.

3.2.5 Theorem

Let $f: V \to W$ be a linear transformation. Then

$$V/(\operatorname{Ker} f) \approx \operatorname{Im} f.$$

Proof There is little to do. The mapping

$$\bar{f}: V/(\operatorname{Ker} f) \to \operatorname{Im} f: \bar{f}(v + \operatorname{Ker} f) = f(v)$$

is a ***group*** isomorphism. All we need is that $\bar{f}$ preserves scalar multiplication. But

$$\bar{f}(a(v + \operatorname{Ker} f)) = \bar{f}(av + \operatorname{Ker} f) = f(av) = af(v) = a\bar{f}(v + \operatorname{Ker} f).$$

We actually proved more than is stated in 3.2.5. The map $V/(\operatorname{Ker} f) \to \operatorname{Im} f$ *induced by* f is an isomorphism. As in the case of groups, 3.2.5 immediately yields the following isomorphism theorems, which are very useful.

3.2.6 Theorem

Let S and T be subspaces of V with $S \subset T$. Then

$$(V/S)/(T/S) \approx V/T.$$

Proof $V/S \to V/T: v + S \to v + T$ is an epimorphism with kernel T/S. Apply 3.2.5.

Let $V = \mathbb{R}^3$, let S be a line through the origin, and let T be a plane through the origin containing S. Then T/S is a subspace of V/S, so we may form the quotient space $(V/S)/(T/S)$. But that vector space is isomorphic to V/T by 3.2.6. They both are isomorphic to $\mathbb{R}$, in fact. (See Problem 9.)

3.2.7 Theorem

Let S and T be subspaces of V. Then

$$S/(S \cap T) \approx (S + T)/T.$$

Proof Note that $S + T$ is indeed a subspace of V containing T. (See Problem 17 in Section 3.1). Now $S \to (S + T)/T: s \to s + T$ is an epimorphism with kernel $S \cap T$. Apply 3.2.5.

To illustrate 3.2.7, consider distinct planes S and T through the origin in the vector space $\mathbb{R}^3$. Then $S + T = \mathbb{R}^3$, so $S/(S \cap T) \approx (S + T)/T = \mathbb{R}^3/T \approx R$. (See Problem 11.)

PROBLEMS

1 Prove that the composition of two linear transformations (when defined) is a linear transformation.

2 Prove that a mapping $f: V \to W$ is a linear transformation if and only if $f(av_1 + bv_2) = af(v_1) + bf(v_2)$.

3 Let V be the vector space of all polynomials with real coefficients. For p in V, let $d(p)$ be the derivative of p. Let $f_x(p) = xp$, that is, just multiplication by x. Prove that d and f_x are linear transformations, and that $d \circ f_x - f_x \circ d = 1_V$. What does this problem have to do with real numbers?

4 Prove that there is a linear transformation i on V in Problem 3 such that $d \circ i = 1_V$, and that d does not have a left inverse.

5 Let f be any linear transformation from $\mathbb{R}^2$ onto $\mathbb{R}$. Prove that $\operatorname{Ker} f = \mathbb{R}v$ for some $v \neq 0$.

6 Let v be any nonzero element of $\mathbb{R}^2$. Prove that $\mathbb{R}^2/\mathbb{R}v \approx \mathbb{R}$.

7 Prove that every nontrivial subspace of $\mathbb{R}^3$ is a line through the origin or a plane through the origin.

8 Prove that every nontrivial subspace of F^3 is of the form $Fv + Fw$ for suitable v and w in F^3.

9 Prove that for any nontrivial subspace S of F^3, F^3/S is isomorphic to either F or F^2.

10 Let $V = F^3$, and let $S = Fv$ for some nonzero v. Prove that there is a subspace T of V strictly between S and V. Prove that there is no subspace strictly between T and V.

11 For the vector spaces in Problem 10, prove that $V/S \approx F^2$ and $V/T \approx F$.

12 In $\mathbb{R}^3$, let

$$S = \mathbb{R}(1, 2, 3) + \mathbb{R}(4, 5, 2),$$

and

$$T = \mathbb{R}(2, 1, -4) + \mathbb{R}(1, 1, 1).$$

Find (a, b, c) in $\mathbb{R}^3$ such that $S \cap T = \mathbb{R}(a, b, c)$.

13 In Problem 12, find a set of representatives for the cosets of S in $\mathbb{R}^3$. Find a set of representatives for the cosets of $S \cap T$ in S.

14 Suppose that v is a nonzero vector in a vector space V over a field F. Prove that the subspace Fv is isomorphic to the vector space F.

15 Prove that F^1 is not isomorphic to F^2. Generalize.

16 Let S and T be subspaces of V, with $S \subset T$. Prove that there is an epimorphism $V/S \to V/T$.

17 Prove that the set of automorphisms of a vector space, with multiplication given by composition of maps, is a group.

18 Let $f\colon V \to W$ be a linear transformation, and suppose that S is a subspace contained in $\operatorname{Ker} f$. Let $\eta\colon V \to V/S$ be the natural linear transformation. Prove that there is a unique linear transformation $g\colon V/S \to W$ such that $g \circ \eta = f$.

19 Let $f\colon V \to W$ be a linear transformation. Suppose that S is a subspace of V such that whenever $g\colon U \to V$ is a linear transformation with $f \circ g = 0$, then $\operatorname{Im} g \subset S$. Prove that $S \supset \operatorname{Ker} f$.

20 Let V be the vector space of all mappings of a set S into a field F. (See Problem 12 in Section 3.1) Let σ be a mapping of S into S. Prove that the mapping

$$V \to V\colon v \to v \circ \sigma$$

is a linear transformation. Prove that this linear transformation is nonsingular if and only if σ is a permutation of S.

3.3 LINEAR INDEPENDENCE AND BASES

Let F be a field, and consider the vector space F^2 of all pairs of elements of F. The set consisting of the two vectors $(1, 0)$ and $(0, 1)$ has the property that every element in F^2 can be written uniquely in the form

$$a(1, 0) + b(0, 1).$$

In fact, $(a, b) = a(1, 0) + b(0, 1)$, and if $(a, b) = c(1, 0) + d(0, 1)$, then $(a, b) = (c, 0) + (0, d) = (c, d)$, so that $a = c$ and $b = d$. One of our principal aims in this section is to show that this phenomenon holds in any finitely generated vector space $\neq \{0\}$. Specifically, we want to show that if V is a finitely generated vector space and $V \neq \{0\}$, then there are vectors $v_1, v_2, \ldots, v_n$ of V such that every element of V can be written uniquely in the form

$$a_1v_1 + a_2v_2 + \cdots + a_nv_n.$$

Such a nonempty family $\{v_1, v_2, \ldots, v_n\}$ of V will be called a ***basis*** of V. It is a very useful concept in the study of vector spaces.

For $a_1, a_2, \ldots, a_n \in F$, the vector $\sum_{i=1}^n a_iv_i \in V$ is a ***linear combination*** of $v_1, v_2, \ldots, v_n$. Thus, the subspace generated by $\{v_1, v_2, \ldots, v_n\}$ is the set of all linear combinations of the vectors $v_1, v_2, \ldots, v_n$. The nonempty family $\{v_1, v_2, \ldots, v_n\}$ is ***linearly independent*** if $\sum_{i=1}^n a_iv_i = 0$ implies that all the $a_i = 0$. This is equivalent to the condition that whenever $\sum_{i=1}^n a_iv_i = \sum_{i=1}^n b_iv_i$, then $a_i = b_i$ for all i.

3.3.1 Definition

The nonempty family $\{v_1, v_2, \ldots, v_n\}$ of vectors in the vector space V is a **basis** *of V if it is linearly independent and generates (or spans) V.*

Thus, $\{v_1, v_2, \ldots, v_n\}$ is a basis of V if and only if every $v \in V$ is a unique linear combination $\sum_{i=1}^n a_iv_i$ of the v_i's; that is, the a_i's are uniquely determined by v. Indeed, if $\{v_1, v_2, \ldots, v_n\}$ is a basis of V, then every $v \in V$ is certainly a linear combination $v = \sum_{i=1}^n a_iv_i$ of the v_i's, and if $v = \sum_{i=1}^n a_iv_i = \sum_{i=1}^n b_iv_i$, then $0 = \sum_{i=1}^n (a_i - b_i)v_i$, whence $a_i = b_i$. Conversely, if every $v \in V$ can be written $\sum_{i=1}^n a_iv_i$ with the a_i's unique, then $0 = \sum_{i=1}^n a_iv_i$ only if each $a_i = 0$. Hence, $\{v_1, v_2, \ldots, v_n\}$ is a basis of V.

Thus, we see that $\{(1, 0), (0, 1)\}$ is a basis of the vector space F^2 of all pairs (a, b) of elements of the field F. More generally, if $F^n = \{(a_1, a_2, \ldots, a_n): a_i \in F\}$ is the vector space of all n-tuples of elements of the field F, then

$$\{(1, 0, \ldots, 0), (0, 1, 0, \ldots, 0), \ldots, (0, 0, \ldots, 1)\}$$

is a basis of F^n. This basis is called the ***natural basis*** of F_n. Two other bases of

F^3 are

$$\{(1,1,0),(0,1,0),(0,0,1)\},$$

and

$$\{(1,1,1),(0,1,1),(0,0,1)\}.$$

At this point, we have the following things on our mind. We want to show that every finitely generated vector space $\neq \{0\}$ has a basis and that any two bases of such a vector space are the same size. The number of elements in a basis of V will be called the ***dimension*** of V, and it will turn out that two vector spaces over a field F are isomorphic if and only if they have the same dimension.

3.3.2 Theorem

Every finitely generated vector space $\neq \{0\}$ has a basis.

We will prove 3.3.2 by proving the following stronger theorem.

3.3.3 Theorem

If $\{v_1, v_2, \ldots, v_n\}$ spans V, then some subfamily is a basis of V.

Proof We may as well suppose that no proper subfamily of $\{v_1, v_2, \ldots, v_n\}$ spans V. It will suffice to show that $\{v_1, v_2, \ldots, v_n\}$ is a basis. All we need is that it is linearly independent. If $\sum_{i=1}^{n} a_i v_i = 0$ and $a_j \neq 0$, then $v_j = -\sum_{i \neq j} (a_i/a_j) v_i$. Therefore, the family of vectors $\{v_1, v_2, \ldots, v_{j-1}, v_{j+1}, \ldots, v_n\}$ spans V, contrary to the fact that no proper part of $\{v_1, v_2, \ldots, v_n\}$ spans V.

3.3.4 Theorem

If V is finitely generated, then any linearly independent family $\{v_1, v_2, \ldots, v_r\}$ can be extended to a basis $\{v_1, v_2, \ldots, v_r, v_{r+1}, \ldots, v_n\}$.

Proof Let $\{w_1, w_2, \ldots, w_m\}$ span V. If some w_i is not in $Fv_1 + Fv_2 + \cdots + Fv_r$, then let v_{r+1} be the first such w_i. Then $\{v_1, v_2, \ldots, v_r, v_{r+1}\}$ is independent. If some w_i is not in $Fv_1 + Fv_2 + \cdots + Fv_{r+1}$, then let v_{r+2} be the first such w_i. Then $\{v_1, v_2, \ldots, v_{r+1}, v_{r+2}\}$ is independent. Continuing this process, we run out of w's in finitely many steps, obtaining an independent family $\{v_1, v_2, \ldots, v_n\}$ such that $Fv_1 + Fv_2 + \cdots + Fv_n$ contains every w_i, and hence V. Thus, $\{v_1, v_2, \ldots, v_n\}$ is a basis of V.

The proof actually yields a little more than the theorem states. If $\{v_1, v_2, \ldots, v_r\}$ is linearly independent, and $\{w_1, w_2, \ldots, w_m\}$ is *any* generating set, then $\{v_1, v_2, \ldots, v_r\}$ can be augmented by some of the w_i's to a basis of V. For example, if $\{v_1, v_2\}$ is linearly independent in F^3, then $\{v_1, v_2\}$ together with one of $(1,0,0)$, $(0,1,0)$, and $(0,0,1)$ is a basis of F^3.

3.3.5 Theorem

If $\{v_1, v_2, \ldots, v_m\}$ spans V and $\{w_1, w_2, \ldots, w_n\}$ is a linearly independent subset of V, then $m \geq n$.

Proof Since $\{v_1, v_2, \ldots, v_m\}$ spans V, then $w_n = \sum_{i=1}^{m} a_i v_i$, and so $0 = -w_n + \sum_{i=1}^{m} a_i v_i$. Hence

$$\{w_n, v_1, v_2, \ldots, v_m\}$$

is dependent. Remove from this family the first member that is a linear combination of the preceding ones. There is such a member, and it is not w_n. This new family

$$\{w_n, v_1, v_2, \ldots, v_{i-1}, v_{i+1}, \ldots, v_m\}$$

still spans V. Do the same thing with this generating family and w_{n-1}. Continue the process. At each step, a w gets added and a v gets removed. When a w is added, a dependent family results. But $\{w_k, w_{k+1}, \ldots, w_n\}$ is independent for every $k \leq n$. Hence there are at least as many v's as w's, and so $m \geq n$.

Letting $\{v_1, v_2, \ldots, v_m\}$ and $\{w_1, w_2, \ldots, w_n\}$ be bases, we conclude that $m = n$, and hence get the following corollary.

3.3.6 Corollary

Any two bases of a finitely generated vector space have the same number of elements.

3.3.7 Definition

The number of elements in a basis of a finitely generated vector space V is called the **dimension** *of V and is denoted* $\dim(V)$. *The dimension of $V = \{0\}$ is defined to be* 0.

Since the number of elements in a basis of a finitely generated vector space is finite, we call such vector spaces ***finite dimensional***.

A couple of facts are worth noting at this point. It is not entirely obvious that a subspace of a finitely generated vector space is finitely generated. It follows from 3.3.5, however. Let V be finitely generated, and let W be a subspace of V. If $\dim(V) = n$, then by 3.3.5, no independent subset of W can have more than n elements. Let $\{w_1, w_2, \ldots, w_m\}$ be an independent subset of W with m as large as possible. Then $\{w_1, w_2, \ldots, w_m\}$ spans W. Otherwise, there is an element $w_{m+1} \in W$ and not in

$$Fw_1 + Fw_2 + \cdots + Fw_m.$$

That is, $\{w_1, w_2, \ldots, w_m, w_{m+1}\}$ is linearly independent, which is an impossibility. Therefore, $\{w_1, w_2, \ldots, w_m\}$ is a basis of W. In particular, W is finitely generated. Furthermore, $\dim(W) \leq \dim(V)$. Of course, by 3.3.4 any basis of W extends to one of V.

Let F be any field, and let F^n be the vector space of all n-tuples of elements of F. Since

$$\{(1, 0, \ldots, 0), (0, 1, 0, \ldots, 0), \ldots, (0, 0, \ldots, 0, 1)\}$$

is a basis of F^n, $\dim(F^n) = n$. We will show that any vector space V over F is isomorphic to F^n if and only if $\dim(V) = n$. This is equivalent to the following theorem.

3.3.8 Theorem

Two finitely generated vector spaces V and W over a field F are isomorphic if and only if $\dim(V) = \dim(W)$.

Proof Suppose that $\alpha: V \to W$ is an isomorphism, and let $\{v_1, v_2, \ldots, v_n\}$ be a basis of V. (If $V = \{0\}$, then the theorem is trivial.) We will show that

$$\{\alpha(v_1), \alpha(v_2), \ldots, \alpha(v_n)\}$$

is a basis of W. If $\sum_{i=1}^n a_i\alpha(v_i) = 0$, then $\alpha(\sum_{i=1}^n a_iv_i) = 0$, whence $\sum_{i=1}^n a_iv_i = 0$ since α is one-to-one. Thus, each $a_i = 0$ and $\{\alpha(v_1), \alpha(v_2), \ldots, \alpha(v_n)\}$ is independent. Let $w \in W$. Then $\alpha(v) = w$ for some $v \in V$. But $v = \sum a_iv_i$ for appropriate $a_i \in F$, and

$$\alpha(v) = w = \alpha(\sum a_iv_i) = \sum \alpha(a_iv_i) = \sum a_i\alpha(v_i).$$

Hence $\dim(V) = \dim(W)$.

The other half follows from the very useful theorem below.

3.3.9 Theorem

Let V and W be vector spaces over F, $\{v_1, v_2, \ldots, v_n\}$ a basis of V, and $w_1, w_2, \ldots, w_n \in W$. Then there is a unique linear transformation $\alpha: V \to W$ such that $\alpha(v_i) = w_i$ for all i. The linear transformation α is one-to-one if and only if $\{w_1, w_2, \ldots, w_n\}$ is linearly independent, and it is onto if and only if $\{w_1, w_2, \ldots, w_n\}$ spans W.

Proof For $v \in V$, $v = \sum a_i v_i$. Define α by $\alpha(v) = \sum a_i w_i$. Now α is well defined since the a_i's are unique. It is easy to check that α is a linear transformation with $\alpha(v_i) = w_i$. Now suppose $\beta: V \to W$ is a linear transformation with $\beta(v_i) = w_i$ for all i. Then $\beta(v) = \beta(\sum a_i v_i) = \sum \beta(a_i v_i) = \sum a_i \beta(v_i) = \sum a_i w_i = \alpha(v)$. Thus $\alpha = \beta$. Now α is one-to-one if and only if $\alpha(\sum a_i v_i) = 0$ implies $\sum a_i v_i = 0$, which holds if and only if $a_i = 0$ for all i. But $\alpha(\sum a_i v_i) = \sum a_i w_i$. Thus, α is one-to-one if and only if $\sum a_i w_i = 0$ implies that $a_i = 0$ for all i. Since

$$\operatorname{Im} \alpha = Fw_1 + Fw_2 + \cdots + Fw_n,$$

α is onto if and only if $\{w_1, w_2, \ldots, w_n\}$ spans W.

To get the missing half of 3.3.8, let $\dim(V) = \dim(W)$, and let $\{v_1, v_2, \ldots, v_n\}$ and $\{w_1, w_2, \ldots, w_n\}$ be bases of V and W, respectively. The linear transformation $\alpha: V \to W$ such that $\alpha(v_i) = w_i$ for all i is one-to-one and onto by 3.3.9. Hence $V \approx W$.

3.3.10 Corollary

The vector space V over the field F is isomorphic to F^n if and only if $\dim(V) = n$. In particular, $F^n \approx F^m$ if and only if $m = n$.

Consider all the finitely generated vector spaces over the field F. By 3.3.8, a complete isomorphism invariant of such a vector space V is $\dim(V)$. That is, V is determined up to isomorphism by the number $\dim(V)$. This does not conclude the study of vector spaces, however. The real thing of interest about vector spaces is linear transformations, which we have hardly begun to study.

Let $V_1, V_2, \ldots, V_n$ be vector spaces over a field F. Let $V_1 \oplus V_2 \oplus \cdots \oplus V_n$ be the vector space whose elements are the elements of $V_1 \times V_2 \times \cdots \times V_n$ and whose vector addition and scalar multiplication are defined by

$$(v_1, v_2, \ldots, v_n) + (w_1, w_2, \ldots, w_n) = (v_1 + w_1, v_2 + w_2, \ldots, v_n + w_n),$$

and

$$a(v_1, v_2, \ldots, v_n) = (av_1, av_2, \ldots, av_n).$$

This construction is the same as that for Abelian groups, except that here we also have scalar multiplication involved. That $V_1 \oplus V_2 \oplus \cdots \oplus V_n$ is indeed a vector space follows easily. This vector space is called the ***external direct sum*** of $V_1, V_2, \ldots, V_n$. For example, F^n is the external direct sum of n copies of the vector space F.

If V_i^* consists of those elements of $V_1 \oplus V_2 \oplus \cdots \oplus V_n$ of the form

$$(0, 0, \ldots, 0, v_i, 0, \ldots, 0)$$

where v_i is in the ith place, then V_i^* is a subspace, and as in the case for Abelian groups, every element in $V_1 \oplus \cdots \oplus V_n$ can be written uniquely in the form $\sum_{i=1}^n v_i^*$, with $v_i^* \in V_i^*$.

3.3.11 Definition

If a vector space V has subspaces $V_1, V_2, \ldots, V_n$ such that every element in V can be written uniquely in the form $v_1 + v_2 + \cdots + v_n$ with $v_i \in V_i$, then we write

$$V = V_1 \oplus V_2 \oplus \cdots \oplus V_n$$

and say that V is the **(internal) direct sum** *of the subspaces $V_1, V_2, \ldots, V_n$.*

For example, in F^n let V_i be all elements of the form $(0, 0, \ldots, 0, a, 0, \ldots, 0)$, where a is in the ith position. Then F_n is the internal direct sum of the V_i.

This procedure is completely analogous to the one followed for Abelian groups (2.7.1). In particular, if V is the internal direct sum of subspaces V_i, then V is isomorphic to the external direct sum of the vector spaces V_i. We want to prove two things. Here they are.

3.3.12 Theorem

Every finite-dimensional vector space $\neq \{0\}$ is the direct sum of one-dimensional subspaces.

3.3.13 Theorem

Every subspace of a finite-dimensional vector space V is a direct summand of V.

Proof Let $\{v_1, v_2, \ldots, v_n\}$ be a basis of V. The subspaces Fv_i are one dimensional, and

$$V = Fv_1 \oplus Fv_2 \oplus \cdots \oplus Fv_n.$$

This proves 3.3.12.

To prove 3.3.13, let W be a subspace of V. Then W is finite dimensional as we have seen. Let $\{w_1, w_2, \ldots, w_m\}$ be a basis of W. By 3.3.4, this extends to a basis

$$\{w_1, w_2, \ldots, w_m, w_{m+1}, \ldots, w_n\}$$

of V. Thus, $V = W \oplus (Fw_{m+1} + Fw_{m+2} + \cdots + Fw_n)$.

We conclude this section with some relationships between the dimensions of various vector spaces and subspaces. Let us summarize what we have learned thus far about such dimensions into one theorem. We will assume throughout that all our vector spaces are finite dimensional.

3.3.14 Theorem

(a) $V \approx W$ *if and only if* $\dim(V) = \dim(W)$.
(b) *If* $\alpha: V \to W$ *is an epimorphism, then* $\dim(W) \leq \dim(V)$.
(c) If $\alpha: V \to W$ *is a monomorphism, then* $\dim(V) \leq \dim(W)$.
(d) *Let* W *be a subspace of* V. *Then* $\dim(V/W) \leq \dim(V)$ *and* $\dim(W) \leq \dim(V)$.
(e) *Let* W *be a subspace of* V. *Then* $W = V$ *if and only if* $\dim(W) = \dim(V)$.

All these have appeared earlier in one form or another. Note that (d) follows from (b) and (c). The exact relationship between dimensions of subspaces and quotient spaces is given next.

3.3.15 Theorem

Let W *be a subspace of* V. *Then*

$$\dim(V) = \dim(W) + \dim(V/W).$$

Proof Let $V = W \oplus X$. If $\{v_1, v_2, \ldots, v_r\}$ is a basis of W and $\{v_{r+1}, v_{r+2}, \ldots, v_n\}$ is a basis of X, then $\{v_1, v_2, \ldots, v_n\}$ is a basis of V. Since $V/W \approx X$, the theorem follows.

Now let W and X be two subspaces of V. We have the subspaces W, X, $W \cap X$, and $W + X$. The relationship between their dimensions is given in the next theorem.

3.3.16 Theorem

Let W *and* X *be subspaces of* V. *Then*

$$\dim(W + X) + \dim(W \cap X) = \dim(W) + \dim(X).$$

Proof $(W+X)/X \approx W/(W \cap X)$, so using 3.3.15 we obtain

$$\dim(W+X) - \dim(X) = \dim(W) - \dim(W \cap X).$$

Also, 3.3.15 gives us the following theorem.

3.3.17 Theorem

Let $\alpha: V \to W$ be a linear transformation. Then

$$\dim(V) = \dim(\operatorname{Ker} \alpha) + \dim(\operatorname{Im} \alpha).$$

Proof $V/\operatorname{Ker} \alpha \approx \operatorname{Im} \alpha$. Apply 3.3.15.

PROBLEMS

1 Prove that if $\{v_1, v_2, \ldots, v_n\}$ is independent, then $v_1 \neq v_2$, and $v_1 \neq 0$.

2 Show by direct computation that in $\mathbb{Q}^3$, the set

$$S = \{(3, 0, -3), (-1, 1, 2), (4, 2, -2), (2, 1, -1)\}$$

is linearly dependent. Does S generate $\mathbb{Q}^3$? Does S contain a basis of $\mathbb{Q}^3$?

3 (a) Show that $\{(1, 0, -1), (1, 2, 1), (0, -3, 2)\}$ is a basis of $\mathbb{Q}^3$.
(b) Is $\{(1, 1, 2, 4), (2, -1, -5, 2), (1, -1, -1, 0), (2, 1, 1, 0)\}$ linearly independent in $\mathbb{Q}^4$?

4 Let $F = \mathbb{Z}/\mathbb{Z}2$. Write down all bases of F^2. Write down all bases of F^3.

5 Let $F = \mathbb{Z}/\mathbb{Z}3$. Write down all bases of F^2.

6 In $\mathbb{R}^4$, let $V = \mathbb{R}(1, 1, 1, 1)$, and let $W = V + \mathbb{R}(1, 1, 1, 0)$. Let $V^\perp = \{(a_1, a_2, a_3, a_4) \in \mathbb{R}^4 \colon \sum a_i b_i = 0 \text{ for all } (b_1, b_2, b_3, b_4) \in V\}$, and define $W^\perp$ similarly. Find a basis for $V^\perp$. Find a basis for $W^\perp$.

7 Suppose that $v_1 \neq 0$. Prove that $\{v_1, v_2, \ldots, v_n\}$ is dependent if and only if some v_{i+1} is a linear combination of $v_1, v_2, \ldots, v_i$.

8 Prove that $\{v_1, v_2, \ldots, v_n\}$ is a basis if and only if it is linearly independent and every family $\{w_1, w_2, \ldots, w_m\}$ with $m > n$ is dependent.

9 Prove that $\{v_1, v_2, \ldots, v_n\}$ is a basis of V if and only if it spans V and no family $\{w_1, w_2, \ldots, w_m\}$ with $m < n$ spans V.

10 Prove that $\{v_1, v_2, \ldots, v_n\}$ is a basis if and only if it is linearly independent and every family properly containing it is dependent.

11 Prove that $\{v_1, v_2, \ldots, v_n\}$ is a basis of V if and only if it spans V and no proper subfamily spans V.

12 Prove that $(V \oplus W)/W \approx V$.

13 Let $\{v_1, v_2, \ldots, v_m\}$ and $\{w_1, w_2, \ldots, w_n\}$ be bases of V. Prove that $m = n$ in the following way. Write $v_1 = \sum_{i=1}^{n} a_i w_i$. We may suppose that $a_1 \neq 0$. Then $V = Fv_1 \oplus Fw_2 \oplus \cdots \oplus Fw_n$, so

$$V/Fv_1 \approx Fv_2 \oplus \cdots \oplus Fv_m \approx Fw_2 \oplus \cdots \oplus Fw_n.$$

Inducting on m yields $m = n$.

14 Prove that if $\alpha: V \to W$ is an epimorphism and V is finite dimensional, then so is W, and $\dim(W) \le \dim(V)$.

15 Prove that if $\alpha: V \to W$ is a monomorphism, V is finite dimensional, and $\dim(V) = \dim(W)$, then α is an isomorphism.

16 Prove that if V is finite dimensional, and W is a subspace of V, then there is an idempotent endomorphism of V with image W. (See Section 2.6.)

17 Let F be any field. As a vector space over F, what are the bases of F? What is $\dim(F)$? Prove that $\{(a_1, a_2), (b_1, b_2)\}$ is a basis of F^2 if and only if $a_1 b_2 - a_2 b_1 \neq 0$.

18 Suppose that F is a field with 9 elements. (There is such a field.) Suppose that P is a subfield with 3 elements. (There is such a subfield.) Consider F as a vector space over P. What is $\dim(F)$?

19 Consider the complex numbers as a vector space over the real numbers. What is its dimension?

20 Consider the real numbers as a vector space over the rational numbers. Prove that it is not finite dimensional.

21 Suppose that $\{u, v, w\}$ is independent. Prove that $\{u, u+v, u+v+w\}$ is independent.

22 Let V be the vector space over the field of real numbers consisting of all functions from the real numbers into the real numbers. Prove that V is not finite dimensional.

23 Let $V_1, V_2, \ldots, V_n$ be subspaces of V. Prove that the map from the external direct sum $V_1 \oplus \cdots \oplus V_n$ into V given by $\alpha(v_1, v_2, \ldots, v_n) = v_1 + v_2 + \cdots + v_n$ is a linear transformation. Prove that V is the direct sum of the subspaces $V_1, V_2, \ldots, V_n$ if and only if α is an isomorphism.

24 Prove that if $V = V_1 \oplus \cdots \oplus V_n$, then $\dim(V) = \sum_{i=1}^{n} \dim(V_i)$.

25 Let V be finite dimensional, and let W and X be subspaces of V. Let $\{v_1, v_2, \ldots, v_r\}$ be a basis of $W \cap X$, let $\{v_1, \ldots, v_r, w_1, \ldots, w_s\}$ be a basis of W, and let $\{v_1, \ldots, v_r, x_1, \ldots, x_t\}$ be a basis of X. Prove that $\{v_1, \ldots, v_r, w_1, \ldots, w_s, x_1, \ldots, x_t\}$ is a basis of $W + X$. (This proves 3.3.16.)

26 Let W be a subspace of V, let $\{w_1, \ldots, w_r\}$ be a basis of W, and let $\{v_1 + W, v_2 + W, \ldots, v_s + W\}$ be a basis of V/W. Prove that $\{w_1, \ldots, w_r, v_1, \ldots, v_s\}$ is a basis of V. (This proves 3.3.15.)

27 Define an infinite set in a vector space to be linearly independent if every nonempty finite subset of it is linearly independent. Define a basis of V to be a linearly independent set that generates V. Prove that the vector space of polynomials over any field has a basis.

28 Let V be the vector space of all sequences $\{a_1, a_2, \ldots .\}$ of real numbers. Let S be the subspace consisting of those sequences with only finitely many nonzero terms. Prove that S has a basis. Try to prove that V has a basis.

3.4 DUAL SPACES

Suppose V and W are vector spaces over the field F. Let $\mathrm{Hom}(V, W)$ denote the set of all linear transformations from V into W. There is a natural way to make $\mathrm{Hom}(V, W)$ into a vector space over F. To make it into an Abelian group, for $f, g \in \mathrm{Hom}(V, W)$, define $f+g$ by the equation

$$(f+g)(v) = f(v) + g(v).$$

That is, $f+g: V \to W: v \to f(v)+g(v)$. First we need to verify that $f+g \in \mathrm{Hom}(V, W)$. It must preserve vector addition and scalar multiplication. The equations

$$\begin{aligned}(f+g)(v_1+v_2) &= f(v_1+v_2) + g(v_1+v_2)\\ &= f(v_1)+f(v_2)+g(v_1)+g(v_2)\\ &= (f(v_1)+g(v_1)) + (f(v_2)+g(v_2))\\ &= (f+g)(v_1) + (f+g)(v_2)\end{aligned}$$

and

$$(f+g)(av) = f(av)+g(av) = af(v)+ag(v) = a(f(v)+g(v)) = a((f+g)(v))$$

show that it does.

At this point, we have an addition defined on $\mathrm{Hom}(V, W)$. To verify that $\mathrm{Hom}(V, W)$ is an Abelian group under this addition is straightforward. For example, the negative of $f \in \mathrm{Hom}(V, W)$ is the map $-f$ defined by $(-f)(v) = -(f(v))$. The zero is the map that takes every $v \in V$ to the $0 \in W$. We leave the details to the reader. There are no difficult points.

Scalar multiplication in $\mathrm{Hom}(V, W)$ is defined by

$$(af)(v) = a(f(v)).$$

We must verify that $af \in \mathrm{Hom}(V, W)$. But

$$\begin{aligned}(af)(v_1+v_2) &= a(f(v_1+v_2)) = a(f(v_1)+f(v_2))\\ &= a(f(v_1)) + a(f(v_2)) = (af)(v_1) + (af)(v_2),\end{aligned}$$

and

$$(af)(bv) = a(b(f(v))) = ab(f(v)) = ba(f(v)) = b(a(f(v))) = b((af)(v)).$$

Now we must ensure that the equations

$$a(f+g)=af+ag,$$

and

$$a(bf)=(ab)f$$

hold. The first holds because for $v \in V$,

$$\begin{aligned}(a(f+g))(v) &= a((f+g)(v)) = a(f(v)+g(v)) \\ &= (af)(v)+(ag)(v) = (af+ag)(v).\end{aligned}$$

The second follows similarly. Note finally that $1 \cdot f = f$.

Thus, from the vector spaces V and W over F we have built the vector space Hom(V, W) over F. There are two very special cases to consider: $W = F$ and $W = V$. Now Hom(V, V) is more than a vector space. It is a ring. Indeed, it is an algebra over F. It will be studied in Chapter 8. Our concern here will be with Hom(V, F).

3.4.1 Definition

The vector space Hom(V, F) *is denoted* V^*, *and is called the* **dual space**, *or the* **conjugate space**, *or the* **adjoint space** *of* V.

Now we will determine the dimension of V^*. It is about as easy to compute dim(*Hom*(V, W)). *We will assume throughout that all our vector spaces are finite dimensional.*

3.4.2 Theorem

$$\dim(\text{Hom}(V, W)) = \dim(V) \cdot \dim(W).$$

Proof If either V or W is $\{0\}$, the theorem is clear. If neither is $\{0\}$, then dim(Hom(V, W)) is the size of any basis of Hom(V, W). We will simply write down a basis of Hom(V, W) in terms of bases of V and W and note its size. Let $\{v_1, v_2, \ldots, v_m\}$ be a basis of V, and let $\{w_1, w_2, \ldots, w_n\}$ be a basis of W. By 3.3.9, there is exactly one linear transformation from V into W that takes v_1 to x_1, v_2 to x_2, . . . , and v_m to x_m, where $x_1, x_2, \ldots, x_m$ are any prescribed vectors in W. Thus, we have defined an element of Hom(V, W) when we prescribe images for $v_1, v_2, \ldots, v_m$. For $i = 1, 2, \ldots, m$, and $j = 1, 2, \ldots, n$, let f_{ij} be the element in Hom(V, W) such that

$$f_{ij}(v_i) = w_j,$$

and

$$f_{ij}(v_k) = 0 \quad \text{if} \quad i \neq k.$$

Thus $f_{23}(v_1) = 0$, $f_{23}(v_2) = w_3$, $f_{23}(v_3) = 0, \ldots, f_{23}(v_m) = 0$.

These functions are more conveniently defined using δ, the Kronecker delta. Let $\delta_{ij} = 0 \in F$ if $i \neq j$, and let $\delta_{ij} = 1 \in F$ if $i = j$. Then for $i = 1, 2, \ldots, m$ and $j = 1, 2, \ldots, n$,

$$f_{ij}(v_k) = \delta_{ik} w_j.$$

We claim that the f_{ij} constitute a basis for Hom(V, W). At least it has the size we want.

Suppose $\sum_{i,j} a_{ij} f_{ij} = 0$. Then

$$\left(\sum_{i,j} a_{ij} f_{ij}\right)(v_k) = 0 = \sum_{i,j} (a_{ij} f_{ij}(v_k)) = \sum_{i,j} a_{ij} \delta_{ik} w_j = \sum_j a_{kj} w_j.$$

Therefore, $a_{kj} = 0$ for $j = 1, 2, \ldots, n$. This holds for $k = 1, 2, \ldots, m$, so that $a_{ij} = 0$ for all i and j. Thus, the f_{ij} are linearly independent.

Let $f \in$ Hom(V, W). Then $f(v_i) = \sum_j a_{ij} w_j$. But

$$\left(\sum_{k,j} a_{kj} f_{kj}\right)(v_i) = \sum_{k,j} (a_{kj} f_{kj}(v_i)) = \sum_{k,j} (a_{kj} \delta_{ki} w_j) = \sum_j a_{ij} w_j.$$

Hence $\sum_{k,j} a_{kj} f_{kj} = f$, so that the f_{ij} span Hom(V, W). The proof is complete.

3.4.3 Corollary

$\dim(V^*) = \dim(V)$.

Of course, 3.4.3 implies that $V \approx V^*$. Now one can wonder why all the fuss about V^*. Up to isomorphism it is just V itself. The reason for the concern is this. The vector space V^* is constructed from V, and there is a natural way to associate subspaces of V with subspaces of V^*. Furthermore, although there is no natural isomorphism $V \to V^*$, there *is* a natural isomorphism $V \to V^{**}$. Also, with each linear transformation $f: V \to W$, there is associated a linear transformation $f^*: W^* \to V^*$. The relationships between all these things, and in general the properties they have, are important and useful. This section develops some of these properties.

In the proof of 3.4.2, we defined a linear transformation $f_{ij} \in$ Hom(V, W) in terms of given bases $\{v_1, v_2, \ldots, v_m\}$ and $\{w_1, w_2, \ldots, w_n\}$ of V and W, respectively. In the case $W = F$, there is a natural basis to pick, namely $\{1\}$. Let us consider that case.

Let $\{v_1, v_2, \ldots, v_n\}$ be a basis of V. Then $v_i^* \in V^*$ defined by

$$v_i^*(v_j) = \delta_{ij}$$

gives a basis $\{v_1^*, v_2^*, \ldots, v_n^*\}$ of Hom(V, F) $= V^*$. This we proved in the course of proving 3.4.2.

3.4.4 Definition

The basis of V^* given by

$$v_i^*(v_j) = \delta_{ij}$$

*is the basis of V^* **dual** to the basis $\{v_1, v_2, \ldots, v_n\}$ of V.*

Let $v \in V$. If $v^* \in V^*$, then $v^*(v) \in F$. Thus, v yields a map $V^* \to F$. This map is linear—that is, it is in V^{**}.

3.4.5 Definition

The map $\eta\colon V \to V^{**}$ defined by

$$\eta(v)(v^*) = v^*(v)$$

*is called the **natural map from V into V^{**}**.*

3.4.6 Theorem

$\eta\colon V \to V^{**}$ is an isomorphism.

Proof We do not know yet that η is even into V^{**}. To see this, we need that for $a \in F$ and $x, y \in V^*$,

$$\eta(v)(x + y) = \eta(v)(x) + \eta(v)(y),$$

and

$$\eta(v)(ax) = a(\eta(v)(x)).$$

But this follows from the equations

$$\eta(v)(x + y) = (x + y)(v) = x(v) + y(v) = \eta(v)(x) + \eta(v)(y),$$

and

$$\eta(v)(ax) = (ax)(v) = a(x(v)) = a(\eta(v)(x)).$$

Thus, η maps V into V^{**}. Now η must preserve vector addition and scalar multiplication. That it does is left to the reader (Problem 2). To get η to be an isomorphism, it suffices to get it to be one-to-one since V is finite dimensional and $\dim(V) = \dim(V^{**})$. Suppose that $\eta(v) = 0$. Then $\eta(v)(v^*) = 0 = v^*(v)$ for all $v^* \in V^*$. But if $v \neq 0$, then v is an element of a basis of V, and the appropriate element of the basis dual to it takes v to $1 \in F$. Thus $v = 0$, and we are done.

Let S be a subset of V. Consider all those $v^* \in V^*$ such that $\text{Ker}(v^*) \supset S$. This subset of V^* is called the ***annihilator of S in V^****. Denote it by $K(S)$.

Thus,

$$K(S) = \{v^* \in V^*: v^*(s) = 0 \text{ for all } s \in S\}.$$

Similarly, if T is a subset of V^*, let

$$K^*(T) = \{v \in V: t(v) = 0 \text{ for all } t \in T\}.$$

Then $K^*(T)$ is called the ***annihilator of T in V***. Now $K(S)$ is defined when S is a subset of a given vector space V. If T is a subset of V^*, then $K(T) \subset V^{**}$. Similar remarks hold for K^*.

The following properties are readily verified.

(a) If $S_1 \subset S_2$, then $K(S_1) \supset K(S_2)$.
(b) If S generates the subspace W of V, then $K(S) = K(W)$.
(c) $K(S)$ is a subspace of V^*.
(d) $K(\{0\}) = V^*$; $K(V) = \{0\}$.
(e) Statements analogous to (a), (b), (c), and (d) hold for K^*.
(f) $K^*(K(S)) \supset S$; $K(K^*(T)) \supset T$.

The map K is from the set of subsets of V to the set of subspaces of V^*. In light of (b), however, the real interest in it will be as a map from the set of subspaces of V to the set of subspaces of V^*. As such, we are going to show that it is one-to-one and onto. By (a), K reverses inclusions, so K will be a ***duality*** between the set of subspaces of V and those of V^*. First we note a useful relationship between η, K, and K^*.

3.4.7 Lemma

Let T be a subspace of V^*. Then $\eta(K^*(T)) = K(T)$.

Proof Let $v^{**} \in K(T)$. Then $v^{**} = \eta(v)$ for some $v \in V$. For $t \in T$, $\eta(v)(t) = v^{**}(t) = 0 = t(v)$, so $v \in K^*(T)$. Let $v \in K^*(T)$. Then $\eta(v)(t) = t(v) = 0$, so $\eta(v) \in K(T)$.

3.4.8 Theorem

Let S be a subspace of V, and let T be a subspace of V^*. Then

$$\dim(V) = \dim(S) + \dim(K(S)),$$

and

$$\dim(V^*) = \dim(T) + \dim(K^*(T)).$$

Proof Let $\{v_1, v_2, \ldots, v_m\}$ be a basis of S. Extend it to a basis $\{v_1, v_2, \ldots, v_m, \ldots, v_n\}$ of V. Let

$$\{v_1^*, v_2^*, \ldots, v_n^*\}$$

be dual to it. Now $\sum_{i=1}^n a_i v_i^*$ is in $K(S)$ if and only if $(\sum a_i v_i^*)(v_j) = 0$ for $j = 1, 2, \ldots, m$. But

$$(\sum a_i v_i^*)(v_j) = \sum a_i(v_i^*(v_j)) = a_j.$$

Hence

$$K(S) = Fv_{m+1}^* + \cdots + Fv_n^*, \quad \text{and} \quad \dim(K(S)) = \dim(V) - \dim(S).$$

If T is a subspace of V^*, then $\dim(V^*) = \dim(T) + \dim(K(T))$. But $\dim(K(T)) = \dim(K^*(T))$ by 3.4.7. Thus,

$$\dim(V^*) = \dim(T) + \dim(K^*(T)).$$

Now we can show that K is a duality.

3.4.9 Theorem

K is a one-to-one correspondence between the set of subspaces of V and the set of subspaces of V^*. Its inverse is K^*.

Proof Let T be a subspace of V^*. Then $K(K^*(T)) \supset T$, but the two have the same dimension by 3.4.8. Therefore, $K(K^*(T)) = T$. Let S be a subspace of V. Then $K^*(K(S)) \supset S$, and by 3.4.8 they have the same dimension. Hence, $K^*(K(S)) = S$. Thus K and K^* are inverses of each other, and the theorem follows.

The maps K and K^* reverse inclusions. That is, if $A \subset B$, then $K(A) \supset K(B)$, and similarly for K^*. A stronger result is that K and K^* interchange $+$ and $\cap$.

3.4.10 Theorem

Let S_1 and S_2 be subspaces of V, and let T_1 and T_2 be subspaces of V^*. Then

(a) $K(S_1 + S_2) = K(S_1) \cap K(S_2)$.
(b) $K(S_1 \cap S_2) = K(S_1) + K(S_2)$,
(c) $K^*(T_1 + T_2) = K^*(T_1) \cap K^*(T_2)$, and
(d) $K^*(T_1 \cap T_2) = K^*(T_1) + K^*(T_2)$.

Proof Let $v^* \in K(S_1 + S_2)$. Then $v^*(s_1 + s_2) = 0$ for all $s_1 \in S_1$ and $s_2 \in S_2$. Letting $s_2 = 0$, we get $v^* \in K(S_1)$. Similarly, $v^* \in K(S_2)$. Thus, $K(S_1 + S_2) \subset K(S_1) \cap K(S_2)$. Now let $v^* \in K(S_1) \cap K(S_2)$. Then $v^*(s_i) = 0$ for all $s_i \in S_i$, $i = 1, 2$. Thus $v^*(s_1 + s_2) = v^*(s_1) + v^*(s_2) = 0$, so (a) holds.

Part (c) is almost identical to (a). Using 3.4.9 and (c), part (b) follows from the equations

$$\begin{aligned} K(S_1) + K(S_2) &= K(K^*(K(S_1) + K(S_2))) \\ &= K((K^*(K(S_1))) \cap (K^*(K(S_2)))) = K(S_1 \cap S_2). \end{aligned}$$

Part (d) follows similarly.

Let $\alpha: V \to W$ be a linear transformation. With any $w^* \in W^*$ we have the map $w^* \circ \alpha$. It is linear, being the composition of two linear maps. Furthermore, $w^* \circ \alpha: V \to F$, so $w^* \circ \alpha \in V^*$. Thus α induces a map $\alpha^*: W^* \to V^*$, namely

$$\alpha^*(w^*) = w^* \circ \alpha.$$

That α^* itself is linear should be no surprise.

3.4.11 Definition

Let $\alpha: V \to W$ be linear. The map $\alpha^*: W^* \to V^*$ induced by α is called the **conjugate** *of α, or the* **adjoint** *of α, or the* **dual** *of α.*

These two facts are formalities:

(a) α^* is a linear transformation, and
(b) the map $\mathrm{Hom}(V, W) \to \mathrm{Hom}(W^*, V^*)$: $\alpha \to \alpha^*$ is a linear transformation (Problem 8).

There are some important relationships between α and α^*. One is that α and α^* have the same ***rank***, where the rank $r(\alpha)$ of a linear transformation α is defined to be $\dim(\mathrm{Im}\,\alpha)$. We will relate this to the rank of matrices at the end of this section.

3.4.12 Theorem

The annihilator of the image of α is the kernel of α^*. The image of α^* is the annihilator of the kernel of α.

Proof We need that $K(\mathrm{Im}\,\alpha) = \mathrm{Ker}\,\alpha^*$. Let $\alpha: V \to W$. Then $w^* \in K(\mathrm{Im}\,\alpha)$ if and only if $w^*(\alpha(v)) = 0$ for all $v \in V$ if and only if

$\alpha^*(w^*)(v) = 0$ for all $v \in V$ if and only if $\alpha^*(w^*) = 0$ if and only if $w^* \in \text{Ker } \alpha^*$.

The second assertion is Problem 11.

3.4.13 Corollary

α and α^* have the same rank.

Proof From 3.4.12, we have

$$\dim(K(\text{Im } \alpha)) = \dim(W) - r(\alpha) = \dim(\text{Ker } \alpha^*) = \dim(W) - r(\alpha^*).$$

3.4.14 Corollary

α is one-to-one if and only if α^* is onto, and α is onto if and only if α^* is one-to-one.

Proof The details are left as an exercise (Problem 9).

3.4.15 Corollary

The linear transformation

$$\text{Hom}(V, W) \rightarrow \text{Hom}(W^*, V^*): \alpha \rightarrow \alpha^*$$

is an isomorphism.

Proof $\text{Hom}(V, W)$ and $\text{Hom}(W^*, V^*)$ both have dimension $\dim(V)\dim(W)$, so it suffices to get the linear transformation to be one-to-one. But if $\alpha^* = 0$, then $K(\text{Im } \alpha) = \text{Ker } \alpha^* = W^*$, so that $\text{Im } \alpha = \{0\}$. Therefore, $\alpha = 0$.

Here is an application to matrices. Consider the $m \times n$ matrix (a_{ij}) with entries from any field F. The ***row space*** of (a_{ij}) is the subspace of F^n generated by the rows $(a_{i1}, a_{i2}, \ldots, a_{in})$ of (a_{ij}), and its dimension is called the ***row rank*** of (a_{ij}). The ***column space*** of (a_{ij}) is the subspace of F^m generated by the columns $(a_{1j}, a_{2j}, \ldots, a_{mj})$ of (a_{ij}), and its dimension is called the ***column rank*** of (a_{ij}). How are these dimensions related?

3.4.16 Theorem

The row rank of a matrix equals its column rank.

Proof This theorem is a reflection of the fact that the rank of a linear transformation is the rank of its dual (3.4.13). Let $\{v_1, v_2, \ldots, v_m\}$ be the natural basis of F^m, and $\{w_1, w_2, \ldots, w_n\}$ be the natural basis of F^n. Let $\{v_1^*, \ldots, v_m^*\}$ and $\{w_1^*, \ldots, w_n^*\}$ be bases dual to these bases. Let $\alpha: F^m \to F^n$ be given by

$$\alpha(v_i) = (a_{i1}, a_{i2}, \ldots, a_{in}).$$

Then $r(\alpha)$ is the row rank of (a_{ij}). Let β be the isomorphism defined by $\beta(v_i^*) = v_i$. Then $\alpha^*(w_i^*) = \sum_j a_{ji}v_j^*$ since

$$(\alpha^*(w_i^*))(v_k) = (w_i^* \circ \alpha)(v_k) = w_i^*(\alpha(v_k))$$

$$= w_i^*(a_{k1}, \ldots, a_{kn}) = w_i^* \sum_j a_{kj}w_j = a_{ki}.$$

Thus,

$$\beta \circ \alpha^*(w_i^*) = \beta(\sum a_{ji}v_j^*) = \sum a_{ji}v_j = (a_{1i}, a_{2i}, \ldots, a_{mi}).$$

Hence, $r(\beta\alpha^*)$ is the column rank of the matrix (a_{ij}). Since β is an isomorphism, $r(\beta \circ \alpha^*) = r(\alpha^*)$, and we know that $r(\alpha^*) = r(\alpha)$. The theorem is proved.

The row rank of a matrix is called simply the ***rank*** of that matrix. Let (a_{ij}) be a matrix and let $(a_{ij})'$ be its ***transpose***. That is, the ijth entry in $(a_{ij})'$ is a_{ji}. Then the rows of (a_{ij}) are the columns of $(a_{ij})'$, and the columns of (a_{ij}) are the rows of $(a_{ij})'$. Sometimes 3.4.16 is expressed by saying that *the rank of a matrix equals the rank of its transpose.*

PROBLEMS

1 Give a complete proof that $\text{Hom}(V, W)$ is a vector space.

2 Prove in detail that $\eta: V \to V^{**}$ is linear.

3 Let $\{v_1, v_2, \ldots, v_n\}$ be a basis of V, let $\{v_1^*, \ldots, v_n^*\}$ be the basis of V^* dual to it, and let $\{v_1^{**}, \ldots, v_n^{**}\}$ be the basis of V^{**} dual to $\{v_1^*, \ldots, v_n^*\}$. Let η be the natural isomorphism from V to V^{**}. Prove that $\eta(v_i) = v_i^{**}$.

4 Let $\eta_V: V \to V^{**}$ and $\eta_W: W \to W^{**}$ be the natural isomorphisms. Let $\alpha \in \text{Hom}(V, W)$. Prove that $\alpha^{**} \circ \eta_V = \eta_W \circ \alpha$.

5 Let S_1 and S_2 be subsets of V. Prove that $K(S_1) = K(S_2)$ if and only if S_1 and S_2 generate the same subspace of V.

6 Let T_1 and T_2 be subsets of V^*. Prove that $K^*(T_1) = K^*(T_2)$ if and only if T_1 and T_2 generate the same subspace of V^*.

7 Prove in detail parts (c) and (d) of 3.4.10.

8 Let α be an element of $\text{Hom}(V, W)$. Prove that α^* is an element of $\text{Hom}(W^*, V^*)$, and that the mapping

$$\text{Hom}(V, W) \to \text{Hom}(W^*, V^*): \alpha \to \alpha^*$$

is a linear transformation.

9 Prove that α is one-to-one if and only if α^* is onto, and prove that α is onto if and only if α^* is one-to-one.

10 Let $x^*, y^* \in V^*$. Prove that if $\text{Ker}\, x^* \subset \text{Ker}\, y^*$, then $ax^* = y^*$ for some $a \in F$.

11 Let $\alpha \in \text{Hom}(V, W)$. Prove that $\text{Im}\, \alpha^* = K(\text{Ker}\, \alpha)$.

12 Let $\alpha \in \text{Hom}(V, W)$, and let $\beta \in \text{Hom}(W, X)$. Prove that

$$(\beta\alpha)^* = \alpha^*\beta^*.$$

13 Let v_i be the vector in F^n with 1 in the ith place and 0 elsewhere. Let $\{v_1^*, \ldots, v_n^*\}$ be the basis dual to $\{v_1, \ldots, v_n\}$. Identify F^{n*} with F^n via the isomorphism α given by $\alpha(v_i^*) = v_i$. Prove that

$$(a_1, a_2, \ldots, a_n)((b_1, b_2, \ldots, b_n)) = \sum a_i b_i.$$

14 Identifying $\mathbb{R}^{3*}$ with $\mathbb{R}^3$ as in Problem 13, find a basis of the annihilator of the subspace generated by $\{(1,1,0), (0,1,2)\}$ and a basis of the annihilator of the subspace generated by $\{(1,1,1)\}$.

4 Rings and modules

4.1 DEFINITIONS AND EXAMPLES

Several algebraic concepts are introduced in this chapter, and the main purpose is to prove one of the fundamental structure theorems in algebra. This theorem generalizes the fundamental theorem of finite Abelian groups (2.7.7) and the theorem that every finitely generated vector space has a basis (3.3.2). It is useful in analyzing linear transformations and will be put to that use in Chapter 5. Before we can even formulate the theorem, however, a number of definitions and facts are necessary. These preliminaries are themselves basic.

4.1.1 Definition

*A **ring** is a set R with two binary operations, called addition and multiplication, and denoted $+$ and $\cdot$, respectively, such that*

(a) *R is an Abelian group under $+$,*
(b) *$\cdot$ is associative, and*
(c) $a \cdot (b + c) = (a \cdot b) + (a \cdot c)$ and
$(a + b) \cdot c = (a \cdot c) + (b \cdot c)$ for all a, b, c in R.

The two equalities in (c) are called the left and right ***distributive laws***, respectively. As usual for additively written Abelian groups, the additive identity of R is denoted by 0, and $-a$ denotes the element of R such that $a + (-a) = 0$.

It may happen that $\cdot$ is commutative, that is, that $a \cdot b = b \cdot a$ for all a, b in R. In that case, R is called a ***commutative ring***. If $R \neq \{0\}$ and there is an element x in R such that $x \cdot a = a \cdot x = a$ for all a in R, then that element is unique, is always denoted by 1, and R is called a ***ring with identity***. As in the case of fields, which are special kinds of rings, we will tend to write $a \cdot b$ as ab.

Before proceeding with examples, some immediate consequences of the axioms will be proved that are needed in performing elementary calculations in rings.

4.1.2 Theorem

Let R be a ring. Then for $a, b, c \in R$, the following hold.

(a) $a \cdot 0 = 0 \cdot a = 0$.
(b) $a(-b) = -(ab) = (-a)b$.
(c) $ab = (-a)(-b)$.
(d) $a(b - c) = ab - ac$ *and* $(a - b)c = ac - bc$.
(e) *If R has an identity, then* $(-1)a = -a$.
(f) $1 \neq 0$.

Proof

(a) $a \cdot 0 = a(0 + 0) = a \cdot 0 + a \cdot 0$. Thus, $a \cdot 0 - a \cdot 0 = 0 = (a \cdot 0 + a \cdot 0) - a \cdot 0 = a \cdot 0 + (a \cdot 0 - a \cdot 0) = a \cdot 0$. Similarly, $0 \cdot a = 0$.
(b) $ab + a(-b) = a(b + (-b)) = a \cdot 0 = 0$. Thus, $a(-b) = -(ab)$. Similarly, $(-a)b = -(ab)$.
(c) $(-a)(-b) = -(a(-b)) = -(-(ab)) = ab$, using part (b).
(d) $a(b - c) = a(b + (-c)) = ab + a(-c) = ab + (-(ac)) = ab - ac$. Similarly, $(a - b)c = ac - bc$.
(e) $(1 - 1)a = 0 \cdot a = 0 = 1 \cdot a + (-1) \cdot a = a + (-1)a$. Thus, $(-1)a = -a$.
(f) If R has an identity, then $R \neq \{0\}$, so there is an a in R such that $a \neq 0$. But $1 \cdot a = a$ and $0 \cdot a = 0$. Thus, $1 \neq 0$.

We have several classifications of rings already—rings with an identity, commutative rings, and fields. There are other general types of rings that should be kept in mind when looking at the examples below. For example, it may happen that $a \cdot b = 0$ without a or b being 0. Such elements are called ***zero divisors.***

4.1.3 Definition

An **integral domain** *is a commutative ring with identity that has no zero divisors.*

Thus, a commutative ring with identity is an integral domain if and only if a, b in R, $ab = 0$, imply that $a = 0$ or $b = 0$. If a ring R has no zero divisors, then its set R^* of nonzero elements is closed under multiplication. It could happen then, that R^* is a group under multiplication.

4.1.4 Definition

A ring R is a **division ring** *if the set R^* of nonzero elements of R is a group under multiplication.*

4.1.5 Theorem

A finite ring with identity and no zero divisors is a division ring.

Proof Let R be such a ring, and let r be a nonzero element of R. We need only to get that r has a right inverse. (See 2.1, Problem 19.) Consider the set $S = \{rx : x \in R\}$. If $rx = ry$, then $r(x - y) = 0$, whence $x - y = 0$ and $x = y$ since R has no zero divisors. Thus the rx are distinct for distinct x, and since R is finite, $S = R$. Therefore some $rx = 1$, and it follows that R is a division ring.

4.1.6 Corollary

A finite integral domain is a field.

Recall that a field is a commutative division ring. Actually, a finite division ring is a field, but that is difficult to prove, and is postponed until 7.5.1.

If R is a ring with identity, the elements in R which have inverses are called ***units***. That is, an element u of R is a unit if there is an element $v \in R$ such that $uv = vu = 1$. If u is a unit, then the element v is unique and is denoted u^{-1} or $1/u$.

If S is a subset of a ring R, then restricting addition and multiplication to S might make S into a ring. If so, S is called a ***subring*** of R. Thus, S is a subring of R if and only if S is a subgroup of R which is closed under multiplication.

4.1.7 Examples

These examples should be examined carefully, and the reader should verify that each example is what it is claimed to be.

(a) Any field is a ring. Thus, the field of rational numbers $\mathbb{Q}$ with the ordinary operations of addition and multiplication is a ring.

Similarly, the real numbers and the complex numbers are rings, since they are fields.

(b) The integers $\mathbb{Z}$ with ordinary addition and multiplication is a ring, in fact, an integral domain.

(c) The set $\mathbb{Z}2$ of even integers is a subring of the ring $\mathbb{Z}$. This ring does not have an identity, but it is commutative and has no zero divisors.

(d) The set $\mathbb{Z}n$ for any integer n is a subring of the ring $\mathbb{Z}$.

(e) In the additive group $\mathbb{Z}/\mathbb{Z}n$ of integers modulo n, multiply modulo n. This makes $\mathbb{Z}/\mathbb{Z}n$ into a ring, which is called the ***ring of integers modulo*** n. [See 2.1.2(e), 2.3.13, and 3.1.4(g).]

(f) Let $\mathbb{Z}_2$ be the set of all 2×2 matrices over $\mathbb{Z}$. That is, $\mathbb{Z}_2$ is the set of all 2×2 matrices whose entries are integers. Define addition and multiplication by the equations

$$\begin{pmatrix} a_{11} & a_{12} \\ a_{21} & a_{22} \end{pmatrix} + \begin{pmatrix} b_{11} & b_{12} \\ b_{21} & b_{22} \end{pmatrix} = \begin{pmatrix} a_{11}+b_{11} & a_{12}+b_{12} \\ a_{21}+b_{21} & a_{22}+b_{22} \end{pmatrix},$$

and

$$\begin{pmatrix} a_{11} & a_{12} \\ a_{21} & a_{22} \end{pmatrix} \cdot \begin{pmatrix} b_{11} & b_{12} \\ b_{21} & b_{22} \end{pmatrix} = \begin{pmatrix} a_{11}b_{11}+a_{12}b_{21} & a_{11}b_{12}+a_{12}b_{22} \\ a_{21}b_{11}+a_{22}b_{21} & a_{21}b_{12}+a_{22}b_{22} \end{pmatrix}.$$

These definitions of $+$ and $\cdot$ make $\mathbb{Z}_2$ into a ring with identity. Since

$$0 = \begin{pmatrix} 0 & 1 \\ 0 & 0 \end{pmatrix}\begin{pmatrix} 1 & 0 \\ 0 & 0 \end{pmatrix} \neq \begin{pmatrix} 1 & 0 \\ 0 & 0 \end{pmatrix}\begin{pmatrix} 0 & 1 \\ 0 & 0 \end{pmatrix},$$

$\mathbb{Z}_2$ is not commutative and has zero divisors.

(g) Let R be any ring, and let n be any positive integer. Let $\mathbb{R}_n$ be the set of all $n \times n$ matrices over R. An element

$$\begin{pmatrix} a_{11} & a_{12} & \cdots & a_{1n} \\ a_{21} & a_{22} & \cdots & a_{2n} \\ \vdots & \vdots & & \vdots \\ a_{n1} & a_{n2} & \cdots & a_{nn} \end{pmatrix}$$

of R_n has all the a_{ij} in R, and it is denoted simply by (a_{ij}). The element a_{ij} denotes the (i, j) entry of (a_{ij}). Addition and multiplication then are defined by the equations

$$(a_{ij}) + (b_{ij}) = (a_{ij} + b_{ij})$$

and

$$(a_{ij}) \cdot (b_{ij}) = \left(\sum_{k=1}^{n} a_{ik} b_{kj} \right).$$

Again, it is straightforward to check that R_n is a ring. It has an identity if and only if R has an identity.

(h) Let $\mathbb{R}$ be the field of real numbers. Let $\mathbb{C}$ be the set of 2×2 matrices with entries from $\mathbb{R}$ of the form

$$\begin{pmatrix} a & b \\ -b & a \end{pmatrix}.$$

Then $\mathbb{C}$ is a subring of $\mathbb{R}_2$. Would you believe that $\mathbb{C}$ is the field of complex numbers?

(i) Let R be any commutative ring. Let $R[x]$ be the set of all polynomials in x with coefficients in R. That is, $R[x]$ is the set of all expressions of the form $a_n x^n + \cdots + a_1 x + a_0$, with a_i in R. Under the usual rules for adding and multiplying polynomials, $R[x]$ is a commutative ring, and it has an identity if and only if R does. (See 4.5.)

(j) Let R be any ring. Let S be the set of all mappings f of the set of nonnegative integers into R such that $f(n) = 0$ for all but finitely many n. Define $(f + g)(n) = f(n) + g(n)$, and define $(f \cdot g)(n) = \sum_{i+j=n} f(i)g(j)$. Then S is a ring with this definition of $+$ and $\cdot$. What is the difference between (i) and (j)? Is commutativity necessary in (i)?

(k) Let G be any Abelian group, and let R be the set of all endomorphisms of G. For $f, g \in R$, define $f + g$ and $f \cdot g$ by

$$(f + g)(x) = f(x) + g(x)$$

and

$$(f \cdot g)(x) = f(g(x)).$$

Then R is a ring with identity and is called the ***endomorphism ring*** of G. If G is the additive group of $\mathbb{Z}$, then R is an integral domain (4.2, Problem 23). If G is the additive group $\mathbb{Z}/p\mathbb{Z}$ of the integers modulo the prime p, then R is a field (4.2, Problem 24).

(l) Let V be any vector space, and let R be the set of all linear transformations from V into V. Define addition and multiplication as in (k). Then R is a ring with identity and is called the ***ring of linear transformations*** of V. We will look at this ring in some detail in Chapter 8.

(m) Let p be a prime, and let

$$R = \{a/b \colon a, b \in Z, (b, p) = 1\}.$$

Then R is a subring of the field $\mathbb{Q}$ of rational numbers. This ring is called the ***ring of integers localized at*** p.

(n) Let P be a nonempty subset of the set of all prime numbers. Let $R = \{a/b \colon a, b \in Z, (b, p) = 1 \text{ for all } p \in P\}$. Then R is a subring of $\mathbb{Q}$.

(o) Let G be a group and let R be a ring. Let $R(G)$ be the set of all maps f from G into R such that $f(x) = 0$ for all but finitely many x in

G. Define $f+g$ and fg by

$$(f+g)(x)=f(x)+g(x)$$

and

$$(fg)(x)=\sum_{uv=x} f(u)g(v).$$

With these definitions, $R(G)$ is a ring. It has an identity if R does, and it is commutative if R and G are commutative. The ring $R(G)$ is called the ***group ring of G over R***.

(p) Let R be any ring. Then $R\times R$ is a ring if one defines addition and multiplication coordinatewise. That is, $(a, b)+(c, d)=(a+c, b+d)$ and $(a, b)(c, d)=(ac, bd)$. It has an identity if and only if R does, and it is commutative if and only if R is commutative. Since $(a, 0)(0, b)=(0, 0)=0$, $R\times R$ has many zero divisors if $R\neq\{0\}$.

(q) Let S be any nonempty set, and let R be any ring. Let R^S be the set of all mappings from S into R. Define addition and multiplication in R^S by

$$(f+g)(s)=f(s)+g(s).$$

and

$$(fg)(s)=f(s)g(s).$$

With these definitions of addition and multiplication, R^S is a ring. It has an identity if and only if R does, and it is commutative if and only if R is commutative.

(r) Let R be any ring, and let $\mathbb{Z}$ be the ring of integers. Let $S=\mathbb{Z}\times R$. Define addition and multiplication in S by $(m, x)+(n, y)=(m+n, x+y)$ and $(m, x)(n, y)=(mn, xy+my+nx)$. Then S is a ring with identity. Identifying R with the subring of elements of S of the form $(0, r)$ imbeds R into a ring with identity.

(s) Let S be any set, and let $P(S)$ be the set of all subsets of S. For $a, b\in P(S)$, define addition and multiplication by

$$a+b=\{s\in S: s\in a\cup b, s\notin a\cap b\} \quad \text{and} \quad ab=a\cap b.$$

Then $P(S)$ is a commutative ring with identity. It is called the ***ring of subsets*** of S.

(t) Let D be the set of all 2×2 matrices over the field $\mathbb{C}$ of complex numbers which have the form

$$\begin{pmatrix} a & b \\ -\bar{b} & \bar{a} \end{pmatrix},$$

where $\bar{x}$ denotes the conjugate of the complex number x. Then D is a subring of the ring $\mathbb{C}_2$ of all 2×2 matrices over $\mathbb{C}$, and D is a division ring. (D is the ring of ***quaternions***. See 2.2, page 39.)

(u) Let $\mathbb{Z}[i] = \{m + ni: m, n \in \mathbb{Z}\}$. Then $\mathbb{Z}[i]$ is an integral domain, in fact, a subdomain of the field $\mathbb{C}$ of complex numbers. Replacing $\mathbb{Z}$ by any subdomain of $\mathbb{C}$ yields a subdomain of $\mathbb{C}$, as does replacing i by $\sqrt{2}$.

PROBLEMS

1 (a) Prove that a ring with identity has just one identity.
(b) Find an example of a ring with identity that has a subring with a different identity.
(c) Find an example of a ring with no identity that has a subring with identity.

2 Let R be a ring, and let $r \in R$. For positive integers n, define r^n inductively by

$$r^1 = r$$

and

$$r^{n+1} = r^n \cdot r.$$

Prove that $r^m r^n = r^{m+n}$ and $(r^m)^n = r^{mn}$. Prove that if R is commutative, then for $r, s \in R$, $(rs)^m = r^m s^m$.

3 Let R be a ring. Since it is an Abelian group, we can "multiply" elements of R by integers. That is, nr is defined for all $n \in \mathbb{Z}$ and $r \in R$, and this "multiplication" has certain properties (2.1.4). Prove that for integers m, n and for elements r and s of R,
(a) $m(rs) = (mr)s$,
(b) $(mn)r = m(nr)$, and
(c) $(mr)(ns) = (mn)(rs)$.

4 Let R be a ring with identity. Let U be the set of units of R. That is, $u \in U$ if and only if there is a $v \in R$ such that $uv = vu = 1$. Prove that U is a group under multiplication. (U is called the ***group of units*** of R.) What is the group of units of $\mathbb{Z}$?

5 (a) Determine which elements in the ring $\mathbb{Q}_2$ of all 2×2 matrices over the field $\mathbb{Q}$ of rational numbers are units.
(b) Determine which elements in the ring $\mathbb{Z}_2$ of all 2×2 matrices over the ring of integers are units.
(c) Let R be a commutative ring with identity. Determine which elements in the ring R_2 of all 2×2 matrices over R are units.

6 Find all subrings of $\mathbb{Z}$.

7 Find all the subrings of $\mathbb{Q}$ which contain $\mathbb{Z}$.

8 Prove that the intersection of any nonempty set of subrings of a ring is a subring.

9 Let S be a subset of the ring R. Prove that there is a subring T such that $S \subset T$ and such that every subring of R containing S contains T.

10 Let S be a nonempty subset of a ring R. Let T be the subring in Problem 9. Determine the elements of T in terms of the elements of S.

11 Let R be a ring. Let

$$Z(R) = \{a \in R: ar = ra \text{ for all } r \in R\}.$$

Prove that $Z(R)$ is a commutative subring of R. [$Z(R)$ is called the ***center*** of R.]

12 Prove that the center of a division ring is a field.

13 Prove that if $r^2 = r$ for all r in a ring R, then R is commutative.

14 Prove that if $r^3 = r$ for all r in a ring R, then R is commutative.

15 Let G be any Abelian group. Prove that the group ring $\mathbb{Z}(G)$ is an integral domain if and only if G is torsion free, that is, if and only if G has no nonzero elements of finite order.

4.2 HOMOMORPHISMS AND QUOTIENT RINGS

In studying algebraic systems such as groups, vector spaces, or rings, a basic concept is that of a homomorphism from one of them to another. A homomorphism from one group to another was a mapping that preserved the operation. A ring has two operations, addition and multiplication, and a homomorphism from one ring to another is simply a mapping that preserves both of these operations.

4.2.1 Definition

Let R and S be rings. A **homomorphism** *from R into S is a mapping $f: R \to S$ such that*

(a) $f(x+y) = f(x) + f(y)$, *and*
(b) $f(xy) = f(x)f(y)$.

As in the case of groups, a homomorphism is an ***epimorphism*** if it is onto, a ***monomorphism*** if it is one-to-one, and an ***isomorphism*** if it is one-to-one and onto. If there exists an isomorphism f from R to S, we say that R is ***isomorphic*** to S, and we write $R \approx S$. As in the case for groups, isomorphism is an equivalence relation on any set of rings. An ***automorphism*** of R is an isomorphism of R with itself.

4.2.2 Definition

Let R and S be rings, and let $f: R \to S$ be a homomorphism. Then the **kernel** *of f is defined as* $\operatorname{Ker} f = \{x \in R: f(x) = 0\}$ and is denoted $\operatorname{Ker} f$.

Thus $\operatorname{Ker} f$ is the kernel of f considered as a homomorphism from the additive group of R into the additive group of S. Since f is a homomorphism from the additive group of R into the additive group of S, we have, for example, that $f(0) = 0$, $f(-x) = -f(x)$, and that f is a monomorphism if and only if $\operatorname{Ker} f = \{0\}$. Since $\operatorname{Im} f = \{f(x): x \in R\}$ is an additive subgroup of S, and since $f(x)f(y) = f(xy)$, $\operatorname{Im} f$ is a subring of S.

For $x \in R$ and $y \in \operatorname{Ker} f$, $f(xy) = f(x)f(y) = f(x) \cdot 0 = 0 = 0 \cdot f(x) = f(y)f(x) = f(yx)$. Thus $\operatorname{Ker} f$ is an additive subgroup of R such that for $x \in R$ and $y \in \operatorname{Ker} f$, xy and yx are in $\operatorname{Ker} f$. That is, $\operatorname{Ker} f$ is an additive subgroup such that for all elements $x \in R$, $x(\operatorname{Ker} f) \subset \operatorname{Ker} f$ and $(\operatorname{Ker} f)x \subset \operatorname{Ker} f$. This latter property of $\operatorname{Ker} f$ is crucial in making $R/\operatorname{Ker} f$ into a ring (4.2.5).

4.2.3 Examples

(a) Let R and S be rings. Then the identity map 1_R of R and the map $f: R \to S$ defined by $f(r) = 0$ for all $r \in R$ are homomorphisms.

(b) Let $\mathbb{C}$ be the field of complex numbers. The mapping from $\mathbb{C}$ to $\mathbb{C}$ given by $f(a + bi) = a - bi$ is an automorphism of $\mathbb{C}$.

(c) Let F be any field, and let $a \in F$. The mapping φ from $F[x]$ to F defined by $\varphi(f) = f(a)$ is a homomorphism. The action of φ on a polynomial is to evaluate that polynomial at a. Note that φ is an epimorphism. $\operatorname{Ker} \varphi = \{f \in F[x]: f(a) = 0\}$. That is, $\operatorname{Ker} \varphi$ is the set of polynomials of which a is a root.

(d) Let R be a ring and let S be a nonempty set. Let R^S be the ring of all mappings from S into R, as in Example 4.1.7(q). Let $s \in S$. Define $\varphi: R^S \to R$ by $\varphi(f) = f(s)$. Then φ is an epimorphism.

(e) Let $R(G)$ be the ring in Example 4.1.7(o). Define the mapping $\varphi: R(G) \to R$ by $\varphi(f) = \sum_{g \in G} f(g)$. Then φ is an epimorphism.

(f) Let $R \times R$ be the ring in Example 4.1.7(p). Then the mapping $\eta: R \times R \to R$ defined by $\eta(x, y) = x$ is an epimorphism. The mapping $\varphi: R \to R \times R$ given by $\varphi(x) = (x, 0)$ is a monomorphism.

(g) Let R_2 be the ring of 2×2 matrices over a ring R. The map

$$R \to R_2: r \to \begin{pmatrix} r & 0 \\ 0 & r \end{pmatrix}$$

is a monomorphism. So is the map

$$R \to R_2: r \to \begin{pmatrix} r & 0 \\ 0 & 0 \end{pmatrix}.$$

Analogous facts hold for $n \times n$ matrices.

Let $f: R \to S$ be a homomorphism. Then $\operatorname{Ker} f$ is a subgroup of R closed under multiplication on either side by any element of R, as we have seen. Such subsets of rings play a role analogous to the role played for groups by normal subgroups.

4.2.4 Definition

An **ideal** *of a ring R is a subgroup I of the additive group of R such that ri and ir are in I for all* $i \in I$ *and all* $r \in R$.

Kernels of homomorphisms are ideals. It will turn out that every ideal of a ring R is the kernel of a homomorphism from R to another ring. In fact, given an ideal I of a ring R, we will now construct a ring R/I, the ***quotient ring of R with respect to I***. There will be a natural homomorphism from R onto R/I whose kernel is I. This construction is fundamental.

4.2.5 The construction of quotient rings

Let R be a ring, and let I be an ideal of R. Since R is an Abelian group under addition and I is a subgroup of that group, we have the ***quotient group*** R/I. Its elements are the cosets $r + I$ of the subgroup I of R. Furthermore, addition in R/I is given by

$$(r + I) + (s + I) = (r + s) + I.$$

In order to make R/I into a ring, we need to define a suitable multiplication on it. The most natural thing to do is to let

$$(r + I)(s + I) = rs + I.$$

However, it is not clear that this defines a binary operation on R/I. It is conceivable that $r + I = r_1 + I$ and $s + I = s_1 + I$, and that $rs + I \neq r_1 s_1 + I$. If this could be ruled out, however, we would have a binary operation on R/I. So suppose $r + I = r_1 + I$ and $s + I = s_1 + I$. Then $r = r_1 + i$ and $s = s_1 + j$ with i and j in I. Thus,

$$rs + I = (r_1 + i)(s_1 + j) + I = r_1 s_1 + r_1 j + i s_1 + ij + I.$$

But $r_1 j$, $i s_1$, and ij are all in I since i and j are in I and I is an ideal. Thus $r_1 j + i s_1 + ij$ is in I, so that $rs + I = r_1 s_1 + I$.

It is easy to verify that with the multiplication

$$(r + I)(s + I) = rs + I,$$

the group R/I becomes a ring. For example,

$$\begin{aligned}(r+I)((s+I)+(t+I)) &= (r+I)((s+t)+I) = r(s+t)+I \\ &= (rs+rt)+I = (rs+I)+(rt+I) \\ &= (r+I)(s+I)+(r+I)(t+I),\end{aligned}$$

so that the left distributive law holds. The other axioms of a ring are just as easy to check. If R has an identity, then $1+I$ is the identity for R/I. If R is commutative, then R/I is commutative. The mapping $\eta: R \to R/I$ defined by $\eta(r) = r + I$ is an epimorphism.

4.2.6 Definition

The ring R/I constructed in 4.2.5 is called the **quotient ring of R with respect to the ideal I,** *or more simply "R modulo I". The homomorphism $\eta: R \to R/I: r \to r + I$ is called the* **natural homomorphism** *from R to R/I.*

Notice that $\operatorname{Ker} \eta = I$.

4.2.7 First isomorphism theorem for rings

Let R and S be rings, and let $f: R \to S$ be a homomorphism. Then

$$R/(\operatorname{Ker} f) \approx \operatorname{Im} f.$$

Proof There is little to do. The mapping

$$\bar{f}: R/(\operatorname{Ker} f) \to \operatorname{Im} f: r + \operatorname{Ker} f \to f(r)$$

establishes an isomorphism between the additive groups of $R/(\operatorname{Ker} f)$ and $\operatorname{Im} f$ by the first isomorphism theorem for groups. We need only show that $\bar{f}$ preserves multiplication. That follows from

$$\begin{aligned}\bar{f}((r+\operatorname{Ker} f)(s+\operatorname{Ker} f)) &= \bar{f}(rs+\operatorname{Ker} f) = f(rs) = f(r)f(s) \\ &= \bar{f}(r+\operatorname{Ker} f)\bar{f}(s+\operatorname{Ker} f).\end{aligned}$$

Other isomorphism theorems are in Problem 27. Let I be an ideal in R. We will determine the ideals in the ring R/I. Ideals of R/I must be of the form J/I for J a subgroup of R since ideals are, in particular, subgroups. If J/I is an ideal, then for any $r \in R$ and $j \in J$, we must have

$$(r+I)(j+I) = rj + I \in J/I.$$

That is, rj must be in J. Similarly, jr must be in J, whence J must be an ideal. On the other hand, if J is an ideal of R containing I, then J/I is an ideal of R/I

since it is a subgroup of R/I and

$$(r+I)(j+I) = rj + I \in J/I$$

and

$$(j+I)(r+I) = jr + I \in J/I$$

for all r in R and j in J. We have the following result.

4.2.8 Theorem

Let I be an ideal in the ring R. Associating each ideal J of R which contains I with J/I is a one-to-one correspondence between the set of ideals containing I and the set of ideals of R/I.

Let F be a field, and let I be an ideal of F. If $i \in I$ and $i \neq 0$, then i has an inverse, and $i \cdot i^{-1} = 1$ is in I. If $a \in F$, then $a \cdot 1 \in I$, so that $I = F$. *Therefore, a field F has only the trivial ideals F and $\{0\}$.*

Now we will look at the ideals and quotient rings of the ring $\mathbb{Z}$ of integers. Let I be an ideal in $\mathbb{Z}$. In particular, I is a subgroup of $\mathbb{Z}$, so it is cyclic. Therefore I consists of all integral multiples of some integer n, that is, $I = \mathbb{Z}n$. Also, for any $n \in \mathbb{Z}$, $\mathbb{Z}n$ is clearly an ideal. The ring $\mathbb{Z}/\mathbb{Z}n$ is called the ***ring of integers modulo*** n. Suppose $\mathbb{Z}n = \mathbb{Z}m$. Then m and n are integral multiples of each other, so that $m = \pm n$. Since $\mathbb{Z}n = \mathbb{Z}(-n)$, we have that an ideal $I = \mathbb{Z}n$ for a unique $n \geq 0$. Thus, *the ideals of $\mathbb{Z}$ are in one-to-one correspondence with the nonnegative integers.*

When does $\mathbb{Z}m \supset \mathbb{Z}n$? If $\mathbb{Z}m \supset \mathbb{Z}n$, then $n \in \mathbb{Z}m$, so $n = km$ for some integer k. On the other hand, if $n = km$, clearly $\mathbb{Z}m \supset \mathbb{Z}n$. Hence, $\mathbb{Z}m \supset \mathbb{Z}n$ if and only if n is a multiple of m. This says that p is a prime if and only if the only ideal which properly contains $\mathbb{Z}p$ is $\mathbb{Z} \cdot 1 = \mathbb{Z}$. To put it another way, p is a prime if and only if $\mathbb{Z}/\mathbb{Z}p$ has only the two trivial ideals $\mathbb{Z}p/\mathbb{Z}p$ and $\mathbb{Z}/\mathbb{Z}p$.

Since fields F have only the two trivial ideals $\{0\}$ and F, if $\mathbb{Z}/\mathbb{Z}n$ is a field, n must be prime. But if $n = p$ is prime, then for any $a + \mathbb{Z}p \neq 0$, a and p are relatively prime. Thus there are integers x and y such that $ax + py = 1$, and

$$ax + \mathbb{Z}p = 1 - py + \mathbb{Z}p = 1 + \mathbb{Z}p,$$

so that $\mathbb{Z}/\mathbb{Z}p$ is a field. We have then that *n is prime if and only if $\mathbb{Z}/\mathbb{Z}n$ is a field.*

4.2.9 Definition

A **maximal ideal** *in a ring R is an ideal M of R such that $M \neq R$, and whenever I is an ideal such that $M \subset I \subset R$, then $I = M$ or $I = R$.*

Thus a maximal ideal is one that is not the whole ring, and such that there are no ideals strictly between it and the whole ring. The maximal ideals of $\mathbb{Z}$ are $\mathbb{Z}p$ with p prime, and these were also the ideals I of $\mathbb{Z}$ such that $\mathbb{Z}/I$ was a field.

4.2.10 Theorem

Let R be a commutative ring with identity. An ideal M of R is maximal if and only if R/M is a field.

Proof If R/M is a field, then it has at least 2 elements. Therefore $M \neq R$. If I were an ideal strictly between M and R, then by 4.2.8, I/M would be an ideal strictly between R/M and 0. But we have observed that fields have no nontrivial ideals. Therefore, M is maximal.

Now suppose that M is maximal. Since R is commutative with identity, so is R/M. We need only show that every nonzero element in R/M has an inverse. Let $a + M \in R/M$ with $a + M \neq 0$. Then

$$Ra + M = \{ra + m: r \in R, m \in M\}$$

is an ideal. If $m \in M$, then $m = 0 \cdot a + m \in Ra + M$, and $a = 1 \cdot a + 0$ is an element of $Ra + M$. Therefore, $Ra + M$ is an ideal properly containing M. Since M is maximal, $Ra + M = R$. Thus for some $r \in R$ and $m \in M$, $ra + m = 1$. Now

$$(r + M)(a + M) = ra + M = (1 - m) + M = 1 + M$$

shows that R/M is a field.

We will use this theorem many times. One half of it can be strengthened. Suppose that R is any ring, and suppose that R/M is a field. Then there are no ideals strictly between M and R, whence M is maximal. Thus if *R/M is a field, then M is maximal.*

PROBLEMS

1. Find all rings with 2 elements. (First, decide what this problem means.)
2. Prove that the rings $\mathbb{Z}2$ and $\mathbb{Z}3$ are not isomorphic. Prove that if m and n are distinct positive integers, then the rings $\mathbb{Z}m$ and $\mathbb{Z}n$ are not isomorphic.
3. Prove that the fields $\mathbb{R}$ and $\mathbb{C}$ are not isomorphic.
4. Let $\mathbb{Q}[\sqrt{2}] = \{a + b\sqrt{2}: a, b \in \mathbb{Q}\}$, and define $\mathbb{Q}[\sqrt{5}]$ similarly. Prove that these are rings, and prove that they are not isomorphic.

5 Let R be the set of 2×2 matrices of the form

$$\begin{pmatrix} a & 2b \\ b & a \end{pmatrix}$$

with a and b elements of $\mathbb{Q}$. Prove that R is a ring and is isomorphic to the ring $\mathbb{Q}[\sqrt{2}]$ of Problem 4.

6 Find all homomorphisms from the ring $\mathbb{Z}/\mathbb{Z}12$ to the ring $\mathbb{Z}/\mathbb{Z}20$.

7 Prove that the quaternions form a division ring [4.1.5(t)].

8 An element x in a ring is called ***idempotent*** if $x^2 = x$. Find all idempotent elements in the rings $\mathbb{Z}/\mathbb{Z}20$ and $\mathbb{Z}/\mathbb{Z}30$.

9 An element x in a ring is called ***nilpotent*** if $x^n = 0$ for some integer n. Find all nilpotent elements in the rings $\mathbb{Z}/\mathbb{Z}30$ and $\mathbb{Z}/\mathbb{Z}40$.

10 Let R be a ring with identity, and suppose that x is a nilpotent element of R. Prove that $1 - x$ is a unit.

11 Prove that the set of nilpotent elements in a commutative ring R forms an ideal N, and prove that R/N has no nonzero nilpotent elements.

12 Suppose that R is a ring with identity and that f is a homomorphism from R into an integral domain S. Prove that $f(1) = 1$ unless $f(1) = 0$.

13 Prove that the composition of ring homomorphisms is a ring homomorphism.

14 Suppose that R and S are rings and that $f: R \to S$ is an epimorphism. Prove the following.

(a) $f(I)$ is an ideal of S if I is an ideal of R.
(b) $f^{-1}(J)$ is an ideal of R if J is an ideal of S.
(c) If R has an identity, then so does S if $S \neq \{0\}$.
(d) If S has an identity, then R might not.

15 Prove that the set $\mathrm{Aut}(R)$ of all automorphisms of a ring R is a group under composition of automorphisms.

16 Let u be a unit in the ring R. Prove that

$$f_u: R \to R: r \to uru^{-1}$$

is an automorphism of R. (f is called an ***inner automorphism*** of R.)

17 Let R be a ring with identity, and let $\mathrm{Inn}(R)$ be the set of all inner automorphisms of R. Prove that $\mathrm{Inn}(R)$ is a subgroup of $\mathrm{Aut}(R)$ and is normal in $\mathrm{Aut}(R)$.

18 Let R be a ring with identity, and let $U(R)$ be the group of units of R. Prove that the map

$$U(R) \to \mathrm{Inn}(R): u \to f_u$$

is a homomorphism. What is its kernel?

19 Determine Aut($\mathbb{Z}$), Inn($\mathbb{Z}$), and $U(\mathbb{Z})$.

20 Determine Aut($\mathbb{Q}$), Inn($\mathbb{Q}$), and $U(\mathbb{Q})$. Determine Aut($\mathbb{Q}[\sqrt{2}]$) and Aut($\mathbb{Q}[i]$).

21 Determine Aut($\mathbb{R}$) for the field $\mathbb{R}$ of real numbers.

22 Find all automorphisms f of $\mathbb{C}$ such that $f(\mathbb{R}) = \mathbb{R}$.

23 Prove that the endomorphism ring of the additive group $\mathbb{Z}$ is isomorphic to the ring $\mathbb{Z}$.

24 Prove that the endomorphism ring of the additive group $\mathbb{Z}/\mathbb{Z}n$ is isomorphic to the ring $\mathbb{Z}/\mathbb{Z}n$.

25 Prove that the intersection of any nonempty set of ideals of a ring R is an ideal of R.

26 Let S be a subset of a ring R. Prove that there is a unique smallest ideal of R containing S. If S is nonempty, what is this ideal in terms of the elements of S?

27 Let I and J be ideals of R. Prove that

$$I + J = \{i + j : i \in I, j \in J\}$$

is an ideal of R. Prove that

$$(I + J)/J \approx I/(I \cap J).$$

Prove that if I, J, and K are ideals of R such that $I \supset K \supset J$, then

$$(I/J)/(K/J) \approx I/K.$$

28 Let $\{I_s\}_{s \in S}$ be a family of ideals of a ring R. Prove that

$$\sum_{s \in S} I_s = \left\{ \sum_{s \in S} i_s : i_s \in I_s,\ i_s = 0 \text{ for all but finitely many } s \right\}$$

is an ideal of R.

29 Let I and J be ideals of R. Prove that

$$IJ = \{i_1 j_1 + i_2 j_2 + \cdots + i_n j_n : i_1, i_2, \ldots, i_n \in I, j_1, j_2, \ldots, j_n \in J, n \in \mathbb{Z}^+\}$$

is an ideal of R. (IJ is called the ***product*** of I and J).

30 Let I and J be ideals of R. What is the relationship between IJ and $I \cap J$?

31 Let P be a set of primes. Let

$$\mathbb{Z}_p = \{a/b : a, b \in \mathbb{Z}, b \neq 0, (b, p) = 1 \text{ for all } p \in P\}.$$

Find all the maximal ideals M of the ring $\mathbb{Z}_p$, and in each case determine the field $\mathbb{Z}_p/M$.

32 Find an example of an integral domain that has exactly one maximal ideal. Now find such an example that is not a field.

33 Find the maximal ideals of $\mathbb{Z}/\mathbb{Z}n$.

34 For which n does $\mathbb{Z}/\mathbb{Z}n$ have exactly one maximal ideal?

35 Let S be any set, and let $P(S)$ be the ring of subsets of S. [See 4.1.7(s).] Let $T \subset S$. Prove that

$$P(S) \to P(T): a \to a \cap T$$

is an epimorphism. What is its kernel?

36 Let S be a nonempty set, and let $P(S)$ be the ring of subsets of S. For $s \in S$, let $I_s = \{a \in P(S): s \notin a\}$. Prove that I_s is a maximal ideal of $P(S)$. What is the field $P(S)/I_s$? Prove that if S is finite, then every maximal ideal of $P(S)$ is an I_s.

37 Let S be a set, and let $P(S)$ be the ring of subsets of S. For any $s \in S$, show that $U = \{u \in P(S): s \in u\}$ satisfies the following.
(a) $S \in U$ and $\varnothing \notin U$.
(b) If $u, v \in U$, then $u \cap v \in U$.
(c) If $u \in U$, $v \in P(S)$, and $u \subset v$, then $v \in U$.
(d) For each $a \in P(S)$, either a or its complement $S \backslash a$ is in U.

38 Let S be a set, and let $P(S)$ be the ring of subsets of S. A subset U of $P(S)$ which satisfies (a)–(d) of Problem 37 is called an ***ultrafilter*** on S. Prove that a subset M of $P(S)$ is a maximal ideal of $P(S)$ if and only if its complement is an ultrafilter on S.

39 Let S be a set, and let F be the field with 2 elements. Prove that the ring F^S of all mappings from S into F is isomorphic to the ring $P(S)$ of all subsets of S.

40 Let R be a ring with identity. Define a new ring S as follows. The elements of S are those of R. If $+$ and $\cdot$ are addition and multiplication in R, respectively, define $\oplus$ and $*$ on S by

$$a \oplus b = a + b + 1$$

and

$$a * b = a \cdot b + a + b.$$

Prove that S is a ring, and prove that it is isomorphic to the ring R.

41 Let $G = \{e, a, b\}$ be the cyclic group of order 3. Calculate the following products in the group ring $\mathbb{Q}(G)$.
(a) $(e + 2a + 3b)(a + b)$.
(b) $(e - a)^3 + (e - a^2)^3$.
(c) $(1/3(e - a) + 1/3(e - a^2))(x(e - a) + y(e - a^2))$, where x and y are any elements of $\mathbb{Q}$.

42 Let $\mathbb{Q}(G)$ be the ring in Problem 41. Prove that

$$\mathbb{Q}(G) = \mathbb{Q}(e + a + a^2) \oplus \mathbb{Q}(e - a) \oplus \mathbb{Q}(e - a^2)$$

is a vector space over $\mathbb{Q}$. Prove that $\mathbb{Q}(e + a + a^2)$ is a field isomorphic to $\mathbb{Q}$. Prove that $\mathbb{Q}(e - a) \oplus \mathbb{Q}(e - a^2)$ is a field. Prove that as a ring, $\mathbb{Q}(G)$ is the direct sum of these two fields.

4.3 FIELD OF QUOTIENTS OF AN INTEGRAL DOMAIN

An integral domain is a commutative ring with identity such that if $ab = 0$, then $a = 0$ or $b = 0$. Fields are integral domains since $ab = 0$ and $a \neq 0$ imply that $a^{-1}ab = a^{-1}0 = b = 0$. An integral domain is not necessarily a field, as the ring $\mathbb{Z}$ of integers testifies. However, the ring $\mathbb{Z}$ is a subring of the field $\mathbb{Q}$ of rational numbers. More than that, every element of $\mathbb{Q}$ can be written in the form ab^{-1} with a, $b \in \mathbb{Z}$. Rational numbers are quotients of integers. This phenomenon holds for any integral domain. That is, any integral domain D is contained in a field F such that for every element a in F, $a = bc^{-1}$ with b, $c \in D$. Such a field is a ***field of quotients*** of D.

4.3.1 Definition

A **field of quotients** *of an integral domain D is a field F and a ring monomorphism $\varphi: D \to F$ with the property that every element a in F can be written in the form $a = \varphi(b)(\varphi(c))^{-1}$ for some b, $c \in D$.*

We have two things in mind. We want to show that every integral domain D has a field of quotients, and we want to show that any two fields of quotients of D are essentially the same. We must decide what this last part should mean.

4.3.2 Definition

Two fields of quotients $\varphi: D \to F$ and $\varphi': D \to F'$ of an integral domain D are **equivalent** *if there is an isomorphism $\beta: F \to F'$ such that $\beta \cdot \varphi = \varphi'$.*

This definition just says that F and F' must be isomorphic via an isomorphism that respects the embeddings $\varphi: D \to F$ and $\varphi': D \to F'$. If we think of D as being contained in F and F', then it says that there must be an isomorphism $\beta: F \to F'$ fixing D elementwise. Here is a picture.

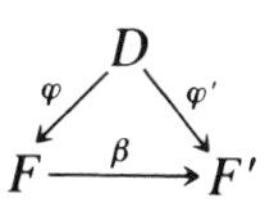

4.3.3 Theorem

Any integral domain D has a field of quotients, and any two fields of quotients of D are equivalent.

Proof This proof is a little long and formal, but conceptually it is easy. A field F will be constructed from D in the same way that $\mathbb{Q}$ is constructed from $\mathbb{Z}$. A rational number is a quotient $a/b = ab^{-1}$ of integers, but two such quotients a/b and c/d may be the same rational number with $a \neq c$ and $b \neq d$. For example, $1/2 = 5/10$. In fact, $a/b = c/d$ if and only if $ad = bc$. We will follow this lead in building our field. Recall that $D^* = D\backslash\{0\}$. In the set $D \times D^*$, let

$$(a, b) \sim (c, d) \quad \text{if } ad = bc.$$

This is an equivalence relation on $D \times D^*$. Only transitivity threatens to be difficult. If $(a, b) \sim (c, d)$ and $(c, d) \sim (e, f)$, then $ad = bc$ and $cf = de$. We need $af = be$. But $ad = bc$ implies $adf = bcf$, and $cf = de$ implies $bcf = bde$. Since D is an integral domain, $af = be$. Thus $\sim$ is transitive.

The equivalence relation $\sim$ on $D \times D^*$ partitions $D \times D^*$ into equivalence classes. Let F be the set of these equivalence classes. We will make F into a field. Denote the equivalence class containing (a, b) by a/b. Then we have $a/b = c/d$ if and only if $ad = bc$. We need to define addition and multiplication on F. Let

$$(a/b)(c/d) = ac/bd$$

and

$$a/b + c/d = (ad + bc)/bd.$$

First, we must make sure that we have really defined binary operations on F. It is conceivable that $a/b = a'/b'$, $c/d = c'/d'$, and $ac/bd \neq a'c'/b'd'$, and similarly for addition. So suppose $a/b = a'/b'$. We will show that $(ad + bc)/bd = (a'd + b'c)/b'd$. This holds if and only if $(ad + bc)b'd = (a'd + b'c)bd$. Since $ab' = ba'$, the left side is $(a'bd + bb'c)d$ and so is the right. It follows that addition is well defined. The proof that multiplication is well defined is similar.

Note that both operations are commutative. Under this addition and multiplication, F is a field. First, $0/1$ is the additive identity since $0/1 + a/b = (0 \cdot b + a \cdot 1)/b = a/b$. The negative of a/b is $(-a)/b$. The associative laws are not difficult. The multiplicative identity is $1/1$. Let us check the distributive law.

$$(a/b)(c/d + e/f) = (a/b)((cf + de)/df) = a(cf + de)/bdf,$$

whereas

$$\begin{aligned}(a/b)(c/d)+(a/b)(e/f) &= ac/bd + ae/bf = (acbf + bdae)/bdbf \\ &= ((acf+ade)/bdf)(b/b) \\ &= ((acf+ade)/bdf)/(1/1) = (acf+ade)/bdf.\end{aligned}$$

Thus, the distributive law holds. If $a/b \neq 0/1$, then $a \neq 0$, and $b/a \in F$. Further, $(a/b)(b/a) = ab/ba = 1/1$. Therefore,

$$(a/b)^{-1} = b/a.$$

In short, one computes with these a/b just as with rational numbers.

We now have a field F. The map

$$\varphi: D \to F: a \to a/1$$

is a ring monomorphism. In fact,

$$\varphi(a+b) = (a+b)/1 = a/1 + b/1 = \varphi(a) + \varphi(b)$$

and

$$\varphi(ab) = ab/1 = (a/1)(b/1) = \varphi(a)\varphi(b).$$

If $\varphi(a) = 0/1$, then $0/1 = a/1$, and so $1 \cdot a = 0 \cdot 1 = a = 0$. We have already observed that $1/b = (b/1)^{-1}$. Thus, for $a/b \in F$, $a/b = (a/1)/(b/1) = \varphi(a)(\varphi(b))^{-1}$.

It remains to show that any two fields of quotients of D are equivalent. Let $\varphi: D \to F$ and $\varphi': D \to F'$ be such. Define

$$\beta: F \to F': \varphi(a)(\varphi(b))^{-1} \to \varphi'(a)(\varphi'(b))^{-1}.$$

It is not entirely obvious that β is well defined. To make sure that it is, suppose that $\varphi(a)(\varphi(b))^{-1} = \varphi(a')(\varphi(b'))^{-1}$. Then $\varphi(a)\varphi(b') = \varphi(a')\varphi(b) = \varphi(ab') = \varphi(a'b)$, and so $ab' = a'b$. Therefore $\varphi'(ab') = \varphi'(a'b)$, and hence $\varphi'(a)(\varphi'(b))^{-1} = \varphi'(a')(\varphi'(b'))^{-1}$. Obviously, $\beta \cdot \varphi = \varphi'$.

We need β to be an isomorphism. First, if $a, b \in D$ with $a \neq 0 \neq b$, then $\varphi(ab) = \varphi(a)\varphi(b)$, so $(\varphi(ab))^{-1} = (\varphi(a)\varphi(b))^{-1} = (\varphi(a))^{-1}(\varphi(b))^{-1}$. Now

$$\begin{aligned}\beta((\varphi(a)(\varphi(b))^{-1})(\varphi(c)(\varphi(d))^{-1})) &= \beta(\varphi(ac)(\varphi(bd))^{-1}) \\ &= \varphi'(ac)(\varphi'(bd))^{-1} \\ &= (\varphi'(a)(\varphi'(b))^{-1})(\varphi'(c)(\varphi'(d))^{-1}) \\ &= \beta(\varphi(a)(\varphi(b))^{-1})\beta(\varphi(c)(\varphi(d))^{-1}).\end{aligned}$$

Thus, β preserves multiplication. We are simply using the definition of β and the fact that every element of F can be written in the form $\varphi(a)(\varphi(b))^{-1}$. If $\beta(\varphi(a)(\varphi(b))^{-1}) = 0$, then $\varphi'(a)(\varphi'(b))^{-1} = 0$, whence $\varphi'(a) = 0$, and so $a = 0$. Thus $\varphi(a)(\varphi(b))^{-1} = 0$, and β is one-to-one. β is onto since every element of F can be written in the form $\varphi'(a)(\varphi'(b))^{-1}$, and so $\beta(\varphi(a)(\varphi(b))^{-1}) = \varphi'(a)(\varphi'(b))^{-1}$. We need that β preserves addition.

$$\begin{aligned}
&\beta(\varphi(a)(\varphi(b))^{-1} + \varphi(c)(\varphi(d))^{-1}) \\
&\quad = \beta(\varphi(a)(\varphi(b))^{-1}(\varphi(d))^{-1}\varphi(d) + \varphi(c)(\varphi(d))^{-1}(\varphi(b))^{-1}\varphi(b)) \\
&\quad = \beta(\varphi(ad)(\varphi(bd))^{-1} + \varphi(bc)(\varphi(bd))^{-1}) \\
&\quad = \beta((\varphi(ad) + \varphi(bc))(\varphi(bd))^{-1}) = \beta(\varphi(ad + bc)(\varphi(bd))^{-1}) \\
&\quad = \varphi'(ad + bc)(\varphi'(bd))^{-1} = \varphi'(a)(\varphi'(b))^{-1} + \varphi'(c)(\varphi'(d))^{-1} \\
&\quad = \beta(\varphi(a)(\varphi(b))^{-1}) + \beta(\varphi(c)(\varphi(d))^{-1}).
\end{aligned}$$

The proof is complete.

In practice, if D is an integral domain, and $\varphi: D \to F$ is a field of quotients of D, then D is identified with its image in F. Then F contains D, and every element in F is of the form ab^{-1} with $a, b \in D$. Also, we write $1/b$ for b^{-1}, or more generally, a/b for ab^{-1}.

This construction has been generalized in many ways. There is an extensive literature on "rings of quotients." One way to generalize it is as follows. Let R be a commutative ring with an identity, and let S be a nonempty multiplicatively closed subset of R containing no zero divisors. If R were an integral domain, then the nonzero elements of R would be such a set. For any commutative ring with identity, the set of elements that are not zero divisors will serve. In $R \times S$, let

$$(a, b) \sim (c, d) \quad \text{if and only if} \quad ad = bc.$$

This defines an equivalence relation, and the set of equivalence classes is a ring under the operations defined as in the integral domain case. This ring is denoted R_S, R is "contained" in R_S, and every element of S has an inverse in R_S. Even if R is an integral domain, S does not have to be taken to be all the nonzero elements of R. For example, in $\mathbb{Z}$ let S be all those integers relatively prime to a fixed prime p. The resulting ring of quotients consists of all those rational numbers whose denominators are relatively prime to p.

There are generalizations to the noncommutative case. If R is a ring with no zero divisors, there is a condition due to Oré which ensures that the construction above can be carried out, yielding an embedding of R in a division ring.

PROBLEMS

1 Find the field of quotients of the integral domain

$$\{a + b\sqrt{3}: a, b \in \mathbb{Z}\}.$$

2 Let $f: R \to S$ be a ring homomorphism, and let R and S be commutative rings with identities. Let G be a subgroup of the group of units of S. Prove that $M = f^{-1}(G)$ is a multiplicatively closed subset of R containing no zero divisors. What is M if f is the natural homomorphism $\mathbb{Z} \to \mathbb{Z}/\mathbb{Z}p$ for a prime p, and $G = (\mathbb{Z}/\mathbb{Z}p)^*$?

3 Prove that if F is a field, then $1_F: F \to F$ is a field of quotients of F.

4 Prove that if $\varphi: D \to F$ is a monomorphism, D is an integral domain, and F is a field, then F contains a subfield F' such that $\varphi: D \to F'$ is a field of quotients of D.

5 Prove that if $\varphi: D \to F$ is a field of quotients of the integral domain D, and F' is a subfield of F containing $\varphi(D)$, then $F' = F$.

6 Let S be a nonempty multiplicatively closed subset of a commutative ring R with identity. Suppose that S contains no zero divisors of R. Define $\sim$ on $R \times S$ by $(a, b) \sim (c, d)$ if $ad = bc$. Let R_S be the set of equivalence classes of $\sim$. Let a/b denote the equivalence class containing (a, b). Prove that

$$(a/b)(c/d) = ac/bd$$

and

$$a/b + c/d = (ad + bc)/bd$$

make R_S into a commutative ring with identity. Let $s \in S$. Prove that

$$\varphi_S: R \to R_S: a \to (as/s)$$

is a monomorphism, and for any $t \in S$, $\varphi_S(t)$ has an inverse in R_S. Prove that every element in R_S is of the form $\varphi_S(a)(\varphi_S(t))^{-1}$ for some $a \in R$ and $t \in S$.

7 Let $R = \mathbb{Z}/\mathbb{Z}n$, and let S be the set of all elements of R that are not zero divisors. Find R_S.

8 Let S be a nonempty multiplicatively closed subset of an integral domain D such that $0 \notin S$. Let $\bar{S} = \{a \in D: ab \in S \text{ for some } b \in D\}$. S is called ***saturated*** if $S = \bar{S}$.

(a) Find all saturated subsets of $\mathbb{Z}$.
(b) Find all saturated subsets of $\mathbb{Q}$.
(c) Find all saturated subsets of a field F.
(d) Prove that $(as)/s \in D_S$ has an inverse if and only if $a \in \bar{S}$.

9 Let S and T be two nonempty multiplicatively closed subsets of an integral domain D such that $0 \notin S$, $0 \notin T$. Prove that there is an isomorphism $\beta: D_S \to D_T$ such that $\beta \cdot \varphi_S = \varphi_T$ if and only if $\bar{S} = \bar{T}$.

4.4 PRINCIPAL IDEAL DOMAINS

For any element in a commutative ring R, the set

$$Ra = \{ra: r \in R\}$$

is an ideal of R. Such an ideal is called ***principal***. Thus if R has an identity, Ra is the smallest ideal containing the element a. The ideal Ra is generally denoted (a).

4.4.1 Definition

A **principal ideal domain** *is an integral domain in which every ideal is principal.*

Principal ideal domain will be abbreviated PID. The ring $\mathbb{Z}$ is a PID. If I is an ideal of $\mathbb{Z}$, it is, in particular, a subgroup of $\mathbb{Z}$, so is cyclic. Thus, $I = \mathbb{Z}n$ for some n. It turns out that the ring $F[x]$ of all polynomials in x with coefficients in F is a PID for any field F (4.5.11). The ring $F[x, y]$ of polynomials in x and y with coefficients in F is not a PID. The ideal generated by the set $\{x, y\}$ is not principal. The ring $\mathbb{Z}[x]$ is not a PID, since the ideal generated by $\{2, x\}$ is not principal. We have not studied polynomial rings yet, and these facts may not be entirely obvious.

Our purpose in this section is to discuss primes, greatest common divisors, factorization of elements into primes, and the like in PID's. Application of these results will be mainly to $F[x]$ in 4.5.

We will assume throughout that R is an integral domain. A number of definitions are needed.

The element a in R is an ***associate*** of the element b in R, if $a = ub$ for some unit u in R. An element a not a unit and not 0 is ***irreducible*** if whenever $a = bc$, either b or c is a unit. The element a ***divides*** the element b, or b is a ***multiple*** of a, denoted $a|b$, if $b = ac$ for some c. The nonzero element p is a ***prime*** if p is not a unit, and whenever $p|ab$, then $p|a$ or $p|b$. A ***greatest common divisor*** (abbreviated gcd) of the set of elements $\{a_1, a_2, \ldots, a_n\}$ is an element d such that $d|a_i$ for all i, and such that if $e|a_i$ for all i, then $e|d$. Two elements a and b are ***relatively prime*** if any gcd of $\{a, b\}$ is a unit. A ***least common multiple*** (abbreviated lcm) of the set of elements $\{a_1, a_2, \ldots, a_n\}$ with not all $a_i = 0$ is an element a such that a is a multiple of each a_i, and such that if e is a multiple of each a_i, then e is a multiple of a.

Being "an associate of" is an equivalence relation. This follows from the fact that the set U of units of R is a multiplicative group. The set of equivalence classes of this equivalence relation partitions R. The equivalence classes on $\mathbb{Z}$ are $\{\pm n\}$, $n = 0, 1, 2, \ldots$. There are only 2 units, ± 1, in $\mathbb{Z}$, so each equivalence class has at most 2 elements in it. For a field F, $\{0\}$ and F^* are the only equivalence classes. For any integral domain R and element a in R, the equivalence class containing a is $aU = \{au : u \in U\}$, where U is the group of units of R. If F is the field of quotients of R, a coset xU of U in F^* is either contained in R or is disjoint from R, according to whether $x \in R$ or $x \notin R$. Thus, the equivalence classes are the cosets of U which are in R.

Suppose that the element a is irreducible and that u is a unit. If $au = bc$, then $a = b(cu^{-1})$. Therefore, either b or cu^{-1} is a unit, and hence either b or c is a unit. Therefore, if an element a is irreducible and u is a unit, then au is irreducible.

Notice that if $a|b$ and u is a unit, then $au|b$. In fact, $ac = b$ implies that $(au)(u^{-1}c) = b$. Suppose that a is a prime and that u is a unit. If $au\,|bd$, then $a|bc$, so $a|b$ or $a|c$. Thus, $au|b$ or $au|c$. Therefore, au is prime if a is prime. Equivalently, if a is prime, then so is any associate of a.

Suppose d and d' are both gcd's of $\{a_1, a_2, \ldots, a_n\}$. Then $d|d'$ and $d'|d$. Hence, $de = d'$ and $d'e' = d$ for some e and e'. Thus $dee' = d$, so $ee' = 1$, whence e and e' are units. Clearly, if d is a gcd of $\{a_1, a_2, \ldots, a_n\}$, then so is du for any unit u. Therefore, if a set $\{a_1, a_2, \ldots, a_n\}$ has a gcd d, then the set of all gcd's of that set is the equivalence class of associates of d. A similar story is true for lcm's. One class of associates is the set U of units of the ring. This class has a natural representative, namely 1. If the set of gcd's of a set $\{a_1, a_2, \ldots, a_n\}$ is U, then we say that the gcd of the set is 1, and we write $\gcd\{a_1, a_2, \ldots, a_n\} = 1$. Thus, a and b are relatively prime if $\gcd\{a, b\} = 1$.

Suppose that a is a prime. If $a = bc$, then $a|bc$, so $a|b$ or $a|c$. If $ad = b$, then $a = adc$, whence $1 = dc$ and c is a unit. Therefore, *primes are irreducible.*

There are two important facts that we are after. First, if R is a PID, then any finite subset $\{a_1, a_2, \ldots, a_n\}$ of R, with not all the $a_i = 0$, has a gcd d in R, and $d = \sum d_i a_i$ for appropriate d_i in R. Next we want to show that every nonzero element a in a PID can be written in the form $a = up_1p_2 \cdots p_n$ with u a unit and the p_i's prime. Furthermore, this can be done in essentially only one way. We can get the first important fact now.

4.4.2 Theorem

Let R be a PID *and let $\{a_1, a_2, \ldots, a_n\}$ be a subset of R, with $n \geq 1$ and not all the $a_i = 0$. Then this set has a* gcd, *and any* gcd *of it is of the form $\sum r_i a_i$ with $r_i \in R$.*

Proof The set $I = Ra_1 + Ra_2 + \cdots + Ra_n$ is an ideal of R. Therefore, $I = Rd$ for some d in R. Now each a_i is in Rd, so $a_i = s_i d$ for some $s_i \in R$. Thus, $d|a_i$ for all i. The element d is in I, so $d = \sum r_i a_i$ for appropriate $r_i \in R$. Therefore, if $e|a_i$ for all i, then $e|d$. Hence, d is a gcd of $\{a_1, a_2, \ldots, a_n\}$. Any gcd of $\{a_1, a_2, \ldots, a_n\}$ is of the form du with u a unit. Therefore $du = \sum (ur_i)a_i$, and the proof is complete.

The element $\sum r_i a_i$ is called a ***linear combination*** of the a_i's. Thus, any gcd of $\{a_1, a_2, \ldots, a_n\}$ is a linear combination of the a_i's.

4.4.3 Corollary

An element in a PID is a prime if and only if it is irreducible.

Proof We have already seen that a prime is irreducible. Suppose that a is irreducible, and suppose that $a|bc$. Let d be a gcd of $\{a, b\}$. Then $a = de$ and $b = df$. If e is a unit, then $a|b$ since $d|b$. If e is not a unit, then d is a unit, and hence $ar + bs = 1$ for some r and s in R. Hence, $arc + bsc = c$. Since a divides both terms on the left of this equation, $a|c$. Hence a is prime.

In PID's we can now use the words "prime" and "irreducible" interchangeably. It is not true that prime and irreducible are the same for arbitrary integral domains. There is an example in 8.2.

4.4.4 Corollary

A nonzero ideal Rp in a PID R is maximal if and only if p is prime.

Proof Suppose that p is prime. If $Rp \subset Ra$, then $p = ra$ for some r in R. Either r or a is a unit. If r is a unit, then $Rp = Ra$. If a is a unit, then $Ra = R$. Thus, Rp is maximal.

Suppose that Rp is maximal and nonzero. If $p = ab$, then $Rp \subset Ra$, so $Ra = R$ or $Ra = Rp$. If $Ra = R$, then a is a unit. If $Ra = Rp$, then $a = rp$, so $p = ab = rpb$, whence $1 = rb$. Thus, b is a unit. Therefore, p is prime.

To write a nonzero element in a PID as a unit times a product of primes is somewhat tricky. In fact, suppose a is not a unit. Why does a have a prime divisor? It is not obvious at all that it does. Even if every nonunit has a prime divisor, then $a = p_1b_1$ with p_1 prime, $b_1 = p_2b_2$ with p_2 prime, and so on, but conceivably the process will never end. So there is some preliminary work to be done.

4.4.5 Definition

Let R be any ring. Then R satisfies the **ascending chain condition,** *or R is* **Noetherian,** *if every ascending sequence $I_1 \subset I_2 \subset I_3 \cdots$ of ideals of R becomes constant.*

That is, R satisfies the ascending chain condition (abbreviated acc) if R has no infinite chain $I_1 \subset I_2 \subset I_3 \subset \cdots$ of ideals with the inclusions all proper ones.

4.4.6 Theorem

A PID *is Noetherian.*

Proof Let $I_1 \subset I_2 \subset I_3 \subset \cdots$ be a chain of ideals of a PID R. Then $\cup I_i = I$ is an ideal of R. But $I = Ra$ for some a in R, and a must be in I_n for some n. Thus,

$$I_n = I_{n+1} = I_{n+2} = \cdots .$$

4.4.7 Corollary

Let R be a PID. *Every ideal $I \neq R$ is contained in a maximal ideal.*

Proof If $I \neq R$ and I is not maximal, then I is properly contained in an ideal $I_1 \neq R$. If I_1 is not maximal, then I_1 is properly contained in an ideal $I_2 \neq R$. The chain

$$I \subset I_1 \subset I_2 \subset \cdots$$

must stop. It stops at a maximal ideal.

Now we are ready to factor an element into a product of primes.

4.4.8 The fundamental theorem of arithmetic

Let R be a PID. *Then any nonzero element a in R can be written in the form $a = up_1p_2 \cdots p_n$, where u is a unit and each p_i is a prime. If also $a = vq_1q_2 \cdots q_m$ with v a unit and each q_i prime, then $m = n$ and, after suitable renumbering, p_i and q_i are associates for all i.*

Proof If a is a unit, there is nothing to do. Suppose that a is not a unit. If a is not prime, then $a = a_1b_1$ with neither a_1 nor b_1 a unit. Thus, $Ra \subsetneq Rb_1$. If b_1 is not prime, then $b_1 = a_2b_2$ with neither a_2 nor b_2 a unit. Thus, $Ra \subsetneq Rb_1 \subsetneq Rb_2$. Keep going. We get an ascending chain of ideals of R. It must terminate. Thus, we come to a b_n that is prime. Hence $a = (a_1a_2 \cdots a_n)b_n$, so a is divisible by a prime. We have shown that every nonunit nonzero element in R is divisible by a prime.

Write $a = p_1c_1$ with p_1 a prime. If c_1 is not a unit, then $c_1 = p_2c_2$ with p_2 prime. If c_2 is not a unit, then $c_2 = p_3c_3$ with p_3 prime. Thus, we get a chain $Ra \subsetneq Rc_1 \subsetneq Rc_2 \subsetneq Rc_3 \subsetneq \cdots$. It must terminate. Therefore, some c_n is a unit. Hence, $a = up_1p_2 \cdots p_n$ with u a unit and with each p_i a prime.

Now suppose that $a = vq_1q_2 \cdots q_m$ with v a unit and each q_i a prime. We will induct on n. If $n = 0$, then a is a unit and hence $m = 0$. If $n > 0$, then $m > 0$. In this case, $p_1 | a$ so p_1 divides some q_i, and we may as well suppose $i = 1$. This means that $p_1u_1 = q_1$ for some unit u_1. We have $a = up_1p_2 \cdots p_n = (vu_1)p_1q_2q_3 \cdots q_m$. Cancel p_1. We get $up_2p_3 \cdots p_n = (vu_1)q_2q_3 \cdots q_m$. By the induction hypothesis, we are through.

4.4.9 Corollary

Every nonzero element a in a PID *can be written in the form*

$$a = up_1^{n_1}p_2^{n_2} \cdots p_k^{n_k},$$

where u is a unit and the p_i are nonassociate primes. If

$$a = vq_1^{m_1}q_2^{m_2} \cdots q_j^{m_j},$$

where v is a unit and the q_i are nonassociate primes, then $j = k$, and after suitable renumbering, p_i and q_i are associates for all i.

PROBLEMS

1 Prove that being "an associate of" is an equivalence relation.

2 Let p be a prime in $\mathbb{Z}$, and let $R = \{a/p^n \colon a, n \in \mathbb{Z}\}$. Prove that R is a PID. What are the primes, and what are the units in R? Prove that if $p = 3$, then a gcd of $\{5, 7\}$ is 9.

3 Prove that any subring of $\mathbb{Q}$ which contains $\mathbb{Z}$ is a PID.

4 Let S be a nonempty multiplicatively closed subset of $\mathbb{Z}$ with $0 \notin S$. What are the primes, and what are the units of $\mathbb{Z}_S$?

5 Let R be a PID, and let S be a nonempty multiplicatively closed subset of R with $0 \notin S$. Prove that R_S is a PID. What are its units?

6 (a) Describe, in terms of the primes of $\mathbb{Z}$, all the subrings of $\mathbb{Q}$ containing $\mathbb{Z}$.
(b) Let R be a PID, and let F be its field of quotients. Describe, in terms of the primes of R, all the subrings of F containing R.

7 Prove that $\mathbb{Z}[i] = \{a + bi \colon a, b \in \mathbb{Z}\}$ is a PID.

8 Prove that any two lcm's of a set $\{a_1, a_2, \ldots, a_n\}$ are associates.

9 Let R be a PID, with a and b in R, and let

$$a = up_1^{n_1}p_2^{n_2} \cdots p_k^{n_k}, \quad b = vp_1^{m_1}p_2^{m_2} \cdots p_k^{m_k},$$

with u and v units, the p_i's primes, and n_i and $m_i \geq 0$. Write down a gcd and an lcm of $\{a, b\}$.

10 Let a and b be in R. Prove that if R is a PID, and if d is a gcd of $\{a, b\}$, then ab/d is an lcm of $\{a,b\}$.

11 Prove that the ideals of a PID R are in one-to-one correspondence with the associate classes of elements of R.

12 Let I and J be ideals of the PID R. If $I = Ra$ and $J = Rb$, prove that $IJ = Rab$. (See 4.2, Problem 29, for the definition of the product IJ.)

13 Let I and J be ideals of the PID R. Prove that if $I = Ra$ and $J = Rb$, then $I + J = Rd$, where d is any gcd of a and b. Prove that $I \cap J = Rc$, where c is any lcm of a and b.

14 Let I and J be ideals in the PID R. Let $I = Ra$, and let $J = Rb$. Prove that $IJ = I \cap J$ if and only if ab is an lcm of a and b.

15 Call an ideal I in a commutative ring R with identity a ***prime*** if $I \neq R$ and if $ab \in I$ implies that $a \in I$ or $b \in I$. Prove that $I \neq R$ is a prime ideal in R if and only if R/I is an integral domain. Prove that every maximal ideal of R is a prime ideal.

16 Let R be a PID, and let $r \neq 0$. Prove that R/Rr is either a field or has zero divisors.

17 Find those r in a PID R such that R/Rr has no nonzero nilpotent elements.

18 Let R be a PID. Prove that an ideal $I \neq 0$ is a prime ideal if and only if $I = Rp$ for some prime p.

19 Let R be a PID, and let I be an ideal of R such that $0 \neq I \neq R$. Prove that $I = P_1 P_2 \cdots P_n$, where each P_i is a prime ideal. Prove that if $I = Q_1 Q_2 \cdots Q_m$ with each Q_i prime, then $m = n$, and after suitable renumbering, $P_i = Q_i$ for all i. That is, prove that every ideal $\neq R$ or 0 is uniquely a product of prime ideals.

20 Call an ideal I in a commutative ring R with an identity ***primary*** if $I \neq R$ and if $ab \in I$ and $a \notin I$, then $b^n \in I$ for some n. Prove that prime ideals are primary. Prove that if R is a PID, then I is primary if and only if $I = P^n$ for some prime ideal P. Prove that if R is a PID and if $P^m = Q^n$ for prime ideals P and Q of R, then $P = Q$ and $m = n$.

21 Let I be an ideal in the PID R, with $I \neq R$. Prove that

$$I = P_1^{n_1} \cap P_2^{n_2} \cap \cdots \cap P_k^{n_k}$$

where the P_i are distinct prime ideals. Prove that if

$$I = Q_1 \cap Q_2 \cap \cdots \cap Q_m$$

with the Q_i primary and no Q_i contained in Q_j for $j \neq i$, then $m = k$, and after suitable renumbering,

$$Q_i = P_i^{n_i}$$

for all i.

22 Let R be a PID, and let a and b be in R with $(a, b) = 1$. Prove that

$$R/Rab \to (R/Ra) \oplus (R/Rb): r + Rab \to (r + Ra, r + Rb)$$

is a ring isomorphism. Generalize.

23 **(Chinese Remainder Theorem)** Let $a, b, c,$ and d be integers with $\gcd(a, b) = 1$. Prove that $x = c \pmod a$ and $x = d \pmod b$ has a simultaneous integral solution x, and that x is unique mod ab. Generalize.

24 Let I and J be ideals of the commutative ring R with identity such that $I + J = R$. Prove that

$$R/(I \cap J) \to (R/I) \oplus (R/J): r + I \cap J \to (r + I, r + J)$$

is a ring isomorphism. Generalize.

4.5 POLYNOMIAL RINGS

Not only are polynomial rings of interest in their own right, but they will play a key role in our study of linear transformations in Chapter 5, and in our study of fields in Chapter 6. They will also make an appearance in a couple of classical theorems in Chapter 8. We have used them as a source of examples several times already, for example in Examples 4.1.7(i) and 4.2.3(c).

Let R be any ring. We think of a polynomial in x with coefficients in R as a symbol

$$a_0 + a_1x + a_2x^2 + \cdots + a_nx^n$$

with $a_i \in R$. If we add and multiply these symbols in the usual way, we get a new ring, denoted $R[x]$. "Adding and multiplying these symbols in the usual way" covers some sins, however. For example, the distinct symbols

$$1 + 2x + 0x^2$$

and

$$1 + 2x$$

represent the same element in $\mathbb{Z}[x]$. More generally, the symbols

$$a_0 + a_1x + \cdots + a_nx^n$$

and

$$b_0 + b_1x + \cdots + b_mx^m$$

represent the same element in $R[x]$ if and only if $a_i = b_i$ whenever either a_i or $b_i \neq 0$. This defines an equivalence relation, and we could define $R[x]$ to be the set of its equivalence classes, and addition and multiplication accordingly. This would give us what we want, but we prefer another way that is equally precise.

Let R be any ring, and let $R[x]$ be the set of all mappings f from the set $\{x^0, x^1, x^2, \ldots\}$ into the set R such that $f(x^n) = 0$ for all but finitely many n.

The largest n such that $f(x^n) \neq 0$, if such exists, is called the ***degree*** of f. If no such n exists, then f has degree $-\infty$. In $R[x]$, add and multiply by the rules

$$(f+g)(x^n) = f(x^n) + g(x^n)$$

and

$$(fg)(x^n) = \sum_{i+j=n} f(x^i)g(x^j).$$

This makes $R[x]$ into a ring. There are many things to check, but they are all easy. Note first that $f+g$ and fg are actually in $R[x]$. The most difficult thing to check is the associative law for multiplication, which we will do now.

$$((fg)h)(x^n) = \sum_{i+j=n} (fg)(x^i)h(x^j) = \sum_{i+j=n} \left(\sum_{k+m=i} f(x^k)g(x^m) \right) h(x^j)$$

$$= \sum_{i+j=n} \left(\sum_{k+m=i} f(x^k)g(x^m)h(x^j) \right) = \sum_{k+m+j=n} f(x^k)g(x^m)h(x^j).$$

Similarly,

$$(f(gh))(x^n) = \sum_{k+m+j=n} f(x^k)g(x^m)h(x^j),$$

and so the associative law for multiplication holds. The remaining details showing that $R[x]$ is a ring are left to the reader.

4.5.1 Definition

The ring $R[x]$ is the **ring of polynomials in x with coefficients in R**.

Let $a \in R$, and let n be a nonnegative integer. One element of $R[x]$ is the function that sends x^n to a and x^m to 0 if $m \neq n$. Denote this element by ax^n. That is, $(ax^n)(x^m) = \delta_{mn}a$. It is called a ***monomial***. Now let f be any element of $R[x]$. Clearly,

$$f = f(x^0)x^0 + f(x^1)x^1 + \cdots + f(x^n)x^n$$

whenever $f(x^m) = 0$ for all $m > n$. That is, every element of $R[x]$ is a sum of monomials. When are two sums

$$a_0x^0 + a_1x^1 + \cdots + a_nx^n$$

and

$$b_0x^0 + b_1x^1 + \cdots + b_mx^m$$

equal? By the definition of addition in $R[x]$, the first takes x^i to a_i for $i \leq n$ and x^k to 0 for $k > n$, and similarly for the second. Thus these two sums of monomials are equal if and only if $a_i = b_i$, with the convention that $a_i = 0$ for

$i > n$ and $b_i = 0$ for $i > m$. That is, they are equal if and only if they are equal as polynomials as we ordinarily think of polynomials. How do such sums of monomials add and multiply? Noting that $(ax^m)(bx^n) = (ab)x^{m+n}$, and that $ax^m + bx^m = (a + b)x^m$ by their very definition, we see that they add and multiply just as we should add and multiply polynomials.

Some notational changes are in order. If R has an identity, then the monomial $1x^n$ is written simply as x^n. In any case, x^1 is written as x, and the monomial ax^0 is written simply as a. Hence we can write any element in $R[x]$ in the form

$$a_0 + a_1x + a_2x^2 + \cdots + a_nx^n$$

with the $a_i \in R$, and we have already observed how unique such a representation is. The letter x is called the ***indeterminate***, and the a_i are called the ***coefficients***. We could just as well have used y, or z, or anything for the indeterminate.

Having $R[x]$, we can form $R[x][y]$, or more generally, $R[x_1][x_2] \cdots [x_n]$. However, $R[x][y] \approx R[y][x]$ in the following special way. An element in $R[x][y]$ is a polynomial in y with coefficients in $R[x]$, and is the sum of monomials. Let

$$(a_0 + a_1x + \cdots + a_nx^n)y^m$$

be a monomial in $R[x][y]$. Since

$$(a_0 + a_1x + \cdots + a_nx^n)y^m = a_0y^m + a_1xy^m + \cdots + a_nx^ny^m,$$

every element in $R[x][y]$ is a sum of monomials of the form $a_{ij}x^iy^j$ with $a_{ij} \in R$. Associate with $a_{ij}x^iy^j$ the monomial $a_{ij}y^jx^i$ in $R[y][x]$. Now the mapping $R[x][y] \to R[y][x]$ that sends $\sum a_{ij}x^iy^j$ to $\sum a_{ij}y^jx^i$ is an isomorphism. The rings $R[x][y]$ and $R[y][x]$ are usually identified and denoted as $R[x, y]$. Since any element in $R[x][y]$ can be written in the form $\sum a_{ij}x^iy^j$ with $a_{ij} \in R$, the elements in $R[x][y]$ are "polynomials in x and y" with coefficients in R. Similar remarks apply to $R[x_1][x_2] \cdots [x_n]$. We leave the details to the reader and will not further concern ourselves with "polynomials in several variables" at the moment.

We are not going to study general polynomial rings in any depth. Our primary interest is in $F[x]$ where F is a field, and at this point it is convenient to restrict attention to rings $D[x]$ where D is an integral domain. So from now on, D will be an integral domain.

The map $D \to D[x]$: $a \to ax^0$ is clearly a ring monomorphism. In fact, we are already writing ax^0 simply as a. Therefore, we consider D as a subring of $D[x]$. The elements of D in $D[x]$ are called ***constants.*** The reason is this. For any element $a_0 + a_1x + \cdots + a_nx^n \in D[x]$, there is associated a mapping from D into D defined by

$$d \to a_0 + a_1d + \cdots + a_nd^n.$$

The constant polynomial a_0 corresponds to the constant function $D \to D$: $d \to a_0$.

An element of $D[x]$ will typically be denoted $p(x)$. If $p(x)=a_0+a_1x+\cdots+a_nx^n$ and $d\in D$, then $p(d)$ denotes the element $a_0+a_1d+\cdots+a_nd^n$ of D. That is, $p(d)$ is "$p(x)$ evaluated at d." Thus, with each $p(x)\in D[x]$ we have associated a map from D into D. Letting Map(D, D) denote the set of all maps from D into D, we then have a map

$$\varepsilon\colon D[x]\to \mathrm{Map}(D, D).$$

It is given by $(\varepsilon(p(x)))(d)=p(d)$.

4.5.2 Theorem

If $p(x)$ and $q(x)$ are in $D[x]$ and $d\in D$, then $p(x)+q(x)$ evaluated at d is $p(d)+q(d)$, and $p(x)q(x)$ evaluated at d is $p(d)q(d)$.

Proof Let $p(x)=\sum a_ix^i$ and $q(x)=\sum b_ix^i$. Then

$$p(x)q(x)=\sum_i\Bigl(\sum_{j+k=i}a_jb_k\Bigr)x^i.$$

Hence $p(x)q(x)$ evaluated at d is $\sum_i(\sum_{j+k=i}a_jb_k)d^i$. Now

$$p(d)q(d)=\Bigl(\sum_i a_id^i\Bigr)\Bigl(\sum_i b_id^i\Bigr)=\sum_i\Bigl(\sum_{j+k=i}a_jb_k\Bigr)d^i,$$

using commutativity of D. Similarly, $p(x)+q(x)$ evaluated at d is $p(d)+q(d)$.

The image of the map

$$\varepsilon\colon D[x]\to \mathrm{Map}(D, D)$$

is the set of ***polynomial functions*** on D. Furthermore, Map(D, D) is a ring if we add and multiply by the rules

$$(\alpha+\beta)(d)=\alpha(d)+\beta(d)$$

and

$$(\alpha\beta)(d)=\alpha(d)\beta(d).$$

Now observe that 4.5.2 just says that $\varepsilon\colon D[x]\to\mathrm{Map}(D, D)$ is a ring homomorphism. This homomorphism is a monomorphism if and only if D is infinite. If D is finite, then ε cannot be one-to-one since $D[x]$ is infinite and Map(D, D) is finite. However, if D is infinite and $\varepsilon(p(x))=0$, then $p(d)=0$ for all $d\in D$. Thus, $p(x)$ would have infinitely many roots in D. We will see in 4.5.10 below that this is an impossibility unless $p(x)=0$.

The upshot is this. If D is an infinite integral domain, then $D[x]$ is isomorphic to a subring of Map(D, D), namely the subring of polynomial functions on D. These functions can be identified in Map(D, D) without first defining $D[x]$. A polynomial function on D is a function $f: D \to D$ such that there exist

$$a_0, a_1, a_2, \ldots, a_n \in D$$

such that

$$f(d) = a_0 + a_1 d + \cdots + a_n d^n$$

for all $d \in D$. Now one could *define* $D[x]$ as the set of all these polynomial functions on D, and then verify that it is a subring of the ring Map(D, D). In fact, in elementary analysis, we are used to thinking of polynomials (with real coefficients, say) as functions. In any case, if D is an infinite integral domain, the ring $D[x]$ and the ring of polynomial functions on D may be identified—they are isomorphic.

If

$$a_0 x^0 + a_1 x^1 + \cdots + a_n x^n \neq 0,$$

its ***degree*** is the largest m such that $a_m \neq 0$. The ***degree*** of $0 \in D[x]$ is defined to be the symbol $-\infty$, and we make the conventions that

$$-\infty + n = n + -\infty = \infty + -\infty = -\infty$$

for all nonnegative integers n. We write $\deg(p(x))$ for the degree of the polynomial $p(x)$.

4.5.3 Theorem

If $p(x)$ and $q(x)$ are in $D[x]$ and if D is an integral domain, then

$$\deg(p(x)q(x)) = \deg(p(x)) + \deg(q(x)).$$

Proof Write $p(x) = a_0 + a_1 x + \cdots + a_n x^n$ and $q(x) = b_0 + b_1 x^1 + \cdots + b_m x^m$. If either is 0, the desired equality is clear. So take $a_n \neq 0 \neq b_m$. The term of the largest degree in $p(x)q(x)$ is $a_n b_m x^{m+n}$, and it is not zero since D is an integral domain.

4.5.4 Corollary

If D is an integral domain, then $D[x]$ is an integral domain.

4.5.5 Corollary

If D is an integral domain, then the units of $D[x]$ are the units of D.

Proof If $p(x)q(x)=1$, then $\deg(p(x)q(x))=\deg(1)=0=\deg(p(x))+\deg(q(x))$, so that $\deg(p(x))=0=\deg(q(x))$. That is, $p(x)$ and $q(x)$ are in D. Therefore, if $p(x)$ is a unit in $D[x]$, it is not only in D, but is a unit there. Units of D are clearly units of $D[x]$.

4.5.6 Corollary

If F is a field, then the units of $F[x]$ are the nonzero elements of F.

Suppose that D is an integral domain and that

$$g(x)=a_0+a_1x+\cdots+a_mx^m \in D[x]$$

with a_m a unit in D. In particular, $g(x)\neq 0$. For any

$$f(x)=b_0+b_1x+\cdots+b_nx^n \in D[x],$$

we want to write $f(x)=g(x)q(x)+r(x)$ with $q(x)$ and $r(x)$ in $D[x]$ and $\deg(r(x))<\deg(g(x))$. If $\deg(f(x))<\deg(g(x))$, it is easy. Just let $q(x)=0$ and $r(x)=f(x)$. If not, $n\geq m$, and

$$f(x)-a_m^{-1}b_nx^{n-m}g(x)$$

has degree at most $n-1$. Therefore, letting $q_1(x)=a_m^{-1}b_nx^{n-m}$ and $h_1(x)=f(x)-g(x)q_1(x)$, we have

$$f(x)=g(x)q_1(x)+h_1(x)$$

with $\deg(h_1(x))<\deg(f(x))$. If $\deg(h_1(x))\geq\deg(g(x))$, write $h_1(x)=q_2(x)g(x)+h_2(x)$ with $\deg(h_2(x))<\deg(h_1(x))$. Eventually we get

$$f(x)=q_1(x)g(x)+q_2(x)g(x)+\cdots+q_n(x)g(x)+h_n(x)$$

with $\deg(h_n(x))<\deg(g(x))$. Denoting

$$q_1(x)+q_2(x)+\cdots+q_n(x)$$

by $q(x)$ and $h_n(x)$ by $r(x)$, we have shown that

$$f(x)=q(x)g(x)+r(x)$$

with $\deg(r(x))<\deg(g(x))$. If also

$$f(x)=q_1(x)g(x)+r_1(x)$$

with

$$\deg(r_1(x))<\deg(g(x)),$$

then

$$g(x)(q(x)-q_1(x))=r_1(x)-r(x).$$

Unless

$$q(x) - q_1(x) = 0,$$

$$\deg(g(x)(q(x) - q_1(x))) > \deg(r_1(x) - r(x)).$$

Therefore, $q(x) = q_1(x)$ and $r(x) = r_1(x)$. Thus, we have

4.5.7 The division algorithm for polynomials

Let D be an integral domain, let $f(x)$ and $g(x)$ be in $D[x]$, and suppose that $g(x) \neq 0$ and has leading coefficient a unit. Then there exist unique polynomials $q(x)$ and $r(x)$ in $D[x]$ such that

$$f(x) = g(x)q(x) + r(x)$$

and

$$\deg(r(x)) < \deg(g(x)).$$

4.5.8 Remainder theorem

Let D be an integral domain, and let a be an element of D. For any $f(x) \in D[x]$,

$$f(x) = (x - a)q(x) + f(a).$$

Proof Write $f(x) = (x - a)q(x) + r(x)$ with $\deg(r(x)) < \deg(x - a)$. We can do this because the leading coefficient of $x - a$ is 1, which is a unit in D. Since $\deg(x - a) = 1$, $r(x) = d \in D$. Hence, $f(a) = (a - a)q(a) + d = d$.

4.5.9 Corollary

Let D be an integral domain, let $a \in D$, and let $f(x) \in D[x]$. Then $x - a$ divides $f(x)$ in $D[x]$ if and only if $f(a) = 0$.

Proof If $x - a$ divides $f(x)$, then $f(x) = (x - a)q(x)$, and $f(a) = (a - a)q(a) = 0$. Conversely, if $f(a) = 0$, then since $f(x) = (x - a)q(x) + f(a)$ by the Remainder Theorem, we have that $f(x) = (x - a)q(x)$.

4.5.10 Corollary

Let D be an integral domain, and let $f(x)$ be a nonzero element in $D[x]$. Then there are at most $\deg(f(x))$ elements $a \in D$ such that $f(a) = 0$. That is, $f(x)$ has at most $\deg(f(x))$ roots in D.

Proof Suppose that $f(a)=0$. Write $f(x)=(x-a)q(x)$. Now $\deg(q(x))+1=\deg(f(x))$ since $f(x)\neq 0$. By induction, $q(x)$ has at most $\deg(q(x))$ roots in D. Any root of $f(x)$ must be either a or a root of $q(x)$ since $f(b)=(b-a)q(b)$, and D is an integral domain. The corollary follows.

Our real aim at this point is the following consequence of the Division Algorithm for Polynomials. It is this.

4.5.11 Theorem

Let F be a field. Then $F[x]$ is a principal ideal domain.

Proof Let I be any ideal of $F[x]$. If $I=\{0\}$, then I is certainly a principal ideal. So suppose that $I\neq\{0\}$. We must find a polynomial $g(x)$ in $F[x]$ such that $I=F[x]g(x)$. Where can we get hold of such a $g(x)$? Since $I\neq\{0\}$, I has a nonzero element $g(x)$ in it of least degree, and any such $g(x)$ works. In fact, let $f(x)$ be any element in I. Write

$$f(x)=g(x)q(x)+r(x),$$

with $\deg(r(x))<\deg(g(x))$. Since $f(x)$ and $g(x)$ are in the ideal I, so is $r(x)$. But no nonzero element in I has degree less than that of $g(x)$. Therefore, $r(x)=0$. Hence $I=F[x]g(x)$, so that $F[x]$ is a PID.

Now we can apply the results of 4.4 to $F[x]$. Recall that 2 elements are associates if one is a unit times the other. The units of $F[x]$ are the nonzero elements of F, that is, the nonzero constants. Therefore, two polynomials are associates if and only if one is a nonzero constant times the other. If

$$p(x)=a_0+a_1x+\cdots+a_nx^n$$

is in $F[x]$ and $a_n\neq 0$, then

$$q(x)=a_0a_n^{-1}+a_1a_n^{-1}x+\cdots+a_{n-1}a_n^{-1}x^{n-1}+x^n$$

is an associate of $p(x)$. That is, every nonzero polynomial has an associate whose coefficient of its highest power term is 1. Such polynomials are called ***monic.*** Thus every equivalence class of nonzero associates has a monic representative, and clearly two monic polynomials that are associates are equal. So every such equivalence class of associates has a special element in it, namely the monic in it.

When is a polynomial in $F[x]$ prime? A nonprime can be factored into two nonunits, and since the units of $F[x]$ are the polynomials of degree 0, a polynomial $p(x)\in F[x]$ is prime if and only if $p(x)$ cannot be written in the form $q(x)r(x)$ with $\deg(q(x))$ and $\deg(r(x))$ both less than $\deg(p(x))$. Since every nonzero element in $F[x]$ is a unit or a product of primes, we have the following theorems.

4.5.12 Theorem

Let F be a field. Then every nonzero element $p(x) \in F[x]$ can be written in the form

$$p(x) = ap_1(x)p_2(x) \cdots p_n(x),$$

where $a \in F$ and each $p_i(x)$ is a monic prime. If

$$p(x) = bq_1(x)q_2(x) \cdots q_n(x),$$

with $b \in F$ and each $q(x)$ a monic prime, then $a = b$, $m = n$, and after suitable renumbering, $p_i(x) = q_i(x)$ for all i.

4.5.13 Theorem

Let F be a field. Then every nonzero element $p(x) \in F[x]$ can be written in the form

$$p(x) = ap_1(x)^{n_1}p_2(x)^{n_2} \cdots p_k(x)^{n_k},$$

where $a \in F$ and the $p_i(x)$ are distinct monic primes. If

$$p(x) = bq_1(x)^{m_1}q_2(x)^{m_2} \cdots q_j(x)^{m_j}$$

is another such representation, then $a = b$, $j = k$, and after suitable renumbering, $p_i(x) = q_i(x)$ and $m_i = n_i$ for all i.

Let $p(x)$ be in $F[x]$. The ideal $F[x]p(x)$ it generates is usually denoted $(p(x))$. By 4.2.10 and 4.4.4, the quotient ring $F[x]/(p(x))$ is a field if and only if $p(x)$ is a prime. Therefore, $F[x]/(p(x))$ is a field if and only if $p(x)$ cannot be factored into polynomials of smaller degree.

If $p_1(x), p_2(x), \ldots, p_n(x)$ are in $F[x]$, not all zero, then they have a greatest common divisor, and using Theorem 4.4.2 it can be written in the form $\sum r_i(x)p_i(x)$. Since any two greatest common divisors must be associates, there is a unique monic greatest common divisor. Similarly, results in PID's for least common multiples hold for $F[x]$.

The division algorithm for polynomials (4.5.7) yields an algorithm for computing the greatest common divisor of two polynomials $p_1(x)$ and $p_2(x)$, and produces the greatest common divisor as a linear combination of $p_1(x)$ and

$p_2(x)$. It works just as is the case of the integers (1.6, page 22). If $p_2(x) \neq 0$, then writing simply p_i for polynomials $p_i(x)$, we get the following equations.

$$\begin{aligned}
p_1 &= p_2q_1 + p_3 && \text{with } -\infty < \deg(p_3) < \deg(p_2);\\
p_2 &= p_3q_2 + p_4 && \text{with } -\infty < \deg(p_4) < \deg(p_3);\\
p_3 &= p_4q_3 + p_5 && \text{with } -\infty < \deg(p_5) < \deg(p_4);\\
& && \vdots\\
p_{k-3} &= p_{k-2}q_{k-3} + p_{k-1} && \text{with } -\infty < \deg(p_{k-1}) < \deg(p_{k-2});\\
p_{k-2} &= p_{k-1}q_{k-2} + p_k && \text{with } -\infty < \deg(p_k) < \deg(p_{k-1});
\end{aligned}$$

$$p_{k-1} = p_kq_{k-1}.$$

Then p_k is the greatest common divisor of p_1 and p_2, and from the equations above, p_k may be written as a linear combination of p_1 and p_2. The procedure is entirely analogous to that for the integers, and again is called the ***Euclidean Algorithm*** (1.6, page 21).

PROBLEMS

1 Prove that $R[x]$ has an identity if and only if R does.

2 Prove that $R[x]$ is commutative if and only if R is.

3 Prove that $\mathbb{Z}[x]$ is not a PID.

4 Prove that if F is a field, then $F[x, y]$ is not a PID.

5 Prove that $2x + 1$ is a unit in $(\mathbb{Z}/\mathbb{Z}4)[x]$. Compare with 4.5.5.

6 Let f be a homomorphism from R to S. Prove that

$$R[x] \to S[x]: \sum r_ix^i \to \sum f(r_i)x^i$$

is a homomorphism.

7 If D is an integral domain, F is its field of quotients, and K is the field of quotients of $D[x]$, prove that $F(x) \approx K$, where $F(x)$ denotes the field of quotients of the integral domain $F[x]$.

8 Let D be an integral domain, and let $d \in D$. By 4.5.2, the map $D[x] \to D: p(x) \to p(d)$ is a ring homomorphism. Is its kernel a principal ideal?

9 Let F be a field, and let K be the quotient field of $F[x]$. For $a \in F$, let

$$K_a = \{f(x)/g(x): f(x), g(x) \in K, g(a) \neq 0\}.$$

Prove that K_a is a subring of K. Prove that

$$K_a \to F: f(x)/g(x) \to f(a)/g(a)$$

defines a ring epimorphism. What is its kernel?

10 Prove that if F is a field with n elements, and if the nonzero element $p(x) \in F[x]$ has degree m, then $F[x]/(p(x))$ has n^m elements.

11 Prove that $x^4 + x$ and $x^2 + x$ in $(\mathbb{Z}/\mathbb{Z}3)[x]$ determine the same element of $\mathrm{Map}(\mathbb{Z}/\mathbb{Z}3, \mathbb{Z}/\mathbb{Z}3)$.

12 Let p be a prime. Prove that in $(\mathbb{Z}/\mathbb{Z}p)[x]$,

$$x^p - x = \prod_{a \in \mathbb{Z}/\mathbb{Z}p} (x - a).$$

13 Let F be a finite field. Prove that every function $F \rightarrow F$ is a polynomial function. That is, prove that $F[x] \rightarrow \mathrm{Map}(F, F)$ is onto. What is the kernel of this homomorphism?

14 Prove that if F is a field, then $F[x]$ has infinitely many primes.

15 Let F be a finite field. Prove that there are primes in $F[x]$ of arbitrarily large degree. Prove that the hypothesis that F be finite is necessary.

16 Let F and K be fields with $F \subset K$. Let $p(x)$, $q(x) \in F[x]$. Prove that the greatest common divisor of $p(x)$ and $q(x)$ as elements of $F[x]$ is the same as their greatest common divisor as elements of $K[x]$.

17 Let p be any prime, $n \geq 1$, and let F be the field of integers modulo p. Prove that there exists $f(x) \in F[x]$ such that $f(x)$ is irreducible, and $\deg(f(x)) = n$.

18 Let p be any prime, and let $n > 1$. Prove that there is a field with order p^n.

19 Let $X = \{x^0, x^1, x^2, \ldots\}$ and $Y = \{y^0, y^1, y^2, \ldots\}$. Let $R[x, y]$ be the set of all maps f from $X \times Y$ into R such that $f((x^i, y^j)) = 0$ for all but finitely many i and j. Add and multiply in $R[x, y]$ by the rules

$$(f + g)(x^i, y^j) = f(x^i, y^j) + g(x^i, y^j),$$

and

$$(fg)((x^i, y^j) = \sum_{p+q=i} \sum_{m+n=j} f((x^p, y^m))g((x^q, y^n)).$$

Prove that $R[x, y]$ is a ring and that $R[x, y] \approx R[x][y]$.

20 Let G and H be semigroups with identities. (A semigroup is a nonempty set with an associative binary operation on it.) Let R be any ring. Let $R(G)$ be the set of all maps α from G into R such that $\alpha(g) = 0$ for all but finitely many $g \in G$. Add and multiply by the rules

$$(\alpha + \beta)(g) = \alpha(g) + \beta(g)$$

and

$$(\alpha\beta)(g) = \sum_{xy=g} \alpha(x)\beta(y).$$

Let $G \times H$ be the direct product of the semigroups G and H. That is,

$G \times H$ is the semigroup whose elements are all pairs (g, h) with $g \in G$, $h \in H$ and whose multiplication is given by $(g_1, h_1)(g_2, h_2) = (g_1 g_2, h_1 h_2)$. Prove that $R(G)$ is a ring (the ***semigroup ring*** of the semigroup G with coefficients in R), and that the rings $R(G)(H)$ and $R(G \times H)$ are isomorphic.

21 Let X be the semigroup $\{x^0, x^1, x^2, \ldots\}$ with multiplication given by $x^m x^n = x^{m+n}$. Prove that for any ring R, $R[x] \approx R(X)$. Prove that $R[x][y] \approx R(X \times Y)$, where Y is defined analogously. Prove that $R[y][x] \approx R[x][y]$ via an isomorphism which fixes R elementwise, takes x to y and takes y to x.

22 Let F be a field, and let S be the subring of the quotient field of $F[x]$ generated by $1/x$ and $F[x]$. Let G be the infinite cyclic group. Prove that $F(G) \approx S$.

23 Define $R[x_1, x_2, x_3, \ldots]$, and show that it is a ring.

24 Show that $R \approx R[x]$ is possible. ($R \neq 0$.)

25 Let R be any ring. Define the ring $R\{x\}$ of "power series" $a_0 + a_1 x + a_2 x^2 + \cdots$, and show that it is a ring.

26 Prove that if F and K are fields and $F[x] \approx K[x]$ as rings, then $F \approx K$ as rings.

27 Prove that if F is a field and the rings $F[x]$ and $R[x]$ are isomorphic, then so are the rings F and R.

28 Prove that if the rings $\mathbb{Z}[x]$ and $R[x]$ are isomorphic, then so are $\mathbb{Z}$ and R.

29 Prove that $R[x]/(x - r)$ and R are isomorphic as rings.

30 Let F be any field. Prove that the rings $F[x]/(x^2 + 1)$ and $F[x]/(x^2 + 4x + 5)$ are isomorphic.

31 Prove that the rings $R[x, y]/(x)$ and $R[y]$ are isomorphic.

32 Find a nonzero prime ideal in $\mathbb{Z}[x]$ that is not maximal.

33 Let D be an integral domain. Find a nonzero prime ideal in $D[x, y]$ that is not maximal.

34 Use the Euclidean Algorithm to find the greatest common divisor of $x^6 - 1$ and $x^{14} - 1$ in $\mathbb{Q}[x]$. Express that greatest common divisor as a linear combination of the two polynomials.

35 Use the Euclidean Algorithm to find the greatest common divisor of $x^4 + 1$ and $x^6 + x^4 + x^3 + 1$ in $\mathbb{Q}[x]$. Express that greatest common divisor as a linear combination of the two polynomials.

36 Let F be the field of integers modulo 2. Use the Euclidean Algorithm to find the greatest common divisor of $x^2 + 1$ and $x^5 + 1$ in $F[x]$. Express that greatest common divisor as a linear combination of the two polynomials.

4.6 MODULES

The concept of module generalizes both that of an Abelian group and that of a vector space. The purpose of this section is to introduce the reader to this more general notion and to lay the groundwork for the theorems in Section 4.7.

A vector space is an Abelian group V, a field F, and a map $F \times V \to V$ satisfying certain rules (3.1.5). Let G be any Abelian group. We know how to "multiply" elements of $\mathbb{Z}$ times elements of G, obtaining elements of G. In fact, if n is a positive integer and $g \in G$, $ng = g + g + \cdots + g$, the sum of g with itself n times. Letting $(-n)g = -(ng)$ and $0g = 0$ completes the definition. Thus, we have a map $\mathbb{Z} \times G \to G$. This map satisfies the same rules that the map $F \times V \to V$ does. That is, it satisfies

(a) $n(g + h) = ng + nh$ for $n \in \mathbb{Z}$ and $g, h \in G$;
(b) $(m + n)g = mg + ng$ for $m, n \in \mathbb{Z}$ and $g \in G$;
(c) $(mn)g = m(ng)$ for $m, n \in \mathbb{Z}$ and $g \in G$; and
(d) $1g = g$ for all $g \in G$.

The appropriate generalization that encompasses both of these situations is given in the following definition.

4.6.1 Definition

A **module** *is an Abelian group M, a ring R with identity, and a map $\cdot: R \times M \to M$ satisfying* (a)–(d) *below. The image of* (r, m) *under the mapping* $\cdot$ *is denoted* rm.

(a) $r(m + n) = rm + rn$ *for* $r \in R$ *and* $m, n \in M$;
(b) $(r + s)m = rm + sm$ for $r, s \in R$ and $m \in M$;
(c) $(rs)m = r(sm)$ for $r, s \in R$ and $m \in M$; and
(d) $1m = m$ for all $m \in M$.

Thus, a module would be a vector space if the ring involved were a field. In the definition above, the ***scalars***, that is, the elements of R, are written on the left, and we say that M is a ***left module over*** R, or that M is a ***left*** R***-module.*** Right modules are defined in the obvious way. Unless specifically stated otherwise, module will always mean ***left*** module. Thus when we say "Let M be a module over R.", we mean that M is an Abelian group, R is a ring with an identity, and there is a map $R \times M \to M$ on hand satisfying (a)–(d) above. From the remarks made before 4.6.1, we see that every Abelian group is a module over $\mathbb{Z}$, and vector spaces are modules over fields. Another example is this. Let R be any ring with identity. Now R is an Abelian group under $+$, and there is a map $R \times R \to R$ satisfying (a)–(d), namely the ring multiplication of R. Thus, *any ring is a module over itself.*

For the remainder of this section, ***ring*** means *ring with identity,* and by a ***subring*** of R we mean *one that contains the identity of R.*

Let M be a module over R, let r and s be in R, and let m and n be in M. Then the following hold, and the proofs are the same as they were for vector spaces (3.1.6).

(a) $r0 = 0$.
(b) $0m = 0$.
(c) $(-r)m = -(m) - r(-m)$.
(d) $(-r)(-m) = rm$.
(e) $(-1)m = -m$.
(f) $r(m - n) = rm - rn$.
(g) $(r - s)m = rm - sm$.

The notions of submodule, quotient module, homomorphism, direct sum, and so on, are completely analogous to the corresponding notions for vector spaces, and we quickly introduce them.

Let M be a module over R. A ***submodule*** of M is a subset S of M such that restricting addition to elements of S and scalar multiplication to $R \times S$ makes S into a module over R. Thus, S is a submodule of M if and only if S is a subgroup of M closed under multiplication by elements of R. The intersection of any family of submodules is a submodule. If S is a subset of M, there is a smallest submodule of M containing S, namely the intersection of all submodules of M containing S. This submodule will be denoted by $\langle S \rangle$. If S is not empty, then $\langle S \rangle$ consists of all linear combinations of elements of S, that is all elements of the form $\sum_{i=1}^{n} r_i s_i$, with $r_i \in R$ and $s_i \in S$. The module M is ***finitely generated*** if $M = \langle S \rangle$ for some finite subset S of M. If $S = \{s_1, s_2, \ldots, s_n\}$, then $\langle S \rangle = \{\sum_{i=1}^{n} r_i s_i : r_i \in R\}$. If $S = \{s\}$, then $\langle S \rangle = \{rs : r \in R\}$. This latter set is denoted Rs. If $M = Rs$ for some $s \in M$, then M is ***cyclic***. If S and T are submodules of M, then $S + T = \{s + t : s \in S, t \in T\}$ is a submodule of M.

Let M and N be modules over R. *A **homomorphism*** from the module M into the module N is a map $f: M \to N$ such that

$$f(m_1 + m_2) = f(m_1) + f(m_2)$$

and

$$f(rm) = rf(m)$$

for m_1, m_2, and $m \in M$ and $r \in R$. Thus, f is a group homomorphism that preserves scalar multiplication. A homomorphism is a ***monomorphism*** if it is one-to-one, an ***epimorphism*** if it is onto, and an ***isomorphism*** if it is both a monomorphism and an epimorphism. If f is an isomorphism, we say that M is

isomorphic to N, and write $M \approx N$. A homomorphism $f: M \to M$ is called an ***endomorphism***, and an endomorphism that is an isomorphism is called an ***automorphism***. The ***kernel*** of the homomorphism $f: M \to N$ is

$$\operatorname{Ker} f = \{m \in M: f(m) = 0\}.$$

The ***image*** of f is

$$\operatorname{Im} f = \{f(m): m \in M\}.$$

$\operatorname{Ker} f$ is a submodule of M, and $\operatorname{Im} f$ is a submodule of N. We readily see that f is an isomorphism if and only if $\operatorname{Ker} f = \{0\}$ and $\operatorname{Im} f = N$.

Let S be a submodule of M. Since it is a subgroup of M, we have the Abelian group M/S. For $r \in R$, let

$$r(m + S) = rm + S.$$

If $m + S = m_1 + S$, then $(rm + S) - (rm_1 + S) = r(m - m_1) + S = S$ since $m - m_1 \in S$ and S is a submodule. Thus the rule $r(m + S) = rm + S$ defines a map $R \times M/S \to M/S$, and it is routine to check that this makes M/S into a module over R. This module is called the ***quotient module of M with respect to S***, or simply "M over S", or "M modulo S".

The isomorphism theorems now follow as in the case of vector spaces. The proofs are identical.

(a) If $f: M \to N$ is a homomorphism, then

$$M/(\operatorname{Ker} f) \approx \operatorname{Im} f.$$

(b) If S and T are submodules of M, then

$$(S + T)/T \approx S/(S \cap T).$$

(c) If S and T are submodules of M with $S \subset T$, then

$$M/T \approx (M/S)/(T/S).$$

There are a couple of new concepts that we need. Let R be any ring. Recall that R is module over itself. What are the submodules of R? They are subgroups I of R closed under multiplication on the left by elements of R, that is, subgroups I such that $ri \in I$ for all $r \in R$ and $i \in I$.

4.6.2 Definition

Let R be a ring. A **left ideal** *of R is a submodule of R, where R is considered as a (left) module over itself.* **Right ideal** *is defined similarly.*

Thus, an ideal of R is both a left and a right ideal. If R is commutative, then right ideals, left ideals, and ideals are the same. If R is the ring of 2×2

matrices over the integers, and

$$m = \begin{pmatrix} 0 & 1 \\ 0 & 0 \end{pmatrix},$$

then Rm is the left ideal consisting of those matrices with first column all zeroes. However, it is not a right ideal since m is in Rm but $m \cdot m'$ is not, where m' is the transpose of m.

Now let M be a module over R, and let $m \in M$. Consider the set $0{:}m = \{r \in R: rm = 0\}$. Since $r, s \in 0{:}m$ implies $(r - s)m = rm - sm = 0 - 0 = 0$, $0{:}m$ is a subgroup of R. If $r_1 \in R$ and $r \in 0{:}m$, then $(r_1 r)m = r_1(rm) = r_1 0 = 0$ so that $0{:}m$ is a submodule of R, that is, a left ideal of R.

4.6.3 Definition

Let M be a module over R, and let $m \in M$. The left ideal $0{:}m = \{r \in R: rm = 0\}$ is the ***order ideal*** *of m.*

If V is a vector space over the field F, and if $v \in V$, then $0{:}v = \{0\}$ or $0{:}v = F$, depending on whether $v \neq 0$ or $v = 0$. Thus, order ideals in this case are not of interest. However, let G be an Abelian group. Then G is a module over $\mathbb{Z}$. For $g \in G$, $0{:}g = \{n \in \mathbb{Z}: ng = 0\}$ is a principal ideal and is generated by some $k \geq 0$. If $k > 0$, then k is the smallest positive integer such that $kg = 0$, that is, k ***is the order of*** g. If $k = 0$, we say that the order of g is infinite. Hence, if the order of an element in an Abelian group is not infinite, that order is the positive generator of the order ideal of that element. For arbitrary rings R, we cannot associate a specific element of $0{:}m$ with m and reasonably call that element the order of m. We can, however, associate the whole left ideal $0{:}m$ with m. That is the appropriate generalization of the notion of the order of an element of an Abelian group.

4.6.4 Theorem

If M is a cyclic R-module, then $M \approx R/I$ for some left ideal I of R. In fact, if $M = Rm$, then

$$M \approx R/(0{:}m).$$

Proof Let $M = Rm$. The map $f: R \to M: r \to rm$ is an epimorphism with kernel $0{:}m$.

Thus, we know what every cyclic R-module looks like. It is a quotient module of the module R itself. This is a generalization of the fact that every cyclic group is $\mathbb{Z}/\mathbb{Z}n$ for some $n \in \mathbb{Z}$.

Let $M_1, M_2, \ldots, M_k$ be modules over the ring R. Let

$$M_1 \oplus M_2 \oplus \cdots \oplus M_k$$

be the module whose elements are the elements of the set $M_1 \times M_2 \times \cdots \times M_k$ and whose addition and scalar multiplication are given by

$$(m_1, m_2, \ldots, m_k) + (n_1, n_2, \ldots, n_k) = (m_1 + n_1, m_2 + n_2, \ldots, m_k + n_k),$$

and

$$r(m_1, m_2, \ldots, m_k) = (rm_1, rm_2, \ldots, rm_k).$$

This construction is the same as the one we did for vector spaces. It is straightforward to show that $M = M_1 \oplus M_2 \oplus \cdots \oplus M_k$ is a module over R. This module is called the ***external direct sum*** of the modules $M_1, M_2, \ldots, M_k$. If we let M_i^* be those elements of M of the form $(0, 0, \ldots, m_i, 0, 0, \ldots, 0)$, where m_i is in the ith place, then M_i^* is a submodule, and as in the case of vector spaces, every element of M can be written uniquely in the form $m_1^* + m_2^* + \cdots + m_k^*$ with $m_i^* \in M_i^*$. When we have such a situation, namely when M has submodules $N_1, N_2, \ldots, N_k$ such that every element in M can be written uniquely in the form $n_1 + n_2 + \cdots + n_k$ with $n_i \in N_i$, we say that the module M is the ***internal direct sum*** of these submodules. We also write $M = N_1 \oplus N_2 \oplus \cdots \oplus N_k$ when M is the internal direct sum of the submodules $N_1, N_2, \ldots, N_k$. We have seen that with an external direct sum $M_1 \oplus M_2 \oplus \cdots \oplus M_k$, there corresponds an internal direct sum $M_1^* \oplus M_2^* \oplus \cdots \oplus M_k^*$. In practice, we can pretend that all direct sums are internal ones. Thus when we write $M = M_1 \oplus M_2 \oplus \cdots \oplus M_k$, we have in mind an internal direct sum unless we specifically say otherwise.

The following are useful. Suppose A and B are submodules of M. Then $M = A \oplus B$ if and only if $A \cap B = \{0\}$ and $M = A + B$. If $M = A \oplus B$, these conditions certainly hold. But if $A \cap B = \{0\}$ and $M = A + B$, then $a + b = a' + b'$ with a and a' in A, and b and $b' \in B$ implies that $a - a' = b' - b \in A \cap B$, whence $a = a'$, $b = b'$, and $M = A \oplus B$.

If A is a ***summand*** of M, that is if $M = A \oplus B$ for some submodule B of M, then the map $\varphi: M \to M: a + b \to a$ is an endomorphism of M such that $\varphi^2 = \varphi$, Ker $\varphi = B$, and Im $\varphi = A$. Now suppose that φ is any endomorphism of M such that $\varphi^2 = \varphi$. Such an endomorphism is called ***idempotent***. We assert that $M = \text{Ker}\ \varphi \oplus \text{Im}\ \varphi$. We must show that

$$(\text{Ker}\ \varphi) \cap (\text{Im}\ \varphi) = \{0\}$$

and

$$M = \text{Ker}\ \varphi + \text{Im}\ \varphi.$$

Let $m \in (\text{Ker}\ \varphi) \cap (\text{Im}\ \varphi)$. Then $m = \varphi(m')$ for some $m' \in M$, and $\varphi(m) = \varphi^2(m') = \varphi(m') = m = 0$ since $m \in \text{Ker}\ \varphi$ and $\varphi^2 = \varphi$. Let $m \in M$. Then $m = (m - \varphi(m)) + \varphi(m)$. Clearly, $\varphi(m) \in \text{Im}\ \varphi$. Also $\varphi(m - \varphi(m)) = \varphi(m) - \varphi^2(m) = \varphi(m) - \varphi(m) = 0$. Thus, $m - \varphi(m)$ is in Ker φ. We have proved the following theorem.

4.6.5 Theorem

If φ is an idempotent endomorphism of a module M, then $M = (\text{Ker } \varphi) \oplus (\text{Im } \varphi)$.

Good theorems about modules are ones that say that certain modules are direct sums of simpler ones. We have two examples of this already. Every finite Abelian group, that is, every finite module over the ring $\mathbb{Z}$, is a direct sum of cyclic modules (2.7.7). Every finitely generated module over a field is a direct sum of cyclic modules (3.3.12). In the next section (4.7.13) we will prove that a finitely generated module over a PID is a direct sum of cyclic modules, of which both the examples above are special cases.

We pause here to look at a simple example illustrating some of our results so far. Let R be the ring of all 2×2 matrices over the field $\mathbb{Q}$. Let M be the set of all 2×3 matrices over $\mathbb{Q}$. Now M is an Abelian group under matrix addition, and matrix multiplication makes M into a left module over R. Let M_i be the matrices in M with zeroes off the ith column. That is,

$$M_i = \{(a_{jk}) \in M \colon a_{jk} = 0 \text{ if } k \neq i.\}$$

Then M_i is a submodule of M. Furthermore, the mapping that replaces the entries off the ith column by zeroes is an idempotent endomorphism of M with image M_i. The kernel of this endomorphism consists of those matrices all of whose entries in the ith column are zero. Therefore, M_i is a direct summand of M by 4.6.5. However, it should be clear directly from the definition of internal direct sum that

$$M = M_1 \oplus M_2 \oplus M_3.$$

Further, M_i is isomorphic to the module C of 2×1 matrices over $\mathbb{Q}$, and M is isomorphic to the (external) direct sum of three copies of C.

Let

$$m_1 = \begin{pmatrix} 1 & 0 & 0 \\ 0 & 0 & 0 \end{pmatrix}, \quad m_2 = \begin{pmatrix} 0 & 1 & 0 \\ 0 & 0 & 0 \end{pmatrix}, \quad m_3 = \begin{pmatrix} 0 & 0 & 1 \\ 0 & 0 & 0 \end{pmatrix}.$$

Then $M_i = Rm_i$, so that the M_i are cyclic. The left ideal $0{:}m_i$ consists of those matrices in R with first column all zeroes. By 4.6.4, $R/0{:}m_i \approx M_i$, and it follows that the M_i are isomorphic to one another. Using the fact that $\mathbb{Q}$ is a field, it can be shown that M_i has no submodules other than 0 and M_i itself. Thus, $M_i = Rm$ for *any* nonzero element m of M_i. Therefore $R/0{:}m$ is isomorphic to M_j for any m in M_i, $i, j = 1, 2, 3$. The $0{:}m$ are not all the same. For example, if

$$m = \begin{pmatrix} 0 & 0 & 0 \\ 0 & 1 & 0 \end{pmatrix},$$

then $0{:}m$ consists of those matrices of R whose second column have only zero entries. Still $R/0{:}m \approx R/0{:}m_i$ for any i.

Consider the R-module R. It has a special element in it, namely 1. Since $R = R \cdot 1$, R is cyclic. Furthermore, if $r \cdot 1 = s \cdot 1$, then $r = s$. Thus every element $r \in R$ can be written in exactly one way as an element of R times 1, namely $r \cdot 1$. Suppose that M is a cyclic R-module with a generator m such that $rm = sm$ only if $r = s$. Then

$$R \to M: r \to rm$$

is clearly an isomorphism. More generally, suppose that M is an R-module, and that $\{m_1, m_2, \ldots, m_k\}$ is a family of elements of M such that for every element $m \in M$ there exists a unique family $\{r_1, r_2, \ldots, r_k\}$ of elements of R such that

$$m = \sum_{i=1}^{k} r_i m_i.$$

Then $\{m_1, m_2, \ldots, m_k\}$ is called a *basis* of M. Clearly

$$M = Rm_1 \oplus Rm_2 \oplus \cdots \oplus Rm_k,$$

and $R \approx Rm_i$ for each i. Therefore, M is isomorphic to a direct sum of copies of the R-module R. Such a module is called a ***free module of finite rank***. We proved that every finitely generated vector space was free (3.3.12). If $M = M_1 \oplus M_2 \oplus \cdots \oplus M_k$ with each $M_i \approx R$, then M has a basis. In fact, if $M_i = Rm_i$, then $\{m_1, m_2, \ldots, m_k\}$ is a basis. Therefore, M is free of finite rank if and only if it has a basis. Free modules play a key role in the theory of modules. The next three theorems detail some of their properties, and will be needed in 4.7.

4.6.6 Theorem

Let M be a free R-module with basis $\{m_1, m_2, \ldots, m_k\}$. Let N be an R-module, and let $\{n_1, n_2, \ldots, n_k\}$ be any family of elements of N. Then there exists exactly one homomorphism $f: M \to N$ such that $f(m_i) = n_i$ for all i.

Proof Let $m \in M$. Then $m = \sum r_i m_i$ with the r_i unique. Therefore, $f(m) = \sum r_i n_i$ defines a map $f: M \to N$. It is easy to check that f is an R-module homomorphism such that $f(m_i) = n_i$. If $g: M \to N$ is another homomorphism such that $g(m_i) = n_i$ for all i, then

$$g(\sum r_i m_i) = \sum g(r_i m_i) = \sum r_i g(m_i) = \sum r_i n_i = \sum r_i f(m_i) = f(\sum r_i m_i).$$

Therefore, $f = g$.

4.6.7 Theorem

Let M be an R-module, and let N be a submodule of M such that M/N is free of finite rank. Then N is a summand of M. That is, $M = N \oplus F$ for some submodule F of M.

Proof Let $\{m_1 + N, m_2 + N, \ldots, m_k + N\}$ be a basis for M/N. By 4.6.6, there is a homomorphism $M/N \to M$ given by $m_i + N \to m_i$. The composition $M \to M/N \to M$ is an endomorphism φ of M such that $\varphi^2 = \varphi$, and Ker $\varphi = N$. Apply 4.6.5.

4.6.8 Theorem

Let M be any finitely generated R-module. Then there exists an epimorphism $F \to M$ with F a free R-module.

Proof Let $\{m_1, m_2, \ldots, m_k\}$ generate M. Let F be the direct sum of k copies of the R-module R. Then F is free and has a basis $\{a_1, a_2, \ldots, a_k\}$. By 4.6.6, there exists a homomorphism $f: F \to M$ such that $f(a_i) = m_i$ for all i. But $\operatorname{Im} f = M$ since $m_i \in \operatorname{Im} f$ for all i and $\{m_1, m_2, \ldots, m_k\}$ generates M.

We have discussed only free modules of ***finite*** rank. The notion can be generalized to free modules of any rank, and the three theorems above hold with their finiteness assumptions dropped.

Let I be an ideal of R, and suppose that M is a module over the ring R/I. Then

$$rm = (r + I)m$$

makes M into an R-module. That is, every R/I-module is, in a natural way, an R-module. If $i \in I$ and $m \in M$, then $im = (i + I)m = 0m = 0$, so that the submodule

$$IM = \{i_1m_1 + i_2m_2 + \cdots + i_km_k : i_j \in I,\ m_j \in M,\ k > 0\}$$

is zero. Now any R-module N such that $IN = 0$ is, in a natural way, an R/I-module. Just define

$$(r + I)n = rn.$$

This is well defined since $r + I = s + I$ implies that $r - s \in I$, which implies that $(r - s)n = 0$. Thus, $rn = sn$. It is trivial to check that this makes N into an R/I-module. Therefore, an R/I-module A is an R-module such that $IA = 0$, and an R-module A with $IA = 0$ is an R/I-module.

4.6.9 Theorem

Let I be an ideal of R, and let M and N be R/I-modules. Then a map $f: M \to N$ is an R/I-module homomorphism if and only if it is an R-module homomorphism. Furthermore, a subgroup S of M is an R-submodule if and only if it is an R/I-submodule.

Proof Suppose $f: M \to N$ is a homomorphism with M and N considered as R-modules. Then $f((r+I)m) = f(rm) = r(f(m)) = (r+I)f(m)$, whence f is an R/I-homomorphism. Similarly, if f is an R/I-homomorphism, then f is an R-homomorphism.

Since $(r+I)s = rs$, a subgroup S of M is an R-submodule if and only if it is an R/I-submodule.

One can say more than is in 4.6.9. For example, M is finitely generated as an R-module if and only if it is finitely generated as an R/I-module. Also, S is a summand of M as an R-module if and only if it is a summand of M as an R/I-module. These are readily verified.

What good is all this? Suppose that I is a maximal ideal in the commutative ring R, and suppose that M is an R-module such that $IM = 0$. Then M is an R/I-module via $(r+I)m = rm$. But R/I is a field. Thus M is a vector space over the field R/I, and we know a few things about vector spaces. For example, if M is finitely generated as an R-module, then it is a finite-dimensional vector space over R/I, and hence is a direct sum of copies of R/I. But such a decomposition is also an R-module decomposition. Thus, the R-module M is a direct sum of copies of the R-module R/I. Furthermore, this situation can come about readily. In fact, let R be any commutative ring, and let M be any R-module. For any maximal ideal I of R,

$$I(M/IM) = 0,$$

so that M/IM is a vector space over the field R/I. If M is a finitely generated R-module, then M/IM is a finite-dimensional vector space over R/I. We will use all this in the next section.

PROBLEMS

1 Let S be a subring of the ring R. Prove that an R-module is an S-module. In particular, R is an S-module.

2 Let $\varphi: S \to R$ be a ring homomorphism with $\varphi(1) = 1$. Let M be an R-module. Prove that scalar multiplication defined by $s \cdot m = \varphi(s)m$ also makes M into an S-module.

3 In Problem 2, let $R = S$, and let φ be an automorphism of the ring R.

Thus, M is a module over R in two ways. Are these two modules isomorphic?

4 In Problem 2, let $R = S = M$, and let φ be an automorphism of the ring R. Thus, R is a module over R in two ways. Are these two modules isomorphic?

5 Prove that there is only one way to make an Abelian group into a module over $\mathbb{Z}$.

6 Prove that there is at most one way to make an Abelian group into a module over $\mathbb{Q}$.

7 Prove that every finite module over $\mathbb{Q}$ is $\{0\}$.

8 Prove that the additive group of $\mathbb{Z}$ cannot be made into a module over $\mathbb{Q}$.

9 Prove that $R[x]$ is an R-module. Prove that

$$\{p(x) \in R[x]: \deg(p(x)) < n\}$$

is a submodule.

10 Let N be a submodule of M. Prove that if M is finitely generated, then M/N is finitely generated. Prove that M is finitely generated if N and M/N are finitely generated.

11 Show that a submodule of a finitely generated module is not necessarily finitely generated.

12 Prove that if $M = A \oplus B$, and that if C is a submodule of M containing A, then $C = A \oplus (C \cap B)$.

13 Let M be a module over R. Let $0{:}M = \{r \in R: rM = \{0\}\}$. Prove that $0{:}M$ is a (two-sided) ideal of R, and prove that M is a module over the ring $R/0{:}M$.

14 Let M be an R-module, and let $\mathrm{End}_{\mathbb{Z}}(M)$ be the ring of all group endomorphisms of M. For $r \in R$, define φ_r by

$$\varphi_r\colon M \to N\colon m \to rm.$$

Prove that $\varphi_r \in \mathrm{End}_{\mathbb{Z}}(M)$, and prove that

$$R \to \mathrm{End}_{\mathbb{Z}}(M)\colon r \to \varphi_r$$

is a ring homomorphism.

15 Let R be a ring, and let $r \in R$. Let $\varphi_r\colon R \to R\colon s \to sr$. Prove that φ_r is an endomorphism of the R-module R. Prove that every endomorphism of the module R is of this form. Prove that $f\colon R \to \mathrm{End}_R(R)\colon r \to \varphi_r$ is a ring anti-isomorphism. That is, f is one-to-one and onto, preserves addition, and reverses multiplication.

16 Let $\mathrm{Hom}_R(M, N)$ be the set of all homomorphisms from the R-module M into the R-module N. Prove that

$$(f + g)(m) = f(m) + g(m)$$

makes $\mathrm{Hom}_R(M, N)$ into an Abelian group.

17 Prove that $\mathrm{Hom}_R(R, M)$ is a left R module if scalar multiplication is defined by $(rf)(s) = f(sr)$. Prove that $M \approx \mathrm{Hom}_R(R, M)$.

18 Let M be a (left) R-module. For $f \in \mathrm{Hom}_R(M, R)$ and $r \in R$, let $(fr)(m) = f(m)r$. Prove that this makes $\mathrm{Hom}_R(M, R)$ into a *right* R-module.

19 Let S be any ring, and let R be the ring of all $n \times n$ matrices over S. Let M be the set of all $n \times m$ matrices over S.

(a) Prove that M is a left module over R under matrix addition in M and with multiplication of elements of R by those of M given by matrix multiplication.

(b) Let

$$M_k = \{(a_{ij}) \in R: a_{ij} = 0 \text{ if } j \neq k\}.$$

Prove that M_k is a submodule, and prove that $M \approx M_1 \oplus M_2 \oplus \cdots \oplus M_m$.

(c) Let C be the left R module of all $n \times 1$ matrices over S. Prove that $C \approx M_1 \approx M_2 \approx \cdots \approx M_m$.

(d) Let m_{ij} be that element of M_i whose (j, i) entry is 1 and all of whose other entries are 0. Prove that $M_i = Rm_{ij}$.

(e) Find $0{:}m_j$.

(f) Prove that the $R/0{:}m_{ij}$ are all isomorphic.

(g) Prove that $0{:}M_i = \{0\}$.

(h) Prove that if S is a field, then the modules M_i are ***simple***, that is, have no submodules except M_i and $\{0\}$.

20 For I a left ideal of R and $r \in R$, let

$$I{:}r = \{x \in R: xr \in I\}.$$

Prove that $I{:}r$ is a left ideal of R. Prove that the submodule $R(r + I)$ of R/I is isomorphic to $R/(I{:}r)$.

21 Let I and J be left ideals of the ring R. Prove that the R-modules R/I and R/J are isomorphic if and only if $I = J{:}s$ for some generator $s + J$ of R/J.

22 Prove that a cyclic module may have different kinds of generators. That is, prove that there exists a cyclic R-module M with generators x and y, such that no automorphism of M takes x onto y.

4.7 MODULES OVER PRINCIPAL IDEAL DOMAINS

The purpose of this section is to prove that a finitely generated module over a PID is a direct sum of cyclic modules, and to get a nice, simple, complete invariant for such modules. This theorem generalizes both the **Fundamental Theorem of Finite Abelian Groups** (2.7.7) and the fact that a finitely generated vector space is a direct sum of subspaces of dimension 1 (3.3.12). It will be our principal tool in analyzing linear transformations in Chapter 5.

4.7.1 Definition

Let R be a PID, *and let M be an R-module. An element m in M is a* **torsion element** *if* $0{:}m \neq \{0\}$, *and is a* **torsion-free element** if $0{:}m = \{0\}$. *The module M is a* **torsion module** *if every element in it is a torsion element, and M is a* **torsion-free module** *if* 0 *is the only torsion element in it.*

Let m and n be torsion elements of M. Let $r \in 0{:}m$ and $s \in 0{:}n$, with r and s nonzero. Then $rs \in 0{:}(m+n)$, and $rs \neq 0$. Thus, $m+n$ is a torsion element. For any $t \in R$, $r \in 0{:}tm$. Hence, tm is a torsion element. We have proved the following.

4.7.2 Theorem

Let R be a PID, *and let M be an R-module. The set M_t of torsion elements of M is a submodule of M. (It is called the* **torsion submodule** *of M.)*

Observe that M/M_t is torsion free. In fact, if r is in $0{:}(m+M_t)$, then $rm \in M_t$, whence $srm = 0$ for some $s \neq 0$. If $r \neq 0$, then $sr \neq 0$, so that $0 \neq sr \in 0{:}m$, and hence $m \in M_t$. Therefore, if $m + M_t \neq 0$, then $r = 0$.

We will proceed as follows. We will prove that finitely generated torsion-free modules over PID's are direct sums of cyclic modules, and that finitely generated torsion modules over PID's are direct sums of cyclic modules. From these two facts, we will conclude that arbitrary finitely generated modules over PID's are direct sums of cyclic modules by showing that any finitely generated module over a PID is a direct sum of its torsion submodule and a torsion-free module.

Notice that any submodule of R is torsion free. Also, any nonzero submodule of R is isomorphic to R. If I is a nonzero submodule of R, then $I = Ra$ for some $a \in R$ since R is a PID. The map $R \to Ra{:}\, r \to ra$ is an isomorphism. Thus, submodules of R are cyclic, and nonzero ones are all free, in fact, all isomorphic to R. Any quotient module R/I with $I \neq 0$ is a torsion module since $I = Ra$ for some $a \neq 0$, and $a \in 0{:}(r+I)$ for all $r \in R$. Finally, notice that if F is the quotient field of R, then F is a torsion-free R-module.

4.7.3 Theorem

A finitely generated torsion-free module M over a PID *is free of finite rank.*

Proof Let M be a finitely generated torsion-free module over the PID R. If M is cyclic, then $M = Rm$ for some $m \in M$. If $m = 0$, we are done. If

$m \neq 0$, then the map $R \to Rm: r \to rm$ is an isomorphism because $rm \neq 0$ if $r \neq 0$. Thus, $M \approx R$ is free of rank 1. Now induct on the smallest number of elements required to generate M. If some generating set of M has only 1 element in it, we just proved that M is free. Suppose now that M is generated by the $k+1$ elements

$$m_1, m_2, \ldots, m_{k+1},$$

and suppose that any torsion-free module over R generated by k elements is free of rank $\leq k$. We may as well suppose that $m_1 \neq 0$. Let $M_1 = \{m \in M: rm \in Rm_1 \text{ for some nonzero } r \in R\}$. That is, $M_1/Rm_1 = (M/Rm_1)_t$. We have $(M/Rm_1)/(M_1/Rm_1) \approx M/M_1$ and torsion free since $M_1/Rm_1 = (M/Rm_1)_t$. Now M/M_1 is generated by the k elements $m_2 + M_1, m_3 + M_1, \ldots, m_{k+1} + M_1$. Therefore, by the induction hypothesis, M/M_1 is free of rank $\leq k$. By 4.6.7, $M = M_1 \oplus N$ with N free of rank $\leq k$. We need M_1 to be free of rank 1; that is, cyclic. In that case, M would be free. First, note that M_1 is finitely generated since it is a homomorphic image of M. Let $m \in M_1$. Then $rm = sm_1$ for some $r \neq 0$. Suppose that $r_1 m = s_1 m_1$ with $r_1 \neq 0$. Then $rr_1 m = rs_1 m_1 = r_1 s m_1$. Hence $(rs_1 - r_1 s)m_1 = 0$. Since $m_1 \neq 0$ and M_1 is torsion free, $rs_1 = r_1 s$, and $s/r = s_1/r_1$ is in the quotient field F of R. What this says is that associating s/r with $m \in M_1$, where $rm = sm_1$ and $r \neq 0$, is a mapping $M_1 \to F$. Now F is an R-module, and it is routine to check that this mapping is a monomorphism.

The following lemma, which generalizes Problem 28 of Section 2.2, completes the proof.

4.7.4 Lemma

Let R be a PID, *and let F be its quotient field. Then any finitely generated R-submodule of F is cyclic (and hence free).*

Proof Let S be a finitely generated submodule of the R-module F. Let $S = Rs_1 + Rs_2 + \cdots + Rs_n$. Write $s_i = r_i/t_i$ with r_i and t_i in R and $t_i \neq 0$. Let $t = \prod_{i=1}^{n} t_i$. Then S is contained in $R(1/t)$. But the cyclic R-module $R(1/t)$ is isomorphic to R via $R \to R(1/t): r \to r/t$. Therefore, S is isomorphic to a submodule of R. Hence, S is cyclic.

Now we come to an important point. We know that if M is a finitely generated torsion-free module over the PID R, then $M = M_1 \oplus M_2 \oplus \cdots \oplus M_k$, with each $M_i \approx R$. Is k an invariant of M? That is, if $M = N_1 \oplus N_2 \oplus \cdots \oplus N_j$, with each $N_i \approx R$, is $j = k$? We know that the M_i's do not have to be the N_i's. For example, any vector space of dimension 2 is the direct sum of any two distinct one-dimensional subspaces. However, in the case of vector spaces, that is, in case R is a field, then $j = k$. Also notice the following. The R-module R is directly ***indecomposable.*** That is, if $R = R_1 \oplus R_2$, then R_1 or R_2 is $\{0\}$. This is simply because $R_1 \cap R_2 = 0$, and any two nonzero submodules of R have nonzero intersection. In fact, if $0 \neq r_i \in R_i$, then $r_1 r_2 \neq 0$, and $r_1 r_2$ is in $R_1 \cap R_2$. Suppose that $M = A_1 \oplus A_2 \oplus \cdots \oplus A_q$ with each A_i indecomposable. Is $A_i \approx R$? Indeed, if $M = A \oplus B$, is A free, or is A even a direct sum of indecomposable modules? What is the situation if S is a submodule of M? We know the answers to all these questions when R is a field. Any subspace of a finite-dimensional vector space is of no greater dimension, and hence is the direct sum of copies of the field. The answers in the case R is a PID are the best possible, but this is far from obvious. It is not at all obvious that a submodule of a finitely generated R-module is finitely generated, for example. This is really the crucial question, along with whether or not $j = k$ above. Let us get $j = k$ first. There are several ways to do it. One way is to define linear independence in M and mimic vector space methods. We choose another which illustrates the use of our previous results.

4.7.5 Theorem

Let M be a free module over the PID *R. If $M = M_1 \oplus M_2 \oplus \cdots \oplus M_j = N_1 \oplus N_2 \oplus \cdots \oplus N_k$ with each M_i and N_i isomorphic to R, then $j = k$.*

Proof If R is a field, then $j = k = \dim(M)$. So suppose that R is not a field. Then there is a prime p in R. The ideal Rp of R is maximal (4.4.4), whence R/Rp is a field. Furthermore, pM is a submodule of M and M/pM is a vector space over R/Rp via $(r + Rp)(m + pM) = rm + pM$. Now

$$pM = p(M_1 \oplus M_2 \oplus \cdots \oplus M_j) = pM_1 \oplus pM_2 \oplus \cdots \oplus pM_j.$$

Therefore,

$$\begin{aligned} M/pM &= (M_1 \oplus M_2 \oplus \cdots \oplus M_j)/(pM_1 \oplus pM_2 \oplus \cdots \oplus pM_j) \\ &\approx M_1/pM_1 \oplus M_2/pM_2 \oplus \cdots \oplus M_j/pM_j. \end{aligned}$$

This last isomorphism is a vector space isomorphism. Since $M_i \approx R$, then $M_i/pM_i \approx R/Rp$, so that the vector space M_i/pM_i has dimension 1. That is, $\dim(M/pM) = j$. Therefore, $j = k$.

If M is a finitely generated torsion-free module, it is free, and completely determined up to isomorphism by the number of summands in any decomposition of M into a direct sum of submodules isomorphic to R. If R is a field, that number is $\dim(M)$. If R is not a field, that number is $\dim(M/pM)$ for any prime p in R. This number is called the ***rank*** of M, denoted by $r(M)$. The rank of the finitely generated torsion-free module M is j if and only if M is the direct sum of j copies of R. We have

4.7.6 Theorem

Two finitely generated torsion-free modules over a PID *are isomorphic if and only if they have the same rank.*

Let us clarify the situation with submodules of finitely generated R-modules. The key is this. If R is a PID, then any submodule of R is cyclic. That is the definition of PID, in fact.

4.7.7 Theorem

Let M be a free module of rank n over a PID *R. Then any submodule of M is free and of rank $\leq n$.*

Proof Induct on n. If $n = 1$, then $M \approx R$, and any nonzero submodule of R is free of rank 1 since R is a PID. Let $M = M_1 \oplus M_2 \oplus \cdots \oplus M_n$ with $M_i \approx R$, and let S be a submodule of M. Project S into M_n. The image X of this projection is free of rank ≤ 1. Its kernel is $K = S \cap (M_1 \oplus M_2 \oplus \cdots \oplus M_{n-1})$, which is free of rank $\leq n - 1$. By 4.6.7, $S = K \oplus X$, and the theorem follows.

4.7.8 Theorem

Let M be a module over the PID *R. If M is generated by n elements, then any submodule of M is generated by $\leq n$ elements.*

Proof Let $M = Rm_1 + Rm_2 + \cdots + Rm_n$. Let X be a free R-module of rank n with basis $\{x_1, x_2, \ldots, x_n\}$. Then by 4.6.6 $f: X \to M: x_i \to m_i$ defines an epimorphism. Let S be a submodule of M. Then $f^{-1}(S) = \{x \in X: f(x) \in S\}$ is a submodule of X. It is free of rank $\leq n$ by 4.7.7. In particular, it is generated by $\leq n$ elements. Hence, S is generated by $\leq n$ elements.

4.7.9 Corollary

A submodule of a cyclic module over a PID *is cyclic.*

Now let us turn to the case of finitely generated torsion R-modules. Suppose that M is such a module, and suppose that $\{m_1, m_2, \ldots, m_k\}$

generates M. Since M is a torsion module, there exist nonzero r_i's such that $r_i m_i = 0$. Then $(r_1 r_2 \cdots r_k) m_i = 0$ for all i. If $m \in M$, then $m = s_1 m_1 + s_2 m_2 + \cdots + s_k m_k$ for some $s_i \in R$, and it follows that $(r_1 r_2 \cdots r_k) m = 0$. That is, $(r_1 r_2 \cdots r_k) M = 0$. But $r_1 r_2 \cdots r_k \neq 0$. We have that

$$0{:}M = \{r \in R : rM = 0\} \neq \{0\}.$$

This ideal is called the ***annihilator*** of M. Thus, *a finitely generated torsion module M has nonzero annihilator*. For finite Abelian groups, this simply says that there is a positive integer n such that $nx = 0$ for all x in that group.

4.7.10 Lemma

Let M be a module over a PID *R*.

(a) *Let $x \in M$ and $0{:}x = Ra$. If $b \in R$ and d is a greatest common divisor of a and b, then $0{:}bx = Rc$ where $a = dc$.*
(b) *An element $x \in M$ has annihilator $0{:}x = R(\prod_{i=1}^{n} r_i)$ with the r_i relatively prime in pairs if and only if there are elements $x_1, x_2, \ldots, x_n$ in M with $x = x_1 + x_2 + \cdots + x_n$ and $0{:}x_i = Rr_i$.*

Proof For (a), clearly $Rc \subset 0{:}bx$. If $q \in 0{:}bx$, then $qb \in 0{:}x = Ra$, so $qb = ra$ for some $r \in R$. Write $d = sa + tb$. Then $qd = saq + tbq = saq + tra = cd(sq + tr)$, so $q = c(sq + tr) \in Rc$.

For (b), let $q_i = \prod_{j \neq i} r_j$. Then the collection $\{q_1, q_2, \ldots, q_n\}$ has greatest common divisor equal to 1, so there is a linear combination $\sum s_i q_i = 1$. Now r_i and $s_i q_i$ are relatively prime, so $(r_i q_i, s_i q_i) = q_i$. Thus by part (a), $0{:}s_i q_i x = Rr_i$. Let $x_i = s_i q_i x$.

For the converse of (b), we will induct on n. If $n = 1$, the Lemma obviously holds. The crucial case is for $n = 2$, and we do that case now. Let x and y be elements with $0{:}x = Rr$ and $0{:}y = Rs$, with r and s relatively prime. There are elements a and b in R with $ar + bs = 1$. Hence $ar(x + y) = ary = (1 - bs)y = y$, and $bs(x + y) = x$. Suppose that $t(x + y) = 0$. Then $art(x + y) = 0 = ty$, so that $t \in 0{:}y = Rs$. Thus, s divides t. Similarly, r divides t, and since r and s are relatively prime, rs divides t. It follows that $0{:}(x + y) = Rrs$. Now for $n > 2$, letting $y = \sum_{i=1}^{n-1} x_i$, we have $0{:}(\sum_{i=1}^{n} x_i) = 0{:}(y + x_n)$. By the induction hypothesis, $0{:}y = R(\prod_{i=1}^{n-1} r_i)$. Since the r_i are relatively prime in pairs, $\prod_{i=1}^{n-1} r_i$ is relatively prime to r_n, and so $0{:}(y + x_n) = 0{:}(\sum_{i=1}^{n} x_i) = R(\prod_{i=1}^{n} r_i)$.

4.7.11 Lemma

If M is a finitely generated torsion module over a PID, *then $0{:}M = 0{:}m$ for some $m \in M$.*

Proof Since M is finitely generated and torsion, $0{:}M \neq \{0\}$. Let

$$0{:}M = Rp_1^{n_1}p_2^{n_2} \cdots p_t^{n_t},$$

with the p_i's nonassociate primes. Let

$$r = p_1^{n_1}p_2^{n_2} \cdots p_t^{n_t},$$

and let $q_i = r/p_i$. There is an element $m_i \in M$ such that $q_i m_i \neq 0$. Let

$$x_i = (r/p_i^{n_i})m_i.$$

Since

$$p_i^{n_i-1}x_i = (r/p_i)m_i \neq 0,$$

and

$$p_i^{n_i}x_i = rm_i = 0,$$

then $0{:}x_i = Rp_i^{n_i}$. By 4.7.10,

$$0{:}(x_1 + x_2 + \cdots + x_t) = Rp_1^{n_1}p_2^{n_2} \cdots p_t^{n_t},$$

and the lemma is proved.

If G is a finite Abelian group, then there is an element x in G whose order m is a multiple of the order of every other element of G. That is, $my = 0$ for every y in G.

4.7.12 Theorem

Every finitely generated torsion module over a PID *is a direct sum* $C_1 \oplus C_2 \oplus \cdots \oplus C_t$ *of cyclic modules such that* $0{:}C_1 \subset 0{:}C_2 \subset \cdots \subset 0{:}C_t$.

Proof Let M be a finitely generated torsion module over the PID R. By 4.7.11, $0{:}M = Rr$, with $r = p_1^{n_1}p_2^{n_2} \cdots p_t^{n_t}$. Let $\{m_1, m_2, \ldots, m_k\}$ generate M. We proceed by induction on k. If $k = 1$, there is nothing to do. So suppose $k > 1$ and that the theorem is true for every R-module generated by fewer than k elements. Each $0{:}m_i = Rr_i$, where

$$r_i = p_1^{n_{i1}}p_2^{n_{i2}} \cdots p_t^{n_{it}}$$

with $n_{ij} \leq n_j$. For each j, some $n_{ij} = n_j$. Thus,

$$0{:}(r_i/p_j^{n_{ij}})m_i = Rp_j^{n_j}.$$

Hence for each n_j there is a multiple x_j of some m_i such that

$$0{:}x_j = Rp_j^{n_j}.$$

Renumber the m_i's so that $n_{11} = n_1$, $n_{12} = n_2, \ldots, n_{1q} = n_q$, $q > 1$, and

$n_{1i} < n_i$ if $i > q$. Now by 4.7.10, $m_1 = m_{11} + m_{12}$ where

$$0{:}m_{11} = Rp_1^{n_1}p_2^{n_2} \cdots p_q^{n_q},$$

and

$$0{:}m_{12} = Rp_{q+1}^{n_{1,q+1}}p_{q+2}^{n_{1,q+2}} \cdots p_t^{n_{1t}},$$

with $n_{1i} < n_i$ for $i > q$. Thus by Problem 9,

$$0{:}(m_{12} + x_{q+1} + \cdots + x_t) = Rp_{q+1}^{n_{q+1}} \cdots p_t^{n_t}.$$

Let $m = m_1 + x_{q+1} + \cdots + x_t$. Then $0{:}m = Rr$, and clearly $\{m, m_2, \ldots, m_k\}$ generates M. Now

$$M/Rm = (M_1/Rm) \oplus (M_2/Rm) \oplus \cdots \oplus (M_v/Rm)$$

with M_i/Rm cyclic and $0{:}(M_1/Rm) \subset 0{:}(M_2/Rm) \subset \cdots \subset 0{:}(M_v/Rm)$. Let $M_i/Rm = R(w_i + Rm)$, and let $0{:}(w_i + Rm) = Rs_i$. Then $s_iw_i = u_im$ for some $u_i \in R$. Also s_i divides r. We have $s_it_i = r$, $s_it_iw_i = t_iu_im = 0$, so that $t_iu_i = rr_i$, and $s_it_iu_i = s_irr_i = ru_i$. Hence $s_ir_i = u_i$. We have $s_iw_i = u_im = s_ir_im$, so $s_i(w_i - r_im) = 0$. Let $w_i - r_im = y_i$. Then $M_i = Ry_i + Rm$ and $Ry_i \cap Rm = \{0\}$. Therefore, $M_i = Ry_i \oplus Rm$, and hence $M = Rm \oplus Ry_1 \oplus \cdots \oplus Ry_v$. Since $Ry_i \approx M_i/Rm$, then $0{:}Rm \subset 0{:}Ry_1 \subset 0{:}Ry_2 \subset \cdots \subset Ry_v$, and the theorem is proved.

4.7.13 Theorem

A finitely generated module over a PID *is a direct sum* $C_1 \oplus C_2 \oplus \cdots \oplus C_n$ *of cyclic modules such that* $0{:}C_1 \subset 0{:}C_2 \subset \cdots \subset 0{:}C_n$. *If*

$$C_1 \oplus C_2 \oplus \cdots \oplus C_n = D_1 \oplus D_2 \oplus \cdots \oplus D_m$$

with C_i *and* D_i *nonzero cyclic and* $0{:}D_1 \subset 0{:}D_2 \subset \cdots \subset 0{:}D_m$, *then* $m = n$, $0{:}C_i = 0{:}D_i$, *and* $C_i \approx D_i$ *for all i.*

Proof Let M be a finitely generated module over the PID R, and let M_t be the torsion submodule of R. Then M/M_t is a torsion-free finitely generated module. Hence, by 4.6.7, $M = F \oplus M_t$ with F free of finite rank. Now M_t is finitely generated since it is a homomorphic image of M. Since $F = C_1 \oplus C_2 \oplus \cdots \oplus C_k$ with $C_i \approx R$, then $0{:}C_i = 0$, and 4.7.12 yields

$$M = C_1 \oplus C_2 \oplus \cdots \oplus C_n$$

with $0{:}C_1 \subset 0{:}C_2 \subset \cdots \subset 0{:}C_n$. Now suppose that

$$M = D_1 \oplus D_2 \oplus \cdots \oplus D_m$$

with $0{:}D_1 \subset 0{:}D_2 \subset \cdots \subset 0{:}D_m$, and all C_i and D_i are nonzero. We need

$m = n$ and $C_i \approx D_i$ for all i. The number of C_i with $0{:}C_i = 0$ is the rank of M/M_t, and hence is the number of D_i with $0{:}D_i = 0$. Thus, we may as well assume that M is torsion. Let $0{:}C_i = Rr_i$, and let $0{:}D_i = Rs_i$. First we get $m = n$. Let p be a prime, and let $r \in R$. If p divides r, then $pR \supset Rr$. Hence $(R/Rr)/p(R/Rr) = (R/Rr)/(Rp/Rr) \approx R/Rp$, which has dimension 1 as a vector space over R/Rp. If $(p, r) = 1$, then $(R/Rr)/p(R/Rr) = (R/Rr)/((Rp + Rr)/Rr) = (R/Rr)/(R/Rr) = 0$. Suppose now that $n \geq m$. Let p be a prime dividing r_n. There is one since $C_n \neq 0$. Then p divides r_i for all i since $Rr_1 \subset Rr_2 \subset \cdots \subset Rr_n$. Therefore, $(R/Rr_i)/(Rp/Rr_i) \approx R/Rp \approx C_i/pC_i$ has dimension 1 as a vector space over the field R/Rp. Hence,

$$\begin{aligned}\dim(M/pM) &= \dim((C_1 \oplus C_2 \oplus \cdots \oplus C_n)/p(C_1 \oplus C_2 \oplus \cdots \oplus C_n)) \\ &= \dim((C_1/pC_1) \oplus (C_2/pC_2) \oplus \cdots \oplus (C_n/pC_n)) = n.\end{aligned}$$

It follows that $m = n$. We want $C_j \approx D_j$ for each j. Letting $0{:}C_j = Rr_j$ as above, we have

$$r_jM = r_jC_1 \oplus r_jC_2 \oplus \cdots \oplus r_jC_{j-1} = r_jD_1 \oplus r_jD_2 \oplus \cdots \oplus r_jD_n.$$

The r_jC_i and r_jD_i are cyclic with

$$0{:}r_jC_1 \subset 0{:}r_jC_2 \subset \cdots \subset 0{:}r_jC_{j-1}$$

and $0{:}r_jD_1 \subset 0{:}r_jD_2 \subset \cdots \subset 0{:}r_jD_n$. Therefore $r_jD_j = 0$ by the first part of the proof, whence s_j divides r_j. Similarly, r_j divides s_j. Therefore $0{:}C_j = 0{:}D_j$, and so $C_j \approx D_j$. This concludes the proof.

For any commutative ring R and ideals I and J of R, the cyclic R-modules R/I and R/J are isomorphic if and only if $I = J$ (Problem 10). Therefore, cyclic R-modules are, up to isomorphism, in one-to-one correspondence with ideals of R. What 4.7.13 asserts is that if R is a PID, then the nonzero finitely generated R-modules are, up to isomorphism, in one-to-one correspondence with finite increasing chains $I_1 \subset I_2 \subset \cdots \subset I_n$ of ideals of R, where each $I_j \neq R$. In fact, 4.7.13 gives such a chain for each such module. Since $C_i \approx R/(0{:}C_i)$, the module can readily be retrieved, up to isomorphism, from the chain of ideals. It is

$$R/(0{:}C_1) \oplus R/(0{:}C_2) \oplus \cdots \oplus R/(0{:}C_n).$$

Of course, any finite increasing chain $I_1 \subset I_2 \subset \cdots \subset I_n$ gives a finitely generated R-module, namely $R/I_1 \oplus R/I_2 \oplus \cdots \oplus R/I_n$.

4.7.14 Corollary

Nonzero finitely generated modules over a PID R *are, up to isomorphism, in one-to-one correspondence with finite increasing chains* $I_1 \subset I_2 \subset \cdots \subset I_n$

of ideals of R such that each $I_j \neq R$. Such a chain corresponds to the module

$$R/I_1 \oplus R/I_2 \oplus \cdots \oplus R/I_n.$$

If M is torsion and nonzero, then the chain of ideals

$$0{:}C_1 \subset 0{:}C_2 \subset \cdots \subset 0{:}C_n$$

in 4.7.13 is called the chain of ***invariant factors*** of M. Letting the invariant factor of the zero module be R and rephrasing, we have

4.7.15 Theorem

Two finitely generated torsion modules over a PID *are isomorphic if and only if they have the same invariant factors.*

How does one calculate the invariant factors of a finitely generated torsion module without actually decomposing that module as in 4.7.13? In the notation of 4.7.13, if $0{:}C_i = Rr_i$, and if p is a prime, then it is easy to see that $\dim(p^tC_i/p^{t+1}C_i) = 1$ or 0, depending on whether p^t does or does not divide r_i. In fact, we did something similar in the proof of 4.7.13. Therefore, $\dim(p^tM/p^{t+1}M)$ is the number of r_i which p^t divides. Therefore, the r_i, and hence the invariant factors, are computable directly from M, without first writing M as a direct sum of cyclics.

Again, in the notation of 4.7.13, let $0{:}C_i = Rr_i$. Knowing r_i gives $0{:}C_i$, but $0{:}C_i$ gives r_i only up to a unit. That is, associates of r_i will work just as well. That is why we use the ideal $0{:}C_i$ as an invariant rather than the r_i. However, for some PID's, every associate class has a canonical representative. For example, in the ring of integers, every associate class not $\{0\}$ has exactly one positive element. If F is a field, then every associate class not $\{0\}$ in $F[x]$ has exactly one monic element. In these situations, one usually chooses to use families $\{r_1, r_2, \ldots, r_n\}$ with r_{i+1} dividing r_i rather than chains of ideals as invariants. Such families are also called ***invariant factors***. The following illustrates this. Its proof is immediate from 4.7.14.

4.7.16 Theorem

Let F be a field. The finitely generated nonzero torsion $F[x]$ modules are, up to isomorphism, in one-to-one correspondence with the finite families

$$\{f_1(x), f_2(x), \ldots, f_n(x)\}$$

of nonconstant monic polynomials such that $f_{i+1}(x)$ divides $f_i(x)$ for all $i < n$. Such a family corresponds to the module $F[x]/(f_1(x)) \oplus F[x]/(f_2(x)) \oplus \cdots \oplus F[x]/(f_n(x))$.

One should compare 4.7.16 with the situation for Abelian groups. See the discussion preceding 2.7.9, for example.

We have given a "canonical" decomposition of a finitely generated torsion module over a PID. Such a nonzero module can be written $M = C_1 \oplus C_2 \oplus \cdots \oplus C_n$ with C_i nonzero cyclic and $0{:}C_1 \subset 0{:}C_2 \subset \cdots \subset 0{:}C_n$. The C_i are unique up to isomorphism. There is another "standard" decomposition of M into a direct sum of cyclics. We need some preliminaries.

4.7.17 Definition

Let M be a module over a PID *R. If p is a prime in R, then* $M_p = \{m \in M\colon 0{:}m = Rp^n$ for some $n\}$. *M_p is called the p-**component** of M. If $M = M_p$, then M is a p-**module**, or a p-**primary** module. A **primary** module is one that is p-primary for some p.*

Note that if p and q are associates, then $M_p = M_q$. Also if p and q are any primes and $M_p = M_q \neq 0$, then p and q are associates. One should note that M_p is a submodule and that $(M/M_p)_p = 0$. Also, submodules, quotient modules, and direct sums of p-modules are p-modules.

4.7.18 Theorem

Every finitely generated torsion module over a PID *is a direct sum of primary modules.*

Proof It suffices to prove the theorem for cyclic modules. So let $M = Rm$, $0{:}m = Rr$, and

$$r = p_1^{n_1} p_2^{n_2} \cdots p_k^{n_k},$$

where the p_i are nonassociates. Let $q_i = r/p_i^{n_i}$. Then

$$0{:}q_i m = Rp_i^{n_i},$$

so that $Rq_i m$ is a p_i-module. We claim that

$$M = Rq_1 m \oplus Rq_2 m \oplus \cdots \oplus Rq_k m.$$

Since the q_i's are relatively prime, there exist $r_i \in R$ such that $\sum r_i q_i = 1$. Hence $m = \sum r_i q_i m$, so that

$$M = Rq_1 m + Rq_2 m + \cdots + Rq_k m.$$

Suppose that $\sum s_i q_i m = 0$. Then

$$0{:}s_1 q_1 m = 0{:} \sum_{i=2}^{k} s_i q_i m.$$

From 4.7.10, it follows that each $s_i q_i m = 0$. Our theorem is proved.

Suppose that M is a finitely generated torsion module over a PID, and suppose that $0{:}M = Rr$. Write r as a product of powers of nonassociate primes $p_1, p_2, \ldots, p_k$. Writing

$$M = C_1 \oplus C_2 \oplus \cdots \oplus C_n$$

with $0{:}C_1 \subset 0{:}C_2 \subset \cdots \subset 0{:}C_n$, and in turn writing each C_i as a direct sum of primary modules as in the proof of 4.7.18, we see that

$$M = M_{p_1} \oplus M_{p_2} \oplus \cdots \oplus M_{p_k}.$$

4.7.19 Corollary

Let M be a finitely generated torsion module over a PID R. *If $0{:}M = Rr$, and*

$$r = p_1^{n_1} p_2^{n_2} \cdots p_k^{n_k}$$

where the p_i are nonassociate primes, then

$$M = M_{p_1} \oplus M_{p_2} \oplus \cdots \oplus M_{p_k}.$$

If $M = P_1 \oplus P_2 \oplus \cdots \oplus P_t$ where the P_i are nonzero primary modules for nonassociate primes, then $k = t$, and after suitable renumbering, $P_i = M_{p_i}$ for all i.

Proof Only the last assertion needs verifying, and it should be obvious at this point.

4.7.20 Corollary

Two finitely generated torsion modules M and N over a PID R *are isomorphic if and only if $M_p \approx N_p$ for all primes $p \in R$.*

The invariant factors of the various M_p are called the ***elementary divisors*** of M.

4.7.21 Corollary

Two finitely generated torsion modules over a PID *are isomorphic if and only if they have the same elementary divisors.*

PROBLEMS

1 Prove that a finitely generated torsion module over $\mathbb{Z}$ is finite.

2 Prove that a finitely generated torsion module over a PID is not necessarily finite.

3 Prove that quotient modules of torsion modules are torsion modules.

4 Prove that direct sums of torsion modules are torsion modules.

5 Prove that direct sums of torsion-free modules are torsion-free modules.

6 Prove that if M is a free module (of finite rank) over a PID and if S is a submodule of M, then S is contained in a summand of M having the same rank as S.

7 Let R be a ring such that every left ideal is finitely generated as an R-module. Prove that submodules of finitely generated R-modules are finitely generated.

8 Let R be a PID, and let p be a prime in R. Prove that if $R/Rp^i \approx R/Rp^j$, then $i = j$.

9 Prove that if M is a module over a PID R, x and y are in M with $0{:}x = Rp_1^{m_1}p_2^{m_2} \cdots p_t^{m_t}$, $0{:}y = Rp_1^{n_1}p_2^{n_2} \cdots p_t^{n_t}$, and $0 \leq m_i < n_i$ for all i, then $0{:}(x + y) = 0{:}(y)$.

10 Prove that if R is a commutative ring with identity, I and J are ideals of R, and the R-modules R/I and R/J are isomorphic, then $I = J$.

11 Let S be a submodule or a quotient module of the direct sum $C_1 \oplus C_2 \oplus \cdots \oplus C_n$ of cyclic modules C_i over a PID, where $0{:}C_1 \subset 0{:}C_2 \subset \cdots \subset 0{:}C_n$. Prove that if
$$S = S_1 \oplus S_2 \oplus \cdots \oplus S_m$$
with each S_i nonzero cyclic and $0{:}S_1 \subset 0{:}S_2 \subset \cdots \subset 0{:}S_m$, then $m \leq n$ and $0{:}C_i \subset 0{:}S_i$ for all $i \leq m$.

12 Let M be a finitely generated torsion module over a PID. Let $m \in M$ with $0{:}m = 0{:}M$. Prove that Rm is a summand of M. Prove that Rm is not necessarily a summand if M is not torsion.

13 How does one tell from the invariant factors of two finitely generated torsion modules whether or not one is isomorphic to a summand of the other? To a submodule of the other?

14 Let C be a cyclic module over the PID R. Let M and N be finitely generated R-modules. Prove that if $C \oplus M \approx C \oplus N$, then $M \approx N$. (This is true for arbitrary R-modules M and N.)

15 Let p be a prime in a PID R, and let M be an R-module. Prove that M_p is a submodule of M, and prove that $(M/M_p)_p = \{0\}$.

16 Prove that submodules, quotient modules, and direct sums of p-modules are p-modules.

17 Let M be a finitely generated torsion module over a PID R. Let $0{:}M = Rr$ and
$$r = p_1^{n_1}p_2^{n_2} \cdots p_k^{n_k},$$
with the p_i nonassociate primes. Prove that
$$M_{p_i} = (r/p_i^{n_i})M.$$

18 Let p be a prime in the PID R, and let C be a cyclic p-module over R. Prove that $C \approx R/Rp^n$ for some n.

19 Let C be a cyclic p-primary module with $0{:}C = Rp^n$. Prove that the submodules of C are exactly the submodules in the chain

$$C \supset pC \supset \cdots \supset p^nC = \{0\}.$$

20 Let M be a module over the PID R, and let p be a prime in R. Let $M[p] = \{m \in M : pm = 0\}$. Prove that $M[p]$ is a submodule of M, and hence a vector space over the field R/pR.

21 Let M be a finitely generated module over a PID R. Prove that for any nonzero prime p in R,

$$r(M) = \dim(M/pM) - \dim(M[p]).$$

22 Let p be a prime in the PID R, and let C be a cyclic p-module. Prove that $\dim(C[p]) = 1$. Prove that

$$\dim((p^kC)[p]/(p^{k+1}C)[p]) = 1 \quad \text{or} \quad 0$$

depending on whether or not $C \approx R/p^{k+1}R$.

23 Let M be a finitely generated p-primary module over the PID R. Prove that

$$\dim((p^kM)[p]/(p^{k+1}M)[p])$$

is the number of summands which are isomorphic to $R/p^{k+1}R$ in any decomposition of M into a direct sum of cyclic modules.

5 Linear transformations

5.1 LINEAR TRANSFORMATIONS AND MATRICES

The objects of real interest concerning vector spaces are linear transformations, especially linear transformations from a vector space V into itself. We have already noted in 3.4 that for any vector space V over a field F, the set $\mathrm{Hom}(V, V)$ of all linear transformations from V into V is again a vector space over F. Addition and scalar multiplication were given by $(\alpha + \beta)(v) = \alpha(v) + \beta(v)$, and $(a\alpha)(v) = a(\alpha(v))$, respectively. But the elements of $\mathrm{Hom}(V, V)$ may also be multiplied in a natural way. These elements are functions from V into V, and such functions may be composed. Thus, linear transformations in $\mathrm{Hom}(V, V)$ are multiplied by the rule $(\alpha\beta)(v) = \alpha(\beta(v))$. It is easy to verify that $\alpha\beta$ is in $\mathrm{Hom}(V, V)$ whenever α and β are in $\mathrm{Hom}(V, V)$. This addition and multiplication of elements of $\mathrm{Hom}(V, V)$ makes $\mathrm{Hom}(V, V)$ into a ring with identity. Multiplication is associative because it is composition of functions. The left distributive law holds since

$$\begin{aligned}(\alpha(\beta + \gamma))(v) &= \alpha((\beta + \gamma)(v)) = \alpha(\beta(v) + \gamma(v)) \\ &= \alpha(\beta(v)) + \alpha(\gamma(v)) = (\alpha\beta)(v) + (\alpha\gamma)(v) = (\alpha\beta + \alpha\gamma)(v),\end{aligned}$$

where α, β, and γ are in $\mathrm{Hom}(V, V)$ and v is in V. Similarly, the right distributive law holds. The identity map $1_V\colon V \to V$ is the identity of the ring. For $a \in F$ and $v \in V$,

$$((a\alpha)\beta)(v) = (a\alpha)(\beta(v)) = a(\alpha(\beta(v)) = a((\alpha\beta)(v)) = (\alpha\beta)(av)$$
$$\alpha(\beta(av)) = \alpha(a(\beta(v))) = \alpha((a\beta)(v)) = (\alpha(a\beta))(v).$$

Summing up, Hom(V, V) is at the same time a ring and a vector space over F. The ring addition is the same as the vector space addition. Furthermore, scalar multiplication and ring multiplication are connected by the relations

$$a(\alpha\beta) = (a\alpha)\beta = \alpha(a\beta).$$

All this is expressed by saying that Hom(V, V) is an ***algebra*** over F. This particular algebra is called the ***algebra of linear transformations*** of V. It will be denoted by $A(V)$. The algebra $A(V)$ is an algebraic object of great interest. For example, some of the most beautiful theorems in ring theory center on "rings of linear transformations." We will take a hard look at $A(V)$ as a ring in Chapter 8. Our concern here is not so much with $A(V)$ as a ring or as an algebra, but rather with the elements of $A(V)$. We will not forget that $A(V)$ is an algebra, but we will concentrate on analyzing its elements rather than its algebraic structure.

First, we will show how linear transformations are synonymous with matrices. This will afford us a concrete representation of linear transformations. Results about linear transformations will translate into results about matrices, and vice versa. Throughout, V will be a finite-dimensional vector space over a field F.

Let $\{v_1, v_2, \ldots, v_n\}$ be a basis of V, and let $\alpha \in A(V)$. Then for each v_j, $\alpha(v_i)$ is a linear combination

$$\alpha(v_j) = \sum_{i=1}^{n} a_{ij}v_i$$

of the basis vectors. The effect of α on each v_j determines α. The effect of α on v_j is determined by $a_{1j}, a_{2j}, \ldots, a_{nj}$. Therefore, the matrix

$$\begin{pmatrix} a_{11} & a_{12} & \cdots & a_{1n} \\ a_{21} & a_{22} & \cdots & a_{2n} \\ \vdots & \vdots & & \vdots \\ a_{n1} & a_{n2} & \cdots & a_{nn} \end{pmatrix}$$

completely determines α. It depends, of course, on the particular basis $\{v_1, v_2, \ldots, v_n\}$ chosen.

5.1.1 Definition

Let V be a finite-dimensional vector space over the field F. Let $\{v_1, v_2, \ldots, v_n\}$ be a basis of V, let α be a linear transformation from V into V, and let $\alpha(v_j) = \sum_{i=1}^{n} a_{ij}v_i$. The **matrix of α relative to the basis**

$\{v_1, v_2, \ldots, v_n\}$ *is*

$$\begin{pmatrix} a_{11} & a_{12} & \cdots & a_{1n} \\ a_{21} & a_{22} & \cdots & a_{2n} \\ \vdots & \vdots & & \vdots \\ a_{n1} & a_{n2} & \cdots & a_{nn} \end{pmatrix}.$$

Let F_n denote the set of all such matrices. That is, F_n is the set of all $n \times n$ arrays of elements of F. Elements of F_n are typically denoted (a_{ij}). Thus, (a_{ij}) is the matrix in F_n whose entry in the ith row and jth column is the element a_{ij} of F. Given any $(a_{ij}) \in F_n$, we may define the linear transformation α by $\alpha(v_j) = \sum a_{ij}v_i$. This, together with our previous remarks, shows that the map $A(V) \to F_n$ obtained by associating with $\alpha \in A(V)$ its matrix relative to the (fixed) basis $\{v_1, v_2, \ldots, v_n\}$ is a one-to-one correspondence. Again, this correspondence depends on the basis chosen. Even the order of the basis vectors matters. A permutation of the basis vectors permutes the columns of the matrix, for example.

Now F_n is itself a ring. In 4.1.7(g), it served as such an example. It is actually an algebra. Here is the definition we need.

5.1.2 Definition

Let R be a commutative ring with an identity. Let R_n denote the set of all $n \times n$ matrices over R. For (r_{ij}) and (s_{ij}) in R_n and r in R, we define

(a) $(r_{ij}) + (s_{ij}) = (r_{ij} + s_{ij})$,

(b) $(r_{ij})(s_{ij}) = (\sum_k r_{ik}s_{kj})$, *and*

(c) $r(r_{ij}) = (rr_{ij})$.

Thus to add two matrices, just add their corresponding entries. Multiplication is a bit more complicated. Part (c) defines scalar multiplication, that is, how to multiply an element of R by an element of R_n. That multiplication is just entrywise multiplication. It is routine to check that (a) and (b) make R_n into a ring with identity. Further, (c) makes R_n into a left module over R. Thus, R_n would be a vector space over R if R were a field. Also $r((r_{ij})(s_{ij})) = (r(r_{ij}))(s_{ij}) = (r_{ij})(r(s_{ij}))$, so that R_n is an ***algebra*** over R. Checking these statements is routine.

Let V be a vector space of dimension n over the field F. A basis $\{v_1, v_2, \ldots, v_n\}$ of V yields a one-to-one correspondence $A(V) \to F_n$ via $\alpha \to (a_{ij})$, where $\alpha(v_j) = \sum a_{ij}v_i$. This is a vector space isomorphism. For example, if α corresponds to (a_{ij}) and β to (b_{ij}), then

$$(\alpha + \beta)(v_j) = \alpha(v_j) + \beta(v_j) = \sum a_{ij}v_i + \sum b_{ij}v_i = \sum (a_{ij} + b_{ij})(v_i),$$

so that $\alpha+\beta$ corresponds to $(a_{ij}+b_{ij})=(a_{ij})+(b_{ij})$. Similarly, scalar multiplication is preserved. To what matrix does $\alpha\beta$ correspond? To see, we must compute $(\alpha\beta)(v_j)$. But

$$(\alpha\beta)(v_j)=\alpha(\beta(v_j))=\alpha(\sum b_{kj}v_k)=\sum b_{kj}\alpha(v_k)$$

$$=\sum b_{kj}(\sum a_{ik}v_i)=\sum(\sum b_{kj}a_{ik})(v_i)=\sum(\sum a_{ik}b_{kj})(v_i).$$

Thus $\alpha\beta$ corresponds to the matrix $(a_{ij})(b_{ij})$. Hence, multiplication is preserved under our correspondence. The one-to-one correspondence we have set up is therefore an isomorphism of algebras. Summing up, we have

5.1.3 Theorem

Let V be a vector space of dimension n over the field F. Let $A(V)$ be the algebra of all linear transformations on V, and let F_n be the algebra of all $n\times n$ matrices over F. Let $\{v_1, v_2, \ldots, v_n\}$ be a basis of V. For $\alpha\in A(V)$, let $\alpha(v_j)=\sum a_{ij}v_i$. Then the map

$$\Phi: A(V)\to F_n: \alpha\to(a_{ij})$$

is an algebra isomorphism. That is, it is one-to-one and onto, $\Phi(a\alpha)=a\Phi(\alpha)$, $\Phi(\alpha+\beta)=\Phi(\alpha)+\Phi(\beta)$, and $\Phi(\alpha\beta)=\Phi(\alpha)\Phi(\beta)$.

The units of a ring are those elements with inverses. For example, α is a unit in $A(V)$ if there is an element β in $A(V)$ with $\alpha\beta=\beta\alpha=1$, the identity map on V. Elements in $A(V)$ and elements in F_n that are units are called ***nonsingular***. This is special linear algebra terminology.

There are some facts about nonsingular linear transformations and matrices that we need. First, if (a_{ij}) is the matrix of α relative to some basis, it should be clear from 5.1.3 that α is nonsingular if and only if (a_{ij}) is nonsingular. Indeed, if $\alpha\beta=\beta\alpha=1$, then in the notation of 5.1.3,

$$\Phi(\alpha\beta)=\Phi(\beta\alpha)=1=\Phi(\alpha)\Phi(\beta)=\Phi(\beta)\Phi(\alpha).$$

Thus α is a unit in $A(V)$ if and only if $\Phi(\alpha)$ is a unit in F_n. Actually, for α to be nonsingular, it is enough to require that it has a left inverse, or that it has a right inverse. In fact, suppose that $\alpha\beta=1$. By 3.3.17, $\dim(V)=\dim(\operatorname{Im}\alpha)+\dim(\operatorname{Ker}\alpha)$. Since $\alpha\beta=1$, then α is certainly onto, whence $\dim(\operatorname{Ker}\alpha)=0$, and so α is one-to-one. Thus, α has an inverse. If $\beta\alpha=1$, then α must be one-to-one, and the equation $\dim(V)=\dim(\operatorname{Im}\alpha)+\dim(\operatorname{Ker}\alpha)$ yields $\dim(\operatorname{Im}\alpha)=\dim(V)$. In other words, α is also onto, and hence is nonsingular. Note further that if β is a right inverse of α, then $\beta\alpha=1$ also. This is because $\alpha\beta=1$ implies that α is nonsingular, so that there is an element γ in $A(V)$ with

$\gamma\alpha = \alpha\gamma = 1$. Thus, $\gamma(\alpha\beta) = \gamma = (\gamma\alpha)\beta = \beta$. Similarly, any left inverse of α is the unique inverse of α. Using 5.1.3, we get the nonobvious facts that if $(a_{ij}) \in F_n$ has either a left or a right inverse, then it is nonsingular, and any left inverse and any right inverse of (a_{ij}) is the (unique) inverse of (a_{ij}). We have proved the following theorem.

5.1.4 Theorem

Let V be a vector space of dimension n over the field F.

(a) *If (a_{ij}) is the matrix of $\alpha \in A(V)$ relative to some basis of V, then (a_{ij}) is nonsingular if and only if α is nonsingular.*

(b) *For any $\alpha \in A(V)$, the following are equivalent.*
   - (i) *α is nonsingular.*
   - (ii) *α has a right inverse.*
   - (iii) *α has a left inverse.*
   - (iv) *α is onto.*
   - (v) *α is one-to-one.*

(c) *For any $(a_{ij}) \in F_n$, the following are equivalent.*
   - (i) *(a_{ij}) is nonsingular.*
   - (ii) *(a_{ij}) has a right inverse.*
   - (iii) *(a_{ij}) has a left inverse.*

Suppose $\{v_1, v_2, \ldots, v_n\}$ and $\{w_1, w_2, \ldots, w_n\}$ are bases of V. For $\alpha \in A(V)$, how is the matrix of α with respect to $\{v_1, v_2, \ldots, v_n\}$ related to the matrix of α with respect to $\{w_1, w_2, \ldots, w_n\}$? To find out, consider the linear transformation γ defined by $\gamma(v_i) = w_i$. Let $\alpha(v_j) = \sum a_{ij}v_i$, $\alpha(w_j) = \sum b_{ij}w_i$, and $\gamma(v_j) = w_j = \sum c_{ij}v_i$. We will compute the matrix of $\alpha\gamma$ relative to the basis $\{v_1, v_2, \ldots, v_n\}$ in two ways. On the one hand, we get

$$\begin{aligned}\alpha\gamma(v_j) &= \alpha(w_j) = \alpha\left(\sum c_{kj}v_k\right)\\ &= \sum c_{kj}\alpha(v_k) = \sum c_{kj}\left(\sum a_{ik}v_i\right) = \sum\left(\sum c_{kj}a_{ik}\right)v_i.\end{aligned}$$

On the other hand,

$$\alpha\gamma(v_j) = \alpha(w_j) = \sum b_{kj}w_k = \sum b_{kj}\left(\sum c_{ik}v_i\right) = \sum\left(\sum b_{kj}c_{ik}\right)v_i.$$

Therefore, $(a_{ij})(c_{ij}) = (c_{ij})(b_{ij})$ since each side is the matrix of $\alpha\gamma$ with respect to $\{v_1, v_2, \ldots, v_n\}$. Now γ is nonsingular, so therefore (c_{ij}) is nonsingular. Hence, the relation $(a_{ij})(c_{ij}) = (c_{ij})(b_{ij})$ can just as well be expressed as

$$(a_{ij}) = (c_{ij})(b_{ij})(c_{ij})^{-1}.$$

This is expressed by saying that (a_{ij}) is ***similar to*** (b_{ij}). Similarity of matrices is an equivalence relation on F_n. In the same vein, two linear transformations α and β in $A(V)$ are called ***similar*** if there is a nonsingular $\gamma \in A(V)$ with $\alpha = \gamma\beta\gamma^{-1}$. Similarlity is an equivalence relation on $A(V)$.

5.1.5 Theorem

Let V be a vector space of dimension n over a field F. Let α and β be in $A(V)$, and let (a_{ij}) and (b_{ij}) be in F_n.

(a) *α and β are similar if and only if they have the same matrix relative to appropriate bases.*

(b) *(a_{ij}) and (b_{ij}) are similar if and only if they are matrices of the same linear transformation relative to appropriate bases.*

Proof What has been shown already is one half of (b), namely that if (a_{ij}) and (b_{ij}) are the matrices of α relative to two bases, then (a_{ij}) and (b_{ij}) are similar. To get the other half of (b), suppose that

$$(a_{ij}) = (c_{ij})(b_{ij})(c_{ij})^{-1}.$$

Let $\{v_1, v_2, \ldots, v_n\}$ be any basis of V. Let (a_{ij}), (b_{ij}), and (c_{ij}) be the matrices of α, β, and γ, respectively, relative to this basis. Since γ is nonsingular,

$$\{\gamma(v_1), \gamma(v_2), \ldots, \gamma(v_n)\}$$

is a basis of V. We have $\gamma\beta\gamma^{-1} = \alpha$ by 5.1.3. Since

$$\alpha(\gamma(v_j)) = \gamma\beta\gamma^{-1}(\gamma(v_j)) = \gamma\beta(v_j) = \gamma(\textstyle\sum b_{ij}v_i) = \textstyle\sum b_{ij}\gamma(v_i),$$

(b_{ij}) is the matrix of α relative to the basis $\{\gamma(v_1), \gamma(v_2), \ldots, \gamma(v_n)\}$. This completes the proof of (b).

The proof of (a) is about the same. Suppose that $\alpha = \gamma\beta\gamma^{-1}$. Let $\{v_1, v_2, \ldots, v_n\}$ be any basis of V. Then $\{\gamma(v_1), \gamma(v_2), \ldots, \gamma(v_n)\}$ is also a basis of V. Let $\beta(v_j) = \sum b_{ij}v_i$. Then

$$\alpha(\gamma(v_j)) = (\alpha\gamma)(v_j) = \gamma\beta(v_j) = \gamma(\textstyle\sum b_{ij}v_i) = \textstyle\sum b_{ij}\gamma(v_i),$$

whence the matrix of α relative to $\{\gamma(v_1), \gamma(v_2), \ldots, \gamma(v_n)\}$ is the same as the matrix of β relative to $\{v_1, v_2, \ldots, v_n\}$.

Now suppose that the matrix (a_{ij}) of α relative to a basis $\{v_1, v_2, \ldots, v_n\}$ is also the matrix of β relative to a basis $\{w_1, w_2, \ldots, w_n\}$. Let γ be the element of $A(V)$ defined by $\gamma(w_j) = v_j$. Then

$$\gamma\beta\gamma^{-1}(v_j) = \gamma\beta(w_j) = \gamma(\textstyle\sum a_{ij}w_i) = \textstyle\sum a_{ij}\gamma(w_i) = \textstyle\sum a_{ij}v_i.$$

Therefore, $\gamma\beta\gamma^{-1} = \alpha$. The proof is complete.

What does all this really say? Given a basis of V, we have an isomorphism

$$A(V) \to F_n.$$

Suppose that we are interested in relating properties of elements of $A(V)$ with those of F_n. That is, we are interested in translating properties of linear transformations into properties of matrices, and vice versa, using this isomorphism. What sort of properties can we expect to so translate? The isomorphism changes if we change basis. But it does not change drastically. If α corresponds to (a_{ij}) via one basis, it corresponds to a matrix similar to (a_{ij}) via any other basis. Further, if (b_{ij}) is similar to (a_{ij}), then there is another basis relative to which α corresponds to (b_{ij}). This is the content of 5.1.5(b). On the other hand, if α corresponds to (a_{ij}) via one basis, and β corresponds to (a_{ij}) via another basis, then α and β are similar, and conversely. This is the content of 5.1.5(a). This may be viewed in the following way. Pick a basis of V. This yields an isomorphism $A(V) \to F_n$. This isomorphism induces a one-to-one correspondence between the similarity classes of $A(V)$ and those of F_n. Now change basis. This results in another isomorphism $A(V) \to F_n$, and it induces a one-to-one correspondence between the similarity classes of $A(V)$ and those of F_n. *The one-to-one correspondences between these similarity classes in these two cases are the same.* Thus, the one-to-one correspondence obtained between the similarity classes of matrices in F_n is independent of the basis chosen. This makes it eminently plausible to try to characterize matrices in F_n, or equivalently, linear transformations in $A(V)$, up to similarity.

What properties must two matrices in F_n share in order to be similar? Are there some simple invariants of matrices that determine them up to similarity? In each similarity class of matrices in F_n, is there a special one that really displays the features of any linear transformation corresponding to it? For example, suppose that the matrix (a_{ij}) is diagonal. That is, suppose that $a_{ij} = 0$ if $i \neq j$. If α corresponds to (a_{ij}) via the basis $\{v_1, v_2, \ldots, v_n\}$, then $\alpha(v_i) = a_{ii}v_i$. Thus α is just a scalar multiplication on each of the subspaces Fv_i, and the action of α on V becomes transparent. Relative to some other basis, the matrix of α may not be so simple. If every similarity class of matrices had a diagonal matrix in it, then given any α, there would be a basis as above. This is too much to hope for. Linear transformations are just not that simple. Equivalently, not every matrix in F_n is similar to a diagonal one. This is actually easy to see geometrically. Rotation of the plane about the origin through 30 degrees, say, in the counterclockwise direction, is a linear transformation. Clearly, no basis exists relative to which the matrix of that linear transformation is diagonal. In fact, the action of α on no vector, except 0, is multiplication by a scalar.

We are going to spend a fair amount of effort finding special representatives, or "canonical forms," of similarity classes of matrices in F_n. The general idea is to find in each similarity class of matrices one special one that displays the essential features of any linear transformation corresponding to it. Roughly speaking, we will look for a representative as "diagonal as possible." There are several such canonical forms.

Remember, our purpose is to analyze the elements of the algebra $A(V)$. The discussion above has centered on characterizing those elements up to similarity. Can we expect to do any better? To what extent can we expect even to distinguish between the elements of the algebra $A(V)$? To what extent do we wish to do so? Suppose, for instance, that $\gamma: V \to W$ is an isomorphism between the finite-dimensional vector spaces V and W. Then

$$\Phi: A(V) \to A(W): \alpha \to \gamma\alpha\gamma^{-1}$$

is an isomorphism between the algebras $A(V)$ and $A(W)$. Surely α acts on V in "the same way" as $\gamma\alpha\gamma^{-1}$ acts on W. Surely, we do not care about any properties of α not shared by $\gamma\alpha\gamma^{-1}$. As linear transformations, they must be just alike. A remarkable fact is that if $\Phi: A(V) \to A(W)$ is any isomorphism between the algebras $A(V)$ and $A(W)$, then it is induced in the way indicated above by an isomorphism $\gamma: V \to W$. This is the Noether–Skolem Theorem, which we will not prove. In particular, if $V = W$, then any automorphism φ of the algebra $A(V)$ is induced by an automorphism $\gamma: V \to V$. This means that $\Phi(\alpha) = \gamma\alpha\gamma^{-1}$ for all α in $A(V)$. Therefore, if Φ is any automorphism of $A(V)$, then the image of any α in $A(V)$ under φ is similar to α. In short, characterizing the elements of $A(V)$ up to similarity is exactly what we should do. We cannot expect to do any better.

PROBLEMS

1 Prove that $\alpha(a + bi) = a - bi$ is a linear transformation of the complex numbers regarded as a vector space over the field of real numbers. Find the matrix of α relative to the basis $\{1, i\}$.

2 Let $\mathbb{R}$ be the field of real numbers. Let α be the linear transformation on $\mathbb{R}^2$ that rotates the plane about the origin through an angle θ in the counterclockwise direction. Find the matrix of α relative to $\{(1, 0), (0, 1)\}$.

3 Let (a_{ij}) be in F_2. Prove that

$$(a_{ij})^2 - (a_{11} + a_{22})(a_{ij}) + (a_{11}a_{22} - a_{12}a_{21}) \cdot 1 = 0.$$

4 Let V be of dimension 2 over the field F, and let α be an element of $A(V)$. Prove that there is a polynomial $f(x)$ in $F[x]$ of degree 2 with $f(\alpha) = 0$.

5 Let F be a field, and let $f(x)$ in $F[x]$ have degree n. Regard $F[x]/(f(x))$ as a vector space over F. Prove that

$$\{1 + (f(x)), x + (f(x)), x^2 + (f(x)), \ldots, x^{n-1} + (f(x))\}$$

is a basis of $F[x]/(f(x))$. Let α be the linear transformation defined by $\alpha(x^i + (f(x))) = x^{i+1} + (f(x))$. Find the matrix of α relative to the basis above.

6 Prove that the ring $A(V)$ is noncommutative if $\dim(V) \geq 2$.

7 Let V be a vector space over F, and suppose that $V \neq 0$. Prove that F is the center of $A(V)$. That is, prove that $\alpha\beta = \beta\alpha$ for all β in $A(V)$ if and only if there is an element a in F such that $\alpha(v) = av$ for all v in V.

8 Prove that (a_{ij}) is in the center of F_n if and only if (a_{ij}) is of the form

$$\begin{pmatrix} a & 0 & \cdots & 0 \\ 0 & a & \cdots & 0 \\ \vdots & \vdots & & \vdots \\ 0 & 0 & \cdots & a \end{pmatrix},$$

that is, if and only if (a_{ij}) is a scalar matrix.

9 Let (a_{ij}) be a nilpotent element of F_n, that is, let $(a_{ij})^m = 0$ for some positive integer m. Prove that $(a_{ij})^n = 0$.

10 Prove that if α is nilpotent, then $1 + \alpha$ is nonsingular.

11 Prove that if α is nilpotent and $a_0, a_1, \ldots, a_m$ are elements of F with $a_0 \neq 0$, then $a_0 + a_1\alpha + \cdots + a_m\alpha^m$ is nonsingular.

12 Prove that a matrix in F_n with exactly one 1 in each row and column and 0 elsewhere is nonsingular. (These matrices are called ***permutation*** matrices.)

13 Suppose that (a_{ij}) is upper triangular. That is, suppose that $a_{ij} = 0$ if $j < i$. Prove that (a_{ij}) is nonsingular if and only if no $a_{ii} = 0$.

14 Suppose that α is in $A(V)$ and that α is idempotent, that is, that $\alpha^2 = \alpha$. Prove that $V = V_1 \oplus V_2$ with $\alpha(v) = v$ for all $v \in V_1$ and $\alpha(V_2) = 0$.

15 Suppose that (a_{ij}) is idempotent. Prove that (a_{ij}) is similar to a matrix of the form

$$\begin{bmatrix} 1 & 0 & 0 & \cdots & 0 & 0 & \cdots & 0 \\ 0 & 1 & 0 & \cdots & 0 & 0 & \cdots & 0 \\ \vdots & & & & \vdots & & & \vdots \\ 0 & 0 & 0 & \cdots & 1 & 0 & \cdots & 0 \\ 0 & 0 & 0 & \cdots & 0 & 0 & \cdots & 0 \\ \vdots & & & & \vdots & & & \vdots \\ 0 & 0 & 0 & \cdots & 0 & 0 & \cdots & 0 \end{bmatrix}.$$

16 Prove that if α and β are idempotent elements of $A(V)$, then α and β are similar if and only if they have the same rank. (The rank of α is $\dim(\mathrm{Im}(\alpha))$.)

17 Let $\mathbb{R}$ be the field of real numbers. Prove that the $n \times n$ matrix of all 1's is similar to the $n \times n$ matrix (a_{ij}) with $a_{11} = n$ and all other entries 0.

5.2 THE RATIONAL CANONICAL FORM FOR MATRICES

Let V be a finite-dimensional vector space over the field F. The principal tool we will use in studying elements of $A(V)$ is 4.7.15—a finitely generated module over a principal ideal domain is determined up to isomorphism by its invariant factors. The special case 4.7.16 of 4.7.15 is what we will actually use.

5.2.1 Definition

Let V be a vector space over the field F, and let α be in $A(V)$. For $f(x)$ in $F[x]$ and v in V, let $f(x) \cdot v = f(\alpha)(v)$. This makes V into a module over $F[x]$, called the **module of the linear transformation** *α, and denoted V^{α}.*

Throughout, V will be finite dimensional. It is straightforward to verify that $f(x) \cdot v = f(\alpha)(v)$ really makes V into a module over $F[x]$. We write V^{α} for this module to distinguish it from the vector space V. As a set, $V = V^{\alpha}$.

Suppose that S is a submodule of V^{α}. Then certainly S is a subspace of V, but also $x \cdot S = \alpha(S) \subset S$. Therefore, S is a subspace of V that is invariant under, that is, taken into itself by, α. Conversely, if $\alpha(S) \subset S$ and S is a subspace of V, then S is a submodule of V^{α}. Thus, the submodules of V^{α} are the subspaces of V invariant under α. In particular, a direct sum decomposition $V^{\alpha} = V_1 \oplus V_2$ is the same thing as a decomposition $V = V_1 \oplus V_2$ of the vector space V into the direct sum of subspaces V_1 and V_2 which are invariant under α. Since V is finitely generated as an F-module, it is certainly finitely generated as an $F[x]$-module. Any basis of V generates the module V^{α}, for example.

5.2.2 Theorem

Let V be a finite-dimensional vector space over the field F, and let α be in $A(V)$. Then V^{α} is a finitely generated torsion $F[x]$-module.

Proof We already observed that V^{α} is finitely generated. There are several ways to show that it is torsion. Here is one way that utilizes the fact that V^{α} is a direct sum of cyclic modules. Write $V^{\alpha} = V_1 \oplus V_2 \oplus \cdots \oplus V_k$ with each V_i a cyclic $F[x]$-module. Each V_i is either torsion or isomorphic to $F[x]$ as an $F[x]$-module. But $F[x]$ is an infinite-dimensional vector space over F. Thus if $V_i \approx F[x]$, then V_i is infinite dimensional as a vector space over F. Thus each V_i is torsion, and hence V^{α} is torsion.

Another way to get V^α torsion is this. The vector space $A(V)$ has dimension $(\dim(V))^2 = m$ over F. Thus $\{1, \alpha, \alpha^2, \ldots, \alpha^m\}$ is dependent. Hence there are elements $a_0, a_1, \ldots, a_m$ in F, not all 0, such that $\sum a_i\alpha^i = 0$. It follows that for any v in V, $(\sum a_ix^i)\cdot v = (\sum a_i\alpha^i)(v) = 0$. But $\sum a_ix^i \neq 0$. Thus, V^α is torsion.

By 4.7.16, V^α is determined up to isomorphism by its family $\{f_1(x), f_2(x), \ldots, f_k(x)\}$ of invariant factors. These are monic polynomials in $F[x]$ such that $f_{i+1}(x)$ divides $f_i(x)$ for $i<k$. Further, $V^\alpha \approx F[x]/(f_1(x)) \oplus \cdots \oplus F[x]/(f_k(x))$. Let us examine this decomposition briefly. First, let $f(x)$ be any monic polynomial in $F[x]$. Then $F[x]/(f(x))$ is a cyclic $F[x]$-module, but it is also a vector space over F. We need a bit of information about the vector space $F[x]/(f(x))$.

5.2.3 Lemma

Let $f(x)$ be in $F[x]$, and let $\deg(f(x)) = n$. Then as a vector space over F, $F[x]/(f(x))$ is of dimension n. In fact, $\{1 + (f(x)), x + (f(x)), \ldots, x^{n-1} + (f(x))\}$ is a basis.

Proof If $\sum_{i=0}^{n-1} a_i(x^i + (f(x))) = 0$, then $\sum_{i=0}^{n-1} a_ix^i$ is in $(f(x))$. But $(f(x))$ contains no nonzero polynomials of degree less than n. Hence, $B = \{1 + (f(x)), x + (f(x)), \ldots, x^{n-1} + (f(x))\}$ is independent. If $g(x)$ is in $F[x]$, then $g(x) = f(x)k(x) + r(x)$, with $\deg(r(x)) < n$. Hence $g(x) + (f(x)) = r(x) + (f(x))$. Now it is clear that B generates $F[x]/(f(x))$, and hence is a basis.

5.2.4 Corollary

Let W be a cyclic submodule of V^α, and let $\dim(W) = m$. If w generates the module W, then

$$\{w, \alpha(w), \alpha^2(w), \ldots, \alpha^{m-1}(w)\}$$

is a basis of the subspace W.

Proof The epimorphism given by $F[x] \to W: 1 \to w$ yields an isomorphism $F[x]/(f(x)) \to W$, where $(f(x)) = 0{:}w$. This $F[x]$-isomorphism is also a vector space isomorphism. Since $\dim(W) = m$, then by 5.2.3, $\deg(f(x)) = m$, and $\{1 + (f(x)), x + (f(x)), \ldots, x^{m-1} + (f(x))\}$ is a basis of $F[x]/(f(x))$. Its image in W is

$$\{w, x\cdot w, \ldots, x^{m-1}\cdot w\} = \{w, \alpha(w), \ldots, \alpha^{m-1}(w)\}.$$

5.2.5 Corollary

Let α be in $A(V)$, and let

$$\{f_1(x), f_2(x), \ldots, f_k(x)\}$$

be the invariant factors of V^α. Then

$$\dim(V) = \deg(f_1(x)) + \cdots + \deg(f_k(x)).$$

Proof $V^\alpha \approx F[x]/(f_1(x)) \oplus \cdots \oplus F[x]/(f_k(x))$. Apply 5.2.3.

Consider again the decomposition

$$V^\alpha \approx F[x]/(f_1(x)) \oplus \cdots \oplus F[x]/(f_k(x)),$$

where $\{f_1(x), \ldots, f_k(x)\}$ are the invariant factors of V^α. Since $f_1(x)$ is a multiple of each $f_i(x)$, it follows that $f_1(x)(F[x]/(f_i(x))) = 0$. Therefore, $f_1(x)V^\alpha = 0$. What does this mean? It means that for each v in V^α, $f_1(x) \cdot v = 0$. But $f_1(x) \cdot v = f_1(\alpha)(v) = 0$. Therefore, $f_1(\alpha) = 0$. The polynomials $g(x)$ in $F[x]$ such that $g(\alpha) = 0$ form an ideal. This ideal is nonzero. Any nonzero ideal of $F[x]$ is generated by a unique monic polynomial. Clearly, no monic polynomial $g(x)$ of degree smaller than $\deg(f_1(x))$ can satisfy $g(\alpha) = 0$. If it did, then $g(x)V^\alpha = 0 = g(x)(F[x]/(f_1(x)))$, which is impossible. Therefore, $f_1(x)$ is the monic generator of the ideal of polynomials $g(x)$ such that $g(\alpha) = 0$. In other words, $f_1(x)$ is the monic polynomial of least degree such that $f_1(\alpha) = 0$. But what is the degree of $f_1(x)$? By 5.2.3, or 5.2.5, $\deg(f_1(x)) \leq \dim(V)$. Using our isomorphism $A(V) \rightarrow F_n$, we have

5.2.6 Theorem

Every linear transformation on a vector space of dimension n over a field F satisfies a monic polynomial in $F[x]$ of degree $\leq n$. Every matrix in F_n satisfies a monic polynomial in $F[x]$ of degree $\leq n$.

5.2.7 Definition

Let α be in $A(V)$. The **minimum polynomial of** α, *denoted $m_\alpha(x)$, is the monic polynomial of least degree such that $m_\alpha(\alpha) = 0$. Let A be in F_n. The* **minimum polynomial of** A, *denoted $m_A(x)$, is the monic polynomial of least degree such that $m_A(A) = 0$.*

Note that 5.2.6 just says that these minimum polynomials are of degree $\leq \dim(V)$.

5.2.8 Corollary

Let α be in $A(V)$, and let

$$\{f_1(x), f_2(x), \ldots, f_k(x)\}$$

be the invariant factors of V^α. Then $m_\alpha(x) = f_1(x)$.

Therefore, the first member of the family of invariant factors of V^α is the minimum polynomial of α. There is a pressing question at hand.

5.2.9 Theorem

Let V be a finite-dimensional vector space over a field F, and let α and β be in $A(V)$. Then the $F[x]$-modules V^α and V^β are isomorphic if and only if α and β are similar.

Proof Suppose that $\alpha = \gamma^{-1}\beta\gamma$. Then γ is an F-isomorphism from V^α to V^β. It would be an $F[x]$-isomorphism if $\gamma(f(x) \cdot v) = f(x) \cdot \gamma(v)$ for all $f(x)$ in $F[x]$ and v in V^α. This is true, however, if $\gamma(x \cdot v) = x \cdot \gamma(v)$ for all v in V^α. But $\gamma(x \cdot v) = \gamma(\alpha(v)) = (\gamma\alpha)(v) = (\beta\gamma)(v) = \beta(\gamma(v)) = x \cdot \gamma(v)$. Therefore, $V^\alpha \approx V^\beta$. Now suppose that $\gamma: V^\alpha \rightarrow V^\beta$ is an $F[x]$-isomorphism. Then γ certainly is an F-isomorphism. That is, γ is nonsingular. For v in V, we have $\gamma(x \cdot v) = x \cdot (\gamma(v))$, whence $\gamma(\alpha(v)) = \beta(\gamma(v))$. Therefore, $\gamma\alpha = \beta\gamma$, or $\alpha = \gamma^{-1}\beta\gamma$. This concludes the proof.

Thus, isomorphism invariants for V^α are similarity invariants for a. Hence, the invariant factors of V^α are complete similarity invariants for α.

5.2.10 Definition

Let V be a finite-dimensional vector space over the field F, and let α be in $A(V)$. The invariant factors, and the elementary divisors, of the module V^α are called the **invariant factors**, *and the* **elementary divisors**, *respectively, of the linear transformation α.*

5.2.11 Corollary

Let α and β be in $A(V)$. Then α and β are similar if and only if they have the same invariant factors, or the same elementary divisors.

Therefore, each α in $A(V)$ is associated with a unique family

$\{f_1(x), \ldots, f_k(x)\}$ of monic polynomials with each a multiple of the next, namely the invariant factors of α. The linear transformations α and β correspond to the same family precisely when they are similar. For any vector space V of dimension n, similarity classes of elements of $A(V)$ are in a natural one-to-one correspondence with similarity classes of elements of F_n. The correspondence is induced by sending α in $A(V)$ to the matrix of α relative to any basis. Therefore, the following definition of invariant factors are elementary divisors of matrices is legitimate; it is independent of V.

5.2.12 Definition

Let (a_{ij}) be in F_n, and let α be any element of $A(V)$ whose matrix relative to some basis of V is (a_{ij}). The **invariant factors,** *and* **elementary divisors** *of (a_{ij}) are the invariant factors, and elementary divisors, respectively, of α.*

5.2.13 Corollary

Two matrices in F_n are similar if and only if they have the same invariant factors, or the same elementary divisors.

5.2.14 Corollary

Let α be in $A(V)$, with $\dim(V) = n$. Then α and (a_{ij}) have the same invariant factors, or the same elementary divisors, if and only if (a_{ij}) is the matrix of α relative to some basis of V.

5.2.15 Corollary

Similar linear transformations, and similar matrices, have the same minimum polynomial.

How does one tell whether or not two matrices in F_n have the same invariant factors? This is the same as asking how one tells whether or not two matrices in F_n are similar. It turns out that there is a matrix in each similarity class that puts on display its invariant factors. This particular canonical form (that is, this special matrix in each similarity class) is called the ***rational canonical form***. We will get at it through V^{α}. Now V^{α} is a direct sum of cyclic modules. Let us examine the cyclic case first. So suppose that V^{α} is cyclic. This is expressed by saying that α is ***cyclic***. So a linear transformation α on a vector space V is cyclic if the associated module V^α is a cyclic module. Suppose that α is cyclic. Then $V^\alpha \approx F[x]/(m_\alpha(x))$, where $m_\alpha(x)$ is the minimum polynomial of α. Let v generate V^α. By 5.2.4,

$$B = \{v, \alpha(v), \ldots, \alpha^{n-1}(v)\}$$

is a basis of V, where $\dim(V) = n$. What is the matrix of α relative to this basis? We need to write each $\alpha(\alpha^i(v))$ as a linear combination of elements of B. Let

$$m_\alpha(x) = x^n + a_{n-1}x^{n-1} + \cdots + a_0.$$

Then

$$\alpha^n = -(a_0 + a_1\alpha + \cdots + a_{n-1}\alpha^{n-1})$$

since $m_\alpha(\alpha) = 0$. Thus,

$$\alpha^n(v) = -(a_0 v + a_1(\alpha(v)) + \cdots + a_{n-1}(\alpha^{n-1}(v))).$$

For $i < n-1$, $\alpha(\alpha^i(v)) = \alpha^{i+1}(v)$ is certainly a linear combination of elements of B. Therefore, the matrix of α relative to the basis B is

$$\begin{pmatrix} 0 & 0 & 0 & \cdots & 0 & 0 & -a_0 \\ 1 & 0 & 0 & \cdots & 0 & 0 & -a_1 \\ 0 & 1 & 0 & \cdots & 0 & 0 & -a_2 \\ \vdots & & & & & & \vdots \\ 0 & 0 & 0 & \cdots & 1 & 0 & -a_{n-2} \\ 0 & 0 & 0 & \cdots & 0 & 1 & -a_{n-1} \end{pmatrix}.$$

This matrix is, of course, uniquely determined by the polynomial $m_\alpha(x)$, and one can read off $m_\alpha(x)$ from it.

5.2.16 Definition

Let $f(x)$ be a monic polynomial in $F[x]$. If $f(x) = x^n + a_{n-1}x^{n-1} + \cdots + a_0$, then the matrix

$$\begin{pmatrix} 0 & 0 & 0 & \cdots & 0 & 0 & -a_0 \\ 1 & 0 & 0 & \cdots & 0 & 0 & -a_1 \\ 0 & 1 & 0 & \cdots & 0 & 0 & -a_2 \\ \vdots & & & & & & \vdots \\ 0 & 0 & 0 & \cdots & 1 & 0 & -a_{n-2} \\ 0 & 0 & 0 & \cdots & 0 & 1 & -a_{n-1} \end{pmatrix}$$

is called the **companion matrix** *of $f(x)$, and it is denoted $C(f(x))$.*

Thus, one matrix of a cyclic linear transformation is the companion matrix of its minimum polynomial.

5.2.17 Corollary

The minimum polynomial of the companion matrix $C(f(x))$ of the monic polynomial $f(x)$ is $f(x)$. That is, $m_{C(f(x))}(x) = f(x)$.

Proof Let $\deg(f(x)) = n$. Multiplication by x is a linear transformation on the vector space $F[x]/(f(x))$. The matrix of this linear transformation relative to the basis $\{1 + (f(x)), x + (f(x)), \ldots, x^{n-1} + (f(x))\}$ is $C(f(x))$. A matrix of a linear transformation satisfies the minimum polynomial of that linear transformation. The minimum polynomial of the linear transformation in question is $f(x)$ since $f(x)$ is the monic polynomial of least degree such that $f(x)(F[x]/(f(x))) = 0$.

Suppose that α is in $A(V)$ and that $V^{\alpha} = V_1 \oplus V_2$. Let $\{v_1, v_2, \ldots, v_m\}$ be a basis of V_1, and let $\{v_{m+1}, v_{m+2}, \ldots, v_n\}$ be a basis of V_2. Since V_i is invariant under α, the matrix of α relative to $\{v_1, v_2, \ldots, v_n\}$ has the form

$$\begin{pmatrix} a_{11} & \cdots & a_{1m} & 0 & \cdots & 0 \\ \vdots & & & & & \\ a_{m1} & \cdots & a_{mm} & 0 & \cdots & 0 \\ 0 & \cdots & 0 & a_{m+1\,m+1} & \cdots & a_{m+1\,n} \\ \vdots & & & & & \vdots \\ 0 & \cdots & 0 & a_{n\,m+1} & \cdots & a_{nn} \end{pmatrix}$$

In short, writing V^{α} as a direct sum $V_1 \oplus V_2 \oplus \cdots \oplus V_k$ of submodules and stringing together bases of $V_1, V_2, \ldots,$ and V_k yields a basis such that the matrix of α relative to it has the form

$$\begin{pmatrix} M_1 & & & & \\ & M_2 & & 0 & \\ & & M_3 & & \\ & 0 & & \ddots & \\ & & & & M_k \end{pmatrix}$$

where each M_i is a $\dim(V_i) \times \dim(V_i)$ array. This is a principal tool in finding special matrices for linear transformations.

5.2.18 Theorem

Let V be a vector space of dimension n over the field F. Let α be a linear transformation on V, and let $\{f_1(x), f_2(x), \ldots, f_k(x)\}$ be the invariant factors of α. Then V has a basis such that the matrix of α relative to it is

$$R(\alpha) = \begin{pmatrix} C(f_1(x)) & & & 0 \\ & C(f_2(x)) & & \\ & 0 & \ddots & \\ & & & C(f_k(x)) \end{pmatrix}.$$

Proof $V^\alpha \approx F[x]/(f_1(x)) \oplus \cdots \oplus F[x]/(f_k(x))$, so $V^\alpha = V_1 \oplus \cdots \oplus V_k$ with $V_i \approx F[x]/(f_i(x))$. Let v_i generate V_i. Then by 5.2.4,

$$\{v_i, \alpha(v_i), \ldots, \alpha^{n_i-1}(v_i)\}$$

is a basis of V_i, where n_i is the degree of $f_i(x)$. Therefore,

$$\{v_1, \alpha(v_1), \ldots, \alpha^{n_1-1}(v_1), v_2, \alpha(v_2), \ldots,$$
$$\alpha^{n_2-1}(v_2), \ldots, v_k, \alpha(v_k), \ldots, \alpha^{n_k-1}(v_k)\}$$

is a basis of V. The matrix of α relative to it is $R(\alpha)$.

5.2.19 Definition

The matrix $R(\alpha)$ in 5.2.18 is called the **rational canonical** *matrix of a. Any matrix in F_n in the form $R(\alpha)$, where $f_{i+1}(x)$ divides $f_i(x)$ for $i < k$, is said to be in* **rational canonical form**. *Such a matrix is denoted $(C(f_i(x)))$.*

5.2.20 Theorem

Every matrix in F_n is similar to exactly one matrix in F_n in rational canonical form.

Proof Let (a_{ij}) be in F_n. Then (a_{ij}) is the matrix of some α in $A(V)$ with $\dim(V) = n$. The matrix (a_{ij}) is similar to the rational canonical form of α. Suppose that (a_{ij}) is also similar to another matrix $(C(g_i(x)))$ in rational canonical form. Then this matrix is the matrix of the linear transformation α relative to some basis. But $V^\alpha \approx V_1 \oplus V_2 \oplus \cdots \oplus V_k$ with $V_i \approx F[x]/(g_i(x))$. Therefore, $\{g_1(x), g_2(x), \ldots, g_k(x)\}$ is the family of invariant factors of α. The theorem follows.

There is an important point in connection with 5.2.19 and 5.2.20. Let (a_{ij}) be in F_n. Complete similarity invariants for (a_{ij}) are the invariant factors of (a_{ij}). That is, two matrices in F_n are similar if and only if they have the same invariant factors. This is the same as having the same rational canonical form in F_n. Now suppose that K is a field containing F. They any (a_{ij}) in F_n is also in K_n. For example, if F is the field $\mathbb{Q}$ of rational numbers, then any matrix in F_n is also an $n \times n$ real matrix. As an element of F_n, (a_{ij}) has invariant factors in $F[x]$ and has a rational canonical form in F_n. But as an element of K_n, (a_{ij}) has invariant factors in $K[x]$ and a canonical form in K_n. By 5.2.19, a matrix in F_n in rational canonical form is in rational canonical form as a matrix in K_n. Thus, 5.2.20 yields the following theorem.

5.2.21 Theorem

Let F be a subfield of the field K. The invariant factors and the rational canonical form of a matrix in F_n are the same as the invariant factors and the rational canonical form of that matrix as an element of K_n.

In particular, a matrix in F_n has the same minimum polynomial as an element of F_n as it does as an element of K_n. Further, two matrices in F_n are similar in F_n if and only if they are similar in K_n.

The analogue of 5.2.21 for other canonical forms is not necessarily true. In fact, 5.2.21 does not hold for elementary divisors. For example, the matrix

$$\begin{pmatrix} 0 & 1 \\ -1 & 0 \end{pmatrix}$$

as a real matrix has as invariant factors only the polynomial x^2+1. The polynomial x^2+1 is also, of course, its only invariant factor as a complex matrix. But its elementary divisors as a complex matrix are $x+i$ and $x-i$, whereas it has only the single elementary divisor x^2+1 as a real matrix. Its rational canonical form both as a real and as a complex matrix is the matrix itself. In the next section, we will get a canonical form that will display the elementary divisors. That canonical form of a matrix might change if the field is enlarged, because the elementary divisors might change.

PROBLEMS

1. Find the minimum polynomial of an idempotent linear transformation.
2. Find the minimum polynomial of a nilpotent linear transformation.

3 Find the minimum polynomial of the 3×3 matrix whose ij entry is $i \cdot \delta_{ij}$.

4 Find the rational canonical form of the matrix

$$\begin{pmatrix} 0 & 1 & 0 & 0 \\ 0 & 0 & 2 & 0 \\ 0 & 0 & 0 & 3 \\ 0 & 0 & 0 & 0 \end{pmatrix}.$$

5 Find all possible rational canonical matrices of α in $A(V)$ when $m_\alpha(x) = (x^2+1)(x+1)^2$ and $\dim(V) = 7$.

6 Find the invariant factors and the rational canonical form of the matrix

$$\begin{pmatrix} 1 & 0 & 0 & 0 & 0 & 0 \\ 0 & 2 & 0 & 0 & 0 & 0 \\ 0 & 0 & 2 & 0 & 0 & 0 \\ 0 & 0 & 0 & 3 & 0 & 0 \\ 0 & 0 & 0 & 0 & 3 & 0 \\ 0 & 0 & 0 & 0 & 0 & 3 \end{pmatrix}.$$

7 Let F be the field of complex numbers, and let $a_1, a_2, \ldots, a_n$ be the n distinct nth roots of 1. Prove that in F_n the matrix $C(x^n - 1)$ is similar to the diagonal matrix (a_{ij}) wtih $a_{ij} = a_i \cdot \delta_{ij}$.

8 Prove that the matrix

$$\begin{pmatrix} 0 & 0 & 0 & 0 & 1 \\ 1 & 0 & 0 & 0 & -5 \\ 0 & 1 & 0 & 0 & 10 \\ 0 & 0 & 1 & 0 & -10 \\ 0 & 0 & 0 & 1 & 5 \end{pmatrix}$$

is not similar to a diagonal matrix.

9 Let α be in $A(V)$, and suppose that α is nilpotent. That is, $\alpha^m = 0$ for some $m > 0$. Prove that α has a matrix of the form

$$\begin{pmatrix} N_1 & & & & \\ & N_2 & & 0 & \\ & & N_3 & & \\ & 0 & & \ddots & \\ & & & & N_k \end{pmatrix}$$

where N_i is an $n_i \times n_i$ matrix of the form

$$\begin{pmatrix} 0 & 0 & 0 & \cdots & 0 & 0 & 0 \\ 1 & 0 & 0 & \cdots & 0 & 0 & 0 \\ \vdots & & & & & \vdots & \\ 0 & 0 & 0 & \cdots & 1 & 0 & 0 \\ 0 & 0 & 0 & \cdots & 0 & 1 & 0 \end{pmatrix}$$

and $n_1 \geq n_2 \geq \cdots \geq n_k$.

10 Prove that the set of similarity classes of nilpotent matrices in F_n is in one-to-one correspondence with the set of partitions $n_1 + n_2 + \cdots + n_k = n$ of n into the sum of positive integers with $n_1 \geq n_2 \geq \cdots \geq n_k$.

11 Prove that if the matrices M_i are nonsingular in F_{n_i}, then the matrix

$$(M_i) = \begin{pmatrix} M_1 & & & \\ & M_2 & 0 & \\ & 0 & \ddots & \\ & & & M_k \end{pmatrix}$$

is nonsingular in $F_{n_1+n_2+\cdots+n_k}$.

12 Prove that if M_i is similar to N_i in F_{n_i}, then (M_i) in Problem 11 is similar to (N_i) in $F_{n_1+n_2+\cdots+n_k}$.

13 Let $\{v_1, v_2, \ldots, v_n\}$ be a basis of V, and let α be defined by $\alpha(v_1) = v_2$, $\alpha(v_2) = v_3, \ldots, \alpha(v_{n-1}) = v_n$, and $\alpha(v_n) = -a_0v_1 - a_1v_2 - \cdots - a_{n-1}v_n$. Compute the matrix of α relative to the basis

$$\left\{v_n + \sum_{i=1}^{n-1} a_i v_i,\ v_{n-1} + \sum_{i=1}^{n-2} a_{i+1} v_i,\ \ldots,\ v_2 + a_{n-1}v_1,\ v_1\right\}.$$

14 Prove that a matrix in F_n is similar to its transpose.

15 Let $\{f_1(x), f_2(x), \ldots, f_k(x)\}$ be the family of invariant factors of α in $A(V)$. Let $V^\alpha = V_1 \oplus V_2 \oplus \cdots \oplus V_k$ with $V_i \approx F[x]/(f_i(x))$. Let α_i be the restriction of α to V_i. Prove that the minimum polynomial of α_i is $f_i(x)$, and prove that α_i is cyclic.

16 Let $\gamma: V \to W$ be an isomorphism between the vector spaces V and W. Let α be in $A(V)$. Prove that the invariant factors of α in $A(V)$ and $\gamma\alpha\gamma^{-1}$ in $A(W)$ are the same.

17 Let K be a field containing the field F. Prove that the similarity classes of F_n are the intersections of the similarity classes of K_n with F_n.

18 Let α be in $A(V)$. Call α ***irreducible*** if V^α is simple, that is, if V^α has no nontrivial submodules. Prove that α is irreducible if and only if $m_\alpha(x)$ is prime and of degree $\dim(V)$.

19 Let α be in $A(V)$. Call α ***indecomposable*** if V^α is indecomposable. Prove that α is indecomposable if and only if for every pair V_1 and V_2 of submodules of V^α, either $V_1 \subset V_2$ or $V_2 \subset V_1$.

20 Let α be in $A(V)$. Prove that α is indecomposable if and only if α is cyclic and $m_\alpha(x)$ is a power of a prime.

21 Let α be in $A(V)$. Call α ***completely reducible*** if V^α is a direct sum of simple submodules. Prove that α is completely reducible if and only if $m_\alpha(x)$ is a product of distinct primes.

22 Let α and β be in $A(V)$ with α nonsingular. Prove that $\alpha + a\beta$ is nonsingular for all but finitely many a in F.

23 Let α and β be in $A(V)$. Prove that β is an endomorphism of the module V^α if and only if $\alpha\beta = \beta\alpha$.

24 Suppose that α in $A(V)$ is cyclic. Prove that the only linear transformations that commute with α are the polynomials in α with coefficients in F.

25 Let α be in $A(V)$, and let

$$V^\alpha = F[x]v_1 \oplus F[x]v_2 \oplus \cdots \oplus F[x]v_k$$

with $0{:}\, v_i = (f_i(x))$, and with $f_{i+1}(x)$ dividing $f_i(x)$. Let $\gamma_j(\sum_{i=1}^k g_i(x)v_i) = g_1(x)v_j$, and let $\pi_j(\sum_{i=1}^k g_i(x)v_i) = g_j(x)v_j$, $j = 1, 2, \ldots, k$. Prove that γ_j and π_j are endomorphisms of V^α. That is, prove that γ_j and π_j are linear transformations of V which commute with α.

26 Let α be in $A(V)$. Prove that every β in $A(V)$ which commutes with every γ in $A(V)$ which commutes with α is a polynomial in α with coefficients in F.

27 Let V be a vector space over the field F, with $V \neq \{0\}$. Prove that the endomorphism ring of V considered as a module over $A(V)$ is F.

28 Let $V \neq \{0\}$. Let R be a subring of $A(V)$ such that 1 is in R and such that no subspace except $\{0\}$ and V is invariant under every element of R. Prove that the endomorphism ring of V considered as an R-module is a division ring.

5.3 EIGENVECTORS AND EIGENVALUES; THE JORDAN CANONICAL FORM

Let α be a linear transformation on a vector space V of dimension n over a field F. In the similarity class of matrices in F_n corresponding to α is the rational canonical form $R(\alpha)$ of α. The matrix $R(\alpha)$ exhibits the invariant factors of α. The linear transformation α may, however, have a matrix that is diagonal, that is, of the form (a_{ij}) with $a_{ij} = 0$ unless $i = j$. Such a matrix of α is particularly revealing. There is a basis $\{v_1, v_2, \ldots, v_n\}$ of V such that $\alpha(v_i) = a_i v_i$. Equivalently, $V = V_1 \oplus V_2 \oplus \cdots \oplus V_n$ with V_i of dimension 1, and the action of α on V_i is just multiplication by the fixed scalar a_i. Now $R(\alpha)$ is never diagonal unless $\dim(V_i) = 1$ for all i. There is, however, a canonical form $J(\alpha)$ for α such that $J(\alpha)$ is diagonal if there is a diagonal matrix corresponding to α. In some sense, $J(\alpha)$ is as diagonal as possible. The matrix $J(\alpha)$ has the further property of putting on display the elementary divisors of α. $J(\alpha)$ is called the ***Jordan canonical form*** of α. This section is mainly concerned with this particular canonical form.

The first item of business is to find out just when there are bases $\{v_1, v_2, \ldots, v_n\}$ of V such that $\alpha(v_i) = a_i v_i$. In particular, we need to know when nonzero vectors v exist such that $\alpha(v) = av$ for some a in F. Some terminology is in order.

5.3.1 Definition

Let α be in $A(V)$. A nonzero v in V is called an **eigenvector** *of α if $\alpha(v) = av$ for some a in F. The scalar a is called an* **eigenvalue** *of α, and v is an eigenvector of α* **belonging** *to the eigenvalue a.*

Thus, the scalar a in F is an eigenvalue of α if there is a nonzero v in V such that $\alpha(v) = av$. Equivalently, a is an eigenvalue of α if $\alpha - a$ is singular. Eigenvalues are also called ***proper values*** and ***characteristic values***. Similarly, eigenvectors are also called ***proper vectors*** and ***characteristic vectors***. The following theorem is basic.

5.3.2 Theorem

Let V be a finite-dimensional vector space over the field F. Let α be a linear transformation on V, and let $m_\alpha(x)$ be the minimum polynomial of α. Then the scalar a is an eigenvalue of α if and only if $x - a$ divides $m_\alpha(x)$ in $F[x]$.

Proof Suppose that $x - a$ divides $m_\alpha(x)$ in $F[x]$. Then $m_\alpha(x) = (x - a)f(x)$, with $f(x)$ in $F[x]$. Since $\deg(f(x)) < \deg(m_\alpha(x))$, there is a vector v in V such that $f(\alpha)(v) \neq 0$. But $m_\alpha(a)(v) = 0 = (\alpha - a)f(\alpha)(v)$. Thus, $\alpha(f(\alpha)(v)) = a(f(\alpha)(v))$. Therefore, $f(\alpha)(v)$ is an eigenvector corresponding to a.

Now suppose that a is an eigenvalue of α. Write $m_\alpha(x) = (x - a)f(x) + b$, with $f(x)$ in $F[x]$ and b in F. There is a nonzero v in V such that $\alpha(v) = av$. Hence,

$$m_\alpha(\alpha)(v) = 0 = (\alpha - a)f(\alpha)(v) + bv = f(\alpha)(\alpha(v) - av) + bv.$$

Since $\alpha(v) = av$, it follows that $bv = 0$. Hence $b = 0$, and $x - a$ divides $m_\alpha(x)$ in $F[x]$.

5.3.3 Corollary

Let α be in $A(V)$. Then α has at most $\dim(V)$ *eigenvalues.*

Proof The degree of $m_\alpha(x)$ is at most $\dim(V)$, and hence $m_\alpha(x)$ has at most $\dim(V)$ distinct roots.

5.3.4 Corollary

Similar linear transformations on a vector space have the same eigenvalues.

Proof Similar linear transformations on a vector space have the same minimum polynomial.

Note that 5.3.4 does not say that similar linear transformations on a vector space have the same eigenvectors. That is not true.

There is a short direct proof of 5.3.4. Suppose that $\alpha = \gamma^{-1}\beta\gamma$, and suppose that a is an eigenvalue of α. Then there is a nonzero v in V such that $\alpha(v) = av$. Now $\beta(\gamma(v)) = \gamma\alpha(v) = \gamma(av) = a\gamma(v)$, whence a is an eigenvalue of β. Note that $\gamma(v) \neq 0$ since $v \neq 0$ and γ is nonsingular.

Suppose that v is an eigenvector of α belonging to a. Then any nonzero scalar multiple of v is also an eigenvector of α. Indeed, for a nonzero b in F, $\alpha(bv) = b\alpha(v) = b(av) = ab(v) = a(bv)$. Since $b \neq 0$, $bv \neq 0$, and bv is an eigenvector of α belonging to a. We would like, if possible, to get a basis of eigenvectors. How does one get independent eigenvectors? The following theorem helps.

5.3.5 Theorem

If $v_1, v_2, \ldots, v_k$ are eigenvectors of α belonging to distinct eigenvalues, then $\{v_1, v_2, \ldots, v_k\}$ is independent.

Proof Suppose that $a_1, a_2, \ldots, a_k$ are distinct, with v_i belonging to a_i. If $\sum b_iv_i = 0$, then $\alpha(\sum b_iv_i) = 0 = \sum a_ib_iv_i$, and $a_1 \sum b_iv_i = \sum a_ib_iv_i$. Therefore, $\sum_{i=2}^{k} (a_1b_i - a_ib_i)v_i = 0$. Inducting on k, we get that $a_ib_i = a_1b_i$ for $i = 2, 3, \ldots, k$. If $b_i \neq 0$, then $a_i = a_1$. Thus, $b_i = 0$ for $i \geq 2$. It follows that $b_1 = 0$ also, and hence $\{v_1, v_2, \ldots, v_k\}$ is independent.

5.3.6 Corollary

Let α be in $A(V)$. If α has $\dim(V)$ *distinct eigenvalues, then V has a basis of eigenvectors of α, and there is a matrix of α which is diagonal.*

Suppose that α has a diagonal matrix. By permuting the basis if necessary, α then has a matrix of the form

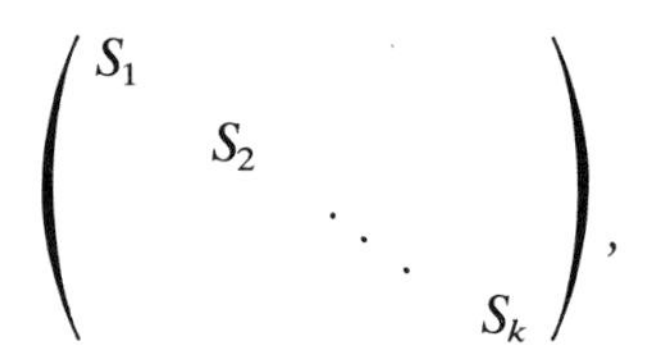

where S_i is an $n_i \times n_i$ scalar matrix

$$\begin{pmatrix} a_i & & & \\ & a_i & & \\ & & \ddots & \\ & & & a_i \end{pmatrix}$$

and $a_i = a_j$ only if $i = j$. The basis involved then has the form

$$\{v_{11}, v_{12}, \ldots, v_{1n_1}, v_{21}, v_{22}, \ldots, v_{2n_2}, \ldots, v_{k1}, v_{k2}, \ldots, v_{kn_k}\}$$

with $\alpha(v_{ij}) = a_i v_{ij}$. The numbers $n_1, n_2, \ldots, n_k$ are invariants of α. Indeed, the elementary divisors of α are

$$\{x - a_1, \ldots, x - a_1, x - a_2, \ldots, x - a_2, \ldots, x - a_k\}$$

with each $x - a_i$ appearing n_i times. The number n_i is also the dimension of the subspace V_i generated by

$$\{v_{ij}: j = 1, 2, \ldots, n_i\}.$$

The subspace V_i is just $\mathrm{Ker}(\alpha - a_i)$. We need some definitions.

5.3.7 Definition

Let V be a vector space over F, and let α be in $A(V)$. Suppose that a is an eigenvalue of α. Then $E(a) = \mathrm{Ker}(\alpha - a)$ is the **eigenspace** *of a. The number* $\dim(E(a))$ *is the* **geometric multiplicity** *of a.*

For a transformation with a diagonal matrix, as above, the multiplicity of

a_i is the number of $x - a_i$ appearing in the family of elementary divisors of α. Thus if we multiply the elementary divisors together, then the multiplicity of a_i is also its multiplicity as a root of that product.

5.3.8 Definition

Let V be a vector space over F, and let α be in $A(V)$. The product of the elementary divisors (or equivalently, the product of the invariant factors) of α is the **characteristic polynomial** *of α, denoted $c_\alpha(x)$.*

Note that $\deg(c_\alpha(x)) = \dim(V)$. Also, similar linear transformations have the same characteristic polynomial. Thus, we define the ***characteristic polynomial of a matrix*** in F_n to be the characteristic polynomial of any α in V corresponding to it. (In 5.4, we will see how to compute the characteristic polynomial of a matrix directly from that matrix.) Finally, note that if α is in $A(V)$ and a is in F, then a is an eigenvalue of α if and only if a is a root of $c_\alpha(x)$. The multiplicity of a as a root of $c_\alpha(x)$ is called its ***algebraic multiplicity***.

How do the two multiplicities of an eigenvalue compare?

5.3.9 Theorem

Let a be an eigenvalue of α in $A(V)$. Then the geometric multiplicity of a is the number of elementary divisors of α which are a power of $x - a$. In particular, the geometric multiplicity of a is no greater than the algebraic multiplicity of a.

Proof Let V_{x-a} be the primary part of V^α corresponding to the prime $x - a$. (See 4.7.17 and 4.7.18.)

$$V_{x-a} = \{v \in V: (x - a)^m \cdot v = 0 \text{ for some } m\}.$$

Now, $\text{Ker}(\alpha - a) \subset V_{x-a}$, and $V_{x-a} = W_1 \oplus W_2 \oplus \cdots \oplus W_k$ with the W_i cyclic modules. Further,

$$W_i \approx F[x]/((x - a)^{n_i}).$$

The kernel of $\alpha - a$ on W_i is $(\alpha - a)^{n_i - 1}W_i$, which is isomorphic to $(x - a)^{n_i - 1}F[x]/((x - a)^{n_i})$. This vector space has dimension 1 since it is cyclic and is annihilated by $x - a$, whence it is isomorphic to $F[x]/(x - a)$.

5.3.10 Theorem

Let V be a vector space over F and let α be in $A(V)$. Then α has a diagonal matrix if and only if the minimum polynomial of α is a product of distinct linear factors in $F[x]$.

Proof Suppose that α has a diagonal matrix. That is, suppose that V has a basis $\{v_1, v_2, \ldots, v_n\}$ consisting of eigenvectors of α. Let $f(x) = \prod(x - a_j)$, where a_j ranges over the distinct eigenvalues of α. Then $f(\alpha)(v_i) = 0$ for all basis vectors v_i, and so $f(\alpha)(v) = 0$ for all v in V. Thus $m_\alpha(x)$ divides $f(x)$, so $m_\alpha(x)$ is the product of distinct linear factors in $F[x]$. [Actually, $m_\alpha(x) = f(x)$.]

Now suppose that the minimum polynomial of α is a product of distinct linear factors in $F[x]$. The module V^α is a direct sum $V_1 \oplus V_2 \oplus \cdots \oplus V_k$ of cyclic modules, and if each V_i has a basis such that the matrix of the restriction of α to V_i is diagonal, then the union of these bases will be a basis of V such that the matrix of α with respect to it is diagonal. Therefore, we may assume that V^α is cyclic, so that $V^\alpha \approx F[x]/(m_\alpha(x))$. Let $m_\alpha(x)$ be the product $\prod(x - a_i)$ of distinct linear factors, and let $f_i(x) = m_\alpha(x)/(x - a_i)$. Then the $f_i(x)$ are relatively prime and there are polynomials $g_i(x)$ in $F[x]$ such that $\sum g_i(x)f_i(x) = 1$. Therefore, $\sum F[x]f_i(x) = F[x]$. Observe that $F[x]f_i(x)/(m_\alpha(x))$ is a submodule of $F[x]/(m_\alpha(x))$, is cyclic, is annihilated by $x - a_i$, and hence is isomorphic to $F[x]/(x - a_i)$. These are primary modules for distinct primes $x - a_i$. It follows that V^α is a direct sum of one-dimensional subspaces, and that V has a basis of eigenvectors.

Recalling that the minimum polynomial of a linear transformation restricted to an invariant subspace is a factor of the original minimum polynomial, we obtain the following result.

5.3.11 Corollary

Suppose that α has a diagonal matrix. Let W be any submodule of V^α. Then W is a direct sum of submodules of dimension 1.

Expressing Theorem 5.3.10 in terms of elementary divisors yields

5.3.12 Corollary

Let α be in $A(V)$. Then α has a diagonal matrix if and only if all the elementary divisors of α are linear.

5.3.13 Corollary

Let (a_{ij}) be in F_n. Then (a_{ij}) is similar in F_n to a diagonal matrix in F_n if and only if the elementary divisors of (a_{ij}) in $F[x]$ are all linear.

One has to be a little careful with 5.3.13. If F is a subfield of a field K, then (a_{ij}) is in K_n, and the elementary divisors of (a_{ij}) as an element of K_n are not necessarily the same as the elementary divisors of (a_{ij}) as an element of F_n. The matrix

$$\begin{pmatrix} 0 & -1 \\ 1 & 0 \end{pmatrix}$$

as a matrix over the reals and as a matrix over the complex numbers is a case in point.

It is easy to write down matrices that are not diagonalizable. For example, let $\mathbb{C}$ be the field of complex numbers. Multiplication by x is a linear transformation α on the vector space $V = \mathbb{C}[x]/((x-1)^2)$. But V^α is cyclic and $(x-1)^2$ is its only elementary divisor. By 5.3.12, no matrix of α is diagonal. One matrix of α is, of course the companion matrix

$$\begin{pmatrix} 0 & -1 \\ 1 & 2 \end{pmatrix}$$

of $(x-1)^2$.

If α does have a diagonal matrix $\operatorname{diag}(a_i)$, then $V^\alpha = V_1 \oplus V_2 \oplus \cdots \oplus V_n$ with $V_i \approx F[x]/(x - a_i)$. Thus, the elementary divisors of α are the members of the family $\{x - a_1, x - a_2, \ldots, x - a_n\}$. The point is that if α does have a diagonal matrix, that matrix displays the elementary divisors of α. Now for any α in $A(V)$, we are going to get a canonical matrix of α that displays the elementary divisors of α and which is diagonal if α has a diagonal matrix.

Let α be in $A(V)$. Then $V^\alpha = V_1 \oplus V_2 \oplus \cdots \oplus V_k$, where V_i is the $p_i(x)$-component of V^α. Each

$$V_i = V_{i1} \oplus V_{i2} \oplus \cdots \oplus V_{ik_i}$$

with V_{ij} cyclic. We want a special basis of V_{ij}. To simplify notation, suppose that V^α is cyclic primary. Thus V^α has a single elementary divisor $p(x)^q$, where $p(x)$ is a prime polynomial in $F[x]$. Of course, $p(x)^q$ is also the minimal polynomial as well as the invariant factor of α in this case. First we will do the special case when $p(x)$ is linear. This is an important special case. For example, if the field F is the field of complex numbers, then every prime polynomial is linear. Also, working through this case may illuminate the general case. So suppose that $p(x) = x - a$. Let v be a generator of the module V^α. Consider the family of vectors

$$v_1 = (\alpha - a)^{q-1}(v), \quad v_2 = (\alpha - a)^{q-2}(v), \ldots, \quad v_q = (\alpha - a)^{q-q}(v).$$

First, note that $\{v_1, v_2, \ldots, v_q\}$ is a basis of V. It is linearly independent since

$\sum_{i=1}^{q} a_i(\alpha - a)^{q-i}(v) = 0$ yields $(\sum_{i=1}^{q} a_i(x - a)^{q-i})V = 0$, whence $m_\alpha(x)$ divides $\sum_{i=1}^{q} a_i(x - a)^{q-i}$. But the degree of the latter polynomial is too small for that unless each $a_i = 0$. Thus, $\{v_1, v_2, \ldots, v_q\}$ is independent. Since $\dim(V) = q$, it is a basis. Now we will compute the matrix of α relative to this basis.

$$\alpha(v_1) = \alpha(\alpha - a)^{q-1}(v) = (\alpha - (\alpha - a))(\alpha - a)^{q-1}(v) = a(\alpha - a)^{q-1}(v) = av_1;$$

$$\begin{aligned}\alpha(v_2) &= \alpha(\alpha - a)^{q-2}(v) = (\alpha - (\alpha - a))(\alpha - a)^{q-2}(v) + (\alpha - a)^{q-1}(v)\\ &= a(\alpha - a)^{q-2}(v) + (\alpha - a)^{q-1}(v) = v_1 + av_2, \ldots;\end{aligned}$$

$$\begin{aligned}\alpha(v_{q-1}) &= \alpha(\alpha - a)(v) = (\alpha - (\alpha - a))(\alpha - a)(v) + (\alpha - a)^2(v)\\ &= a(\alpha - a)(v) + (\alpha - a)^2(v) = v_{q-2} + av_{q-1};\end{aligned}$$

$$\alpha(v_q) = \alpha(v) = (\alpha - (\alpha - a))(v) + (\alpha - a)(v) = av + (\alpha - a)(v) = v_{q-1} + av_q.$$

Therefore, the matrix of α has the form

$$\begin{pmatrix} a & 1 & 0 & 0 & \cdots & 0 & 0 \\ 0 & a & 1 & 0 & \cdots & 0 & 0 \\ \vdots & & & & & & \vdots \\ 0 & 0 & 0 & 0 & \cdots & a & 1 \\ 0 & 0 & 0 & 0 & \cdots & 0 & a \end{pmatrix}$$

We have the following theorem.

5.3.14 Theorem

Let V be a vector space over the field F, and let α be in $A(V)$. Suppose that the minimum polynomial factors into linear factors in $F[x]$. Then V has a basis such that the matrix of α relative to it has the form

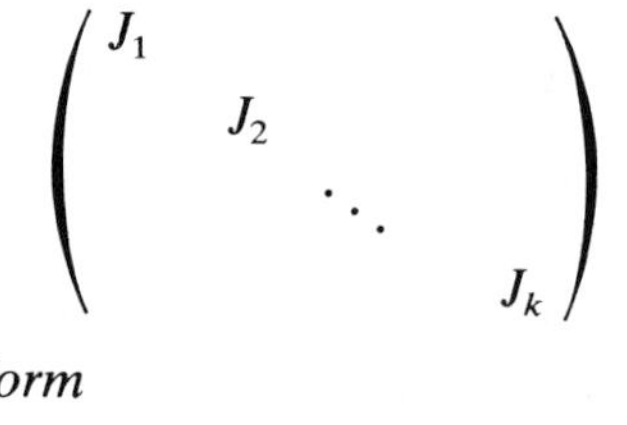

where each J_i has the form

$$\begin{pmatrix} a_i & 1 & 0 & \cdots & 0 & 0 \\ 0 & a_i & 1 & \cdots & 0 & 0 \\ \vdots & & & & & \vdots \\ 0 & 0 & 0 & \cdots & a_i & 1 \\ 0 & 0 & 0 & \cdots & 0 & a_i \end{pmatrix},$$

with the a_i eigenvalues of α.

5.3.15 Definition

$J(\alpha)$ is the **Jordan canonical form for** *α. A matrix in the form of $J(\alpha)$ is said to be* **in Jordan canonical form.**

Note that it is not really proper to say *the* Jordan canonical form for α. There is no particular way in which to order the J_i. However, apart from changing the order of the J_i, the form is unique. If J_i is $n_i \times n_i$, then the elementary divisors are just

$$(x - a_i)^{n_i}, \quad i = 1, 2, \ldots, k.$$

The a_i are not necessarily distinct. However, $J(\alpha)$ does display the elementary divisors of α.

5.3.16 Corollary

Let (a_{ij}) be in F_n, and suppose that the minimum polynomial of (a_{ij}) factors into linear factors in $F[x]$. Then (a_{ij}) is similar to a matrix in Jordan canonical form.

We will extend the definition of Jordan canonical form shortly. Consider again the case where V^α is cyclic primary. Then α has elementary divisor $p(x)^q$. Let $p(x) = a_0 + a_1 x + \cdots + a_{r-1}x^{r-1} + x^r$, and let v be a generator of the module V^α. Consider the family of vectors

$$v_1 = p(\alpha)^{q-1}(v), \quad v_2 = \alpha p(\alpha)^{q-1}(v), \ldots, \quad v_r = \alpha^{r-1}p(\alpha)^{q-1}(v),$$
$$v_{r+1} = p(\alpha)^{q-2}(v), \quad v_{r+2} = \alpha p(\alpha)^{q-2}(v), \ldots, \quad v_{2r} = \alpha^{r-1}p(\alpha)^{q-2}(v), \ldots,$$
$$v_{(q-1)r+1} = v, \quad v_{(q-1)r+2} = \alpha(v), \ldots, \quad v_{qr} = \alpha^{r-1}(v).$$

Each v_i is of the form $f_i(x)\cdot v$. Distinct $f_i(x)$ have distinct degrees, and $\deg(f_i(x)) < qr$. No $f_i(x) = 0$. Since

$$\{v, x\cdot v, \ldots, x^{qr-1}\cdot v\} = \{v, \alpha(v), \ldots, \alpha^{qr-1}(v)\}$$

is a basis, it follows that $\{v_1, v_2, \ldots, v_{qr}\}$ is a basis of V. Now we compute the matrix of α relative to this basis. We get

$$\alpha(v_1) = v_2, \quad \alpha(v_2) = v_3, \ldots, \quad \alpha(v_{r-1}) = v_r,$$
$$\alpha(v_r) = \alpha^r(p(\alpha)^{q-1}(v)) = (\alpha^r - p(\alpha))p(\alpha)^{q-1}(v)$$
$$= -a_0 v_1 - a_1 v_2 - \cdots - a_{r-1}v_r;$$

$$\alpha(v_{r+1}) = v_{r+2}, \quad \alpha(v_{r+2}) = v_{r+3}, \ldots, \quad \alpha(v_{2r-1}) = v_{2r},$$

$$\alpha(v_{2r}) = \alpha^r(p(\alpha)^{q-2})(v) = (\alpha^r - p(\alpha))(p(\alpha)^{q-2})(v) + p(\alpha)^{q-1}(v)$$
$$= -a_0 v_{r+1} - a_1 v_{r+2} - \cdots - a_{r-1} v_{2r} + v_1;$$
$$\vdots$$
$$\alpha(v_{(q-1)r+1}) = v_{(q-1)r+2}, \quad \alpha(v_{(q-1)+2}) = v_{(q-1)+3}, \ldots, \quad \alpha(v_{qr-1}) = v_{qr},$$
$$\alpha(v_{qr}) = \alpha^r(v) = (\alpha^r - p(\alpha))(v) + p(\alpha)(v)$$
$$= -a_0 v_{(q-1)r+1} - a_1 v_{(q-1)r+2} - \cdots - a_{r-1} v_{qr} + v_{(q-2)r+1}.$$

Thus the matrix of α relative to the basis $\{v_1, v_2, \ldots, v_{qr}\}$ is

$$J(p(x)^q) = \begin{pmatrix} C(p(x)) & D & & \\ & C(p(x)) & D & \\ & & \ddots & D \\ & & & C(p(x)) \end{pmatrix},$$

where $C(p(x))$ is the companion matrix of $p(x)$ and D is the $r \times r$ matrix with 1 in the upper right-hand corner and 0 elsewhere. We call $J(p(x)^q)$ the ***Jordan matrix*** of the prime power polynomial $p(x)^q$.

In general, let α be in $A(V)$, and write

$$V^\alpha = V_1 \oplus V_2 \oplus \cdots \oplus V_k,$$

where V_i are cyclic primary modules with elementary divisor $q_i(x)$. Pick a basis of V_i as above. Putting these bases of the V_i together, we get a matrix of α of a certain form.

5.3.16 Theorem

Let α be in $A(V)$, and let

$$q_1(x), q_2(x), \ldots, q_k(x)$$

be the elementary divisors of α. Then V has a basis such that the matrix of α relative to it has the form

$$J(\alpha) = \begin{pmatrix} J(q_1(x)) & & & \\ & J(q_2(x)) & & \\ & & \ddots & \\ & & & J(q_k(x)) \end{pmatrix},$$

where $J(q_i(x))$ is the Jordan matrix of $q_i(x)$.

Again, there is no particular order into which to put the $J(q_i(x))$, and they may of course not all be distinct. But up to such orderings, $J(\alpha)$ is unique.

5.3.17 Definition

$J(\alpha)$ is called the **Jordan canonical form for** *α. A matrix in the form $J(\alpha)$ is said to be* **in Jordan canonical form**.

5.3.18 Corollary

Let (a_{ij}) be in F_n. Then (a_{ij}) is similar to a matrix in Jordan canonical form.

Note that the matrix in 5.3.16 is the matrix in 5.3.14 in case all the elementary divisors are powers of a linear polynomial. Further, if the elementary divisors are linear, then $J(\alpha)$ is indeed diagonal. Thus, $J(\alpha)$ is diagonal if α has a diagonal matrix at all.

A matrix (a_{ij}) is ***upper triangular*** if $a_{ij} = 0$ whenever $j < i$. ***Lower triangular*** is defined analogously. When we say simply ***triangular***, we mean upper triangular. In case $m_\alpha(x)$ factors into linear factors in $F[x]$, then $J(\alpha)$ is triangular. Therefore, every matrix in F_n whose minimum polynomial is a product of linear factors in $F[x]$ is similar in F_n to a triangular matrix. This special fact is easy to prove directly. It does not require information about the modules V^α, for example. Furthermore, the converse holds. That is, a matrix in F_n is similar in F_n to a triangular matrix if and only if its minimal polynomial is a product of linear factors in $F[x]$. The pertinent theorem follows.

5.3.19 Theorem

Let V be an n-dimensional vector space over the field F. Let α be in $A(V)$. Then α has a triangular matrix if and only if its minimal polynomial is a product of linear factors in $F[x]$.

Proof Let α be in $A(V)$, and suppose that $m_\alpha(x)$ is a product of linear factors in $F[x]$. Then α has a triangular matrix by 5.3.16, but as indicated, we will give a direct proof of this fact. Let $\alpha^*: V^* \to V^*$ be the dual of α. Note that $m_\alpha(\alpha^*)(v^*) = v^* m_\alpha(\alpha) = 0$ for all v^* in V^*, so that the minimum polynomial of α^* divides that of α. (They are actually equal.) Thus the minimum polynomial of α^* is also a product of linear factors, and α^* has an eigenvector v^* in V^*. The one-dimensional subspace Fv^* is invariant under α^*, whence the $(n-1)$-dimensional subspace $W = \{w \in V: v^*(w) = 0\}$ is invariant under α. Indeed, $v^*(\alpha(w)) = (\alpha^*(v^*))(w) = (av^*)(w) = 0$, so that $\alpha(w)$ is in W for all w in W. By induction on $\dim(V) = n$, W has a basis

$$\{w_2, w_3, \ldots, w_n\}$$

such that the matrix of the restriction of α to W is triangular. If v is not in W, then $\{v, w_2, w_3, \ldots, w_n\}$ is a basis of V such that the matrix of α relative to it is triangular.

Conversely, if α has a triangular matrix (a_{ij}) with respect to the basis $\{v_1, v_2, \ldots, v_n\}$, then

$$(\alpha - a_{kk})(Fv_1 + Fv_2 + \cdots + Fv_k) \subset Fv_1 + Fv_2 + \cdots + Fv_{k-1},$$

so that $\prod_{i=1}^{n}(\alpha - a_{ii})$ annihilates V. Thus the minimum polynomial of α divides $\prod_{i=1}^{n}(x - a_{ii})$, so it factors into linear factors.

PROBLEMS

1 Suppose that $\dim(V) = n$ and that $\alpha \in A(V)$ has n distinct eigenvalues. Prove that α is cyclic.

2 Prove that if $a \in F$ is an eigenvalue of α, then for $f(x) \in F[x]$, $f(a)$ is an eigenvalue for $f(\alpha)$.

3 Suppose that a is an eigenvalue of the cyclic linear transformation α. Let $m_\alpha(x) = (x - a)^k f(x)$, where $x - a$ does not divide $f(x)$. Prove that the eigenvectors of α belonging to a are precisely the nonzero elements of $(\alpha - a)^{k-1} f(\alpha)(V)$.

4 Prove that if α is cyclic and a is an eigenvalue of α, then $\dim(\text{Ker}(\alpha - a)) = 1$.

5 Let a be an eigenvalue of α, and let

$$\{f_1(x), f_2(x), \ldots, f_k(x)\}$$

be the family of invariant factors of α. Prove that $\dim(\text{Ker}(\alpha - a))$ is the number of the $f_i(x)$ divisible by $x - a$.

6 Let V be a vector space over F and let α be in $A(V)$. Prove that α has a diagonal matrix if and only if the characteristic polynomial $c_\alpha(x)$ of α is a product of linear factors in $F[x]$ and the geometric multiplicity of each eigenvalue a of α is equal to its algebraic multiplicity.

7 Suppose that $V^\alpha = V_1 \oplus V_2$ and that α has a diagonal matrix relative to some basis. Prove that V has a basis

$$\{v_1, v_2, \ldots, v_r, w_1, w_2, \ldots, w_s\}$$

with $v_i \in V_1$, $w_i \in V_2$, and such that the matrix of α relative to this basis is diagonal.

8 Let $\alpha \in A(V)$, and suppose that W is a subspace of V invariant under α. Suppose that α has a diagonal matrix. Prove that the restriction of α to W has a diagonal matrix. Prove that the linear transformation on V/W induced by α has a diagonal matrix.

9 Let V be a vector space over F, and let $\alpha \in A(V)$. Let $f_i(x)$, $i = 1, 2, \ldots, k$, be monic polynomials in $F[x]$ of distinct degrees less than $\deg(m_\alpha(x))$. Prove that

$$\{f_i(\alpha)(v): i = 1, 2, \ldots, k\}$$

is independent for any v such that $0{:}v = (m_\alpha(x))$.

10 Prove that α is nilpotent if and only if $m_\alpha(x) = x^k$ for some k.

11 Prove that a triangular matrix (a_{ij}) with $a_{ii} = 0$ for all i is nilpotent.

12 Prove that if α is nilpotent, then any triangular matrix of α has all its main diagonal entries 0.

13 Prove that if α is nilpotent, then it has a matrix of the form

$$\begin{pmatrix} 0 & a_2 & & & \\ & 0 & a_3 & & \\ & & 0 & & \\ & & & \ddots & a_n \\ & & & & 0 \end{pmatrix}$$

with each a_i either 0 or 1.

14 Suppose that α is nonsingular on the vector space V over the field F, and that the minimum polynomial is a product of linear factors in $F[x]$. Prove that the eigenvalues of α^{-1} are the inverses of those of α with the same multiplicities.

15 Find all possible Jordan canonical forms for the 6×6 matrices in $\mathbb{Q}_6$ whose minimal polynomial is
(a) $(x-1)(x-2)^2(x+4)^3$;
(b) $(x+1)(x^4+x^3+x^2+x+1)$;
(c) $(x^2+1)(x)$;
(d) x^2+1;
(e) $(x^2+x+1)(x^2+5x+7)$.

16 Suppose that $\alpha \in A(V)$ and that W is a submodule of V^α. Prove that if the linear transformations on W and V/W induced by α have triangular matrices, then so does α.

17 Call a subset L of $A(V)$ ***triangulable*** if there is a basis of V relative to which the matrix of each $\alpha \in L$ is triangular. Suppose that $L \subset A(V)$ and suppose that if α and β are in L, then $\alpha\beta = \beta\alpha$. Suppose that the subspace W of V is invariant under each $\alpha \in L$. Let L_1 be the set of linear transformations on W induced by the elements of L, and let L_2 be the set of linear transformations on V/W induced by the elements of L. Prove that if L_1 and L_2 are triangulable, then so is L.

18 Suppose that α and β are in $A(V)$, $\alpha\beta = \beta\alpha$, and α and β have diagonal matrices relative to appropriate bases. Prove that $\alpha\beta$ has a diagonal matrix relative to some basis.

19 Let $L \subset A(V)$, and suppose that if α and β are in L, then $\alpha\beta = \beta\alpha$.
(a) Suppose that each $\alpha \in L$ has a triangular matrix. Prove that L is triangulable.
(b) Suppose that each $\alpha \in L$ has a diagonal matrix. Prove that L is diagonalizable.

20 Suppose that α, β, and γ are in $A(V)$ and all have triangulable matrices. Suppose that $\alpha\beta = \beta\alpha$, $\beta\gamma = \gamma\beta$, and $\alpha\gamma - \gamma\alpha = \beta$. Prove that there is a basis of V relative to which the matrices of α, β, and γ are triangular.

21 Let $\alpha \in A(V)$. Define f_α on $A(V)$ by $f_\alpha(\beta) = \alpha\beta - \beta\alpha$. Prove that f_α is a linear transformation on the vector space $A(V)$ and that f_α has a diagonal matrix if and only if α does.

5.4 DETERMINANTS

In this section, we present an introduction to the theory of determinants. This theory evolved in the eighteenth century, actually preceding the theory of matrices. The reader is likely to be somewhat familiar with determinants, for example, with their use in solving systems of linear equations (***Cramer's Rule***, 5.4.9). Determinants also provide a means to compute the characteristic polynomial of a matrix and hence of a linear transformation on a vector space (the ***Cayley–Hamilton Theorem***, 5.4.14).

There are essentially two ways to define determinant—the classical way, and via alternating forms. We choose the first, but do characterize them in terms of alternating forms.

We shall consider determinants of square matrices with entries from an arbitrary commutative ring with identity. There is a reason for this generality. It will be necessary in any case to consider determinants of matrices with entries from $F[x]$, where F is a field. Once this is done, we may as well let the entries come from any commutative ring with identity.

Let S_n denote the symmetric group on $\{1, 2, \ldots, n\}$. Recall that an element σ in S_n is ***even*** if it is the product of an even number of transpositions, and is ***odd*** otherwise. Let $(-1)^\sigma$ be 1 if σ is even and -1 if σ is odd. Note that $(-1)^\sigma(-1)^\tau = (-1)^{\sigma\tau}$ and $(-1)^\sigma = (-1)^\tau$ if $\sigma^{-1} = \tau$. Here is the definition of determinant.

5.4.1 Definition

Let R be a commutative ring with identity, and let (a_{ij}) be an $n \times n$ matrix with entries from R. Then the **determinant** *of (a_{ij}), denoted* $\det(a_{ij})$, *is the* *element*

$$\sum_\sigma (-1)^\sigma a_{1\sigma(1)} a_{2\sigma(2)} \cdots a_{n\sigma(n)}$$

of R, where the summation is over all σ in S_n.

As usual, let R_n denote the set of all $n \times n$ matrices over R. Then det is a function from R_n to R. This function has a number of fundamental properties that we will now proceed to derive. We have two principal goals in mind—to compute the characteristic polynomial of a matrix directly from that matrix, and to derive Cramer's Rule.

If (a_{ij}) is in R_n, then $\det(a_{ij})$ is the sum of $n!$ terms, each term of which is (neglecting sign) the product of n elements of (a_{ij}) no two of which are from the same row or column. For example,

$$\det\begin{pmatrix} a_{11} & a_{12} & a_{13} \\ a_{21} & a_{22} & a_{23} \\ a_{31} & a_{32} & a_{33} \end{pmatrix} = \begin{matrix} a_{11}a_{22}a_{33} - a_{11}a_{23}a_{32} + a_{12}a_{23}a_{31} \\ - a_{12}a_{21}a_{33} + a_{13}a_{21}a_{32} - a_{13}a_{22}a_{31}. \end{matrix}$$

If any row of (a_{ij}) consists of all zeroes, then $\det(a_{ij}) = 0$. This follows from the fact that each term of $\det(a_{ij})$ contains a factor from each row, and hence each term is zero. A similar argument shows that if a column consists of all zeroes, then $\det(a_{ij}) = 0$. The following theorem eliminates the need for making such similar arguments.

5.4.2 Theorem

The determinant of a matrix is the same as the determinant of its transpose.

Proof Let $(b_{ij}) = (a_{ij})'$, the transpose of (a_{ij}). That is, $b_{ij} = a_{ji}$ for all i and j. From the definition of determinant,

$$\det(a_{ij}) = \sum_\sigma (-1)^\sigma a_{1\sigma(1)} a_{2\sigma(2)} \cdots a_{n\sigma(n)}.$$

Each σ permutes $\{1, 2, \ldots, n\}$. Thus as i ranges through this set, so does $\sigma(i)$. Let $\tau = \sigma^{-1}$. Rearranging the terms of the product $\prod_i a_{i\sigma(i)} = \prod_i a_{\tau(\sigma(i))\sigma(i)}$ in order of increasing $\sigma(i)$ yields $\prod_i a_{i\sigma(i)} = \prod_i a_{\tau(i)i}$. Therefore,

$$\det(a_{ij}) = \sum_\sigma (-1)^\sigma a_{\tau(1)1} a_{\tau(2)2} \cdots a_{\tau(n)n}.$$

But as σ ranges through S_n, so does τ. Since $(-1)^\sigma = (-1)^\tau$, we get

$$\det(a_{ij}) = \sum_\tau (-1)^\tau a_{\tau(1)1} a_{\tau(2)2} \cdots a_{\tau(n)n}.$$

Since $a_{\tau(i)i} = b_{i\tau(i)}$, it follows that $\det(a_{ij}) = \det(b_{ij})$.

In the proof above, the commutativity of R was used. This special property of the ring R will be used many times without particular reference.

5.4.3 Lemma

Interchanging two rows or two columns of a matrix changes the sign of the determinant of that matrix.

Proof The language of 5.4.3 is a little loose. The determinant of a matrix is an element of a commutative ring and does not have a "sign," that is, it is neither "positive" nor "negative." What is meant is that if (b_{ij}) is a matrix gotten from (a_{ij}) by interchanging two rows or two columns of (a_{ij}), then $\det(a_{ij}) = -\det(b_{ij})$. By 5.4.2, it suffices to prove this for rows. Interchange the qth and rth rows of (a_{ij}). Let τ be the transposition (qr). The determinant of the resulting matrix is

$$\sum_{\sigma} (-1)^{\sigma} a_{1\sigma(1)} \cdots a_{(q-1)\sigma(q-1)} a_{r\sigma(q)} a_{(q+1)\sigma(q+1)}$$

$$\cdots a_{(r-1)\sigma(r-1)} a_{q\sigma(r)} a_{(r+1)\sigma(r+1)} \cdots a_{n\sigma(n)}$$

$$= \sum_{\sigma} (-1)^{\sigma} \prod_{i} a_{i\sigma\tau(i)} = -\sum_{\sigma} (-1)^{\sigma\tau} \prod_{i} a_{i\sigma\tau(i)} = -\det(a_{ij}).$$

The interchange of two rows of a matrix is an odd permutation of the rows of that matrix. Further, any permutation of the rows of a matrix is a product of successive interchanges of two rows. This is just another way of saying that a permutation is a product of transpositions. Thus, we see that permuting the rows of a matrix (a_{ij}) by a permutation σ results in a matrix whose determinant is $(-1)^{\sigma} \det(a_{ij})$. By 5.4.2, this holds for columns also.

5.4.4 Lemma

If two rows or two columns of a matrix are equal, then the determinant of that matrix is zero.

Proof Note that 5.4.4 is not implied by 5.4.3. That lemma just says that if two rows of (a_{ij}) are equal, then $\det(a_{ij}) = -\det(a_{ij})$, which does not imply that $\det(a_{ij}) = 0$. For example, in the field of integers modulo 2, $-1 = 1 \neq 0$.

Suppose that the qth and rth columns of (a_{ij}) are the same. Let

$\tau = (qr)$. Then for each i, $a_{i\sigma(i)} = a_{i\sigma\tau(i)}$, and

$$\begin{aligned}\det(a_{ij}) &= \sum_{\sigma \text{ even}} (-1)^\sigma \prod_i a_{i\sigma(i)} + \sum_{\sigma \text{ odd}} (-1)^\sigma \prod_i a_{i\sigma(i)} \\ &= \sum_{\sigma \text{ even}} (-1)^\sigma \prod_i a_{i\sigma(i)} - \sum_{\sigma \text{ odd}} (-1)^{\sigma\tau} \prod_i a_{i\sigma\tau(i)} \\ &= \sum_{\sigma \text{ even}} (-1)^\sigma \prod_i a_{i\sigma(i)} - \sum_{\sigma \text{ even}} (-1)^\sigma \prod_i a_{i\sigma(i)} = 0.\end{aligned}$$

An application of 5.4.2 gets the result for rows.

5.4.5 Theorem

$\det((a_{ij})(b_{ij})) = \det(a_{ij})\det(b_{ij})$.

Proof

$$\det((a_{ij})(b_{ij})) = \det\left(\sum_k a_{ik}b_{kj}\right) = \sum_\sigma (-1)^\sigma \prod_i \left(\sum_k a_{ik}b_{k\sigma(i)}\right),$$

and

$$\det(a_{ij})\det(b_{ij}) = \left(\sum_\sigma (-1)^\sigma \prod_i a_{i\sigma(i)}\right)\left(\sum_\sigma (-1)^\sigma \prod_i b_{i\sigma(i)}\right).$$

This last expression may be rewritten as $\sum_{\sigma,\tau} (-1)^{\sigma\tau}(\prod_i a_{i\sigma(i)}b_{i\tau(i)})$. The expression for $\det((a_{ij})(b_{ij}))$ may be rewritten as

$$\sum_\sigma (-1)^\sigma \sum_\tau \left(\prod_i a_{i\tau(i)}b_{\tau(i)\sigma(i)}\right).$$

To see this, observe that

$$\sum_\sigma (-1)^\sigma \prod_i \left(\sum_k a_{ik}b_{k\sigma(i)}\right) = \sum_\sigma (-1)^\sigma \sum_f \left(\prod_i a_{if(i)}b_{f(i)\sigma(i)}\right)$$

where f ranges over all the functions from the set $\{1, 2, \ldots, n\}$ to itself, not just the permutations. Given one of the functions f which is not a permutation, there must be a pair of integers $1 \le s < t \le n$ with $f(s) = f(t)$. Let τ be the transposition (st). Then $\sigma\tau(i) = \sigma(i)$ if $i \ne s, t$ and $\sigma\tau(s) = \sigma(t)$, $\sigma\tau(t) = \sigma(s)$. Moreover, $(-1)^{\sigma\tau} = -(-1)^\sigma$. Thus,

$$(-1)^\sigma \prod_i a_{if(i)}b_{f(i)\sigma(i)} + (-1)^{\sigma\tau} \prod_i a_{if(i)}b_{f(i)\sigma\tau(i)} = 0.$$

To see this, note that all the factors in each product are the same except possibly when $i = s, t$. But

$$[a_{sf(s)}b_{f(s)\sigma(s)}][a_{tf(t)}b_{f(t)\sigma(t)}] = [a_{sf(s)}b_{f(s)\sigma\tau(s)}][a_{tf(t)}b_{f(t)\sigma\tau(t)}],$$

since $f(s) = f(t)$, $\sigma\tau(s) = \sigma(t)$, and $\sigma\tau(t) = \sigma(s)$. Thus the terms involving functions f which are not permutations cancel out in pairs, so that

$$\sum_{\sigma}(-1)^{\sigma}\sum_{f}\left(\prod_{i} a_{if(i)}b_{f(i)\sigma(i)}\right)=\sum_{\sigma}(-1)^{\sigma}\sum_{\tau}\left(\prod_{i} a_{i\tau(i)}b_{\tau(i)\sigma(i)}\right),$$

where τ ranges over all permutations of $\{1, 2, \ldots, n\}$. Let $\gamma=\tau^{-1}$. Then $\prod_i a_{i\tau(i)}b_{\tau(i)\sigma(i)}=\prod_i a_{i\tau(i)}b_{i\sigma\gamma(i)}$, just noting that the b_{jk} in $\prod_i a_{i\tau(i)}b_{\tau(i)\sigma(i)}$ are exactly those such that $\sigma\gamma(j)=k$. Thus,

$$\det((a_{ij})(b_{ij}))=\sum_{\sigma,\tau}(-1)^{\sigma}\prod_{i} a_{i\tau(i)}b_{i\sigma\gamma(i)}.$$

Replacing $\sigma\gamma$ by σ and holding τ fixed yields

$$\det((a_{ij})(b_{ij}))=\sum_{\sigma,\tau}(-1)^{\sigma\tau}\prod_{i} a_{i\tau(i)}b_{i\sigma(i)},$$

which concludes the proof.

We now note several consequences of this theorem. Let A and B be in R_n, and let I_n be the identity of R_n. If A is a unit in R_n, then $\det(AA^{-1})=\det(I_n)=1=(\det A)(\det A^{-1})$, so that $\det A^{-1}=(\det A)^{-1}$. Further,

$$\det(A^{-1}BA)=(\det A)^{-1}(\det B)(\det A)=\det B.$$

Thus, conjugates have the same determinant. Therefore, we could define $\det(\alpha)$ for α any linear transformation on a finite-dimensional vector space V by letting it be the determinant of any matrix of α. We will return to determinants of linear transformations later. Finally, note that

$$\det(AB)=(\det A)(\det B)=(\det B)(\det A)=\det(BA).$$

Now we come to a result that is useful in the actual computation of determinants. There is a method of such computation known as "expanding along a row (or column)," and we will have need for it presently.

Let (a_{ij}) be in R_n. Let A_{ij} be the matrix gotten from (a_{ij}) by deleting the ith row and the jth column of (a_{ij}). Thus, A_{ij} is an $(n-1)\times(n-1)$ matrix. The matrix A_{ij} is the ***minor*** of the element a_{ij}. For reasons evident below, the quantities $\alpha_{ij}=(-1)^{i+j}\det A_{ij}$ are important. The α_{ij} are the ***cofactors of the matrix*** (a_{ij}), and α_{ij} is the ***cofactor of the entry*** a_{ij}.

5.4.6 Theorem

$$\det(a_{ij})=\sum_{j} a_{ij}\alpha_{ij}=\sum_{i} a_{ij}\alpha_{ij}.$$

Proof

$$\det(a_{ij})=\sum_{\sigma}(-1)^{\sigma}\prod_{i} a_{i\sigma(i)}=\sum_{j}\sum_{\sigma(1)=j}(-1)^{\sigma}\prod_{i} a_{i\sigma(i)}$$

$$=\sum_{j} a_{1j}\sum_{\sigma(1)=j}(-1)^{\sigma}\prod_{i>1} a_{i\sigma(i)}.$$

Let $\tau = (1j)$. Now

$$\sum_{\sigma(1)=j} (-1)^{\sigma} \prod_{i>1} a_{i\sigma(i)} = \sum_{\sigma(1)=j} (-1)^{\sigma} a_{j\sigma(j)} \prod_{1\neq i\neq j} a_{i\sigma(i)}$$

$$= - \sum_{\sigma(1)=j} (-1)^{\sigma\tau} a_{j\sigma\tau(j)} \prod_{1\neq i\neq j} a_{i\sigma\tau(i)}.$$

As σ runs through S_n with $\sigma(1) = j$, $\sigma\tau$ runs through all the permutaions of $\{1, 2, \ldots, j-1, j+1, \ldots, n\}$. Thus, $\sum_{\sigma(1)=j} (-1)^{\sigma\tau} a_{j\sigma\tau(1)} \prod_{1\neq i\neq j} a_{i\sigma\tau(i)}$ is the determinant of the matrix obtained from A_{1j} by moving its jth row to its first row. That moving changes the sign of $\det(A_{ij})$ $j-2$ times. Therefore,

$$\det(a_{ij}) = \sum_{j=1}^{n} (-1)^{j+1} a_{1j} \det A_{1j}.$$

For the general result, change (a_{ij}) by moving its ith row to its first row, obtaining a matrix (b_{ij}). Then

$$\det(a_{ij}) = (-1)^{i-1} \det(b_{ij}) = (-1)^{i-1} \sum_{j=1}^{n} (-1)^{j+1} b_{1j} \det B_{1j}$$

$$= \sum_{j=1}^{n} (-1)^{i+j} a_{ij} \det A_{ij} = \sum_{j=1}^{n} a_{ij}\alpha_{ij}.$$

An application of 5.4.2 completes the proof.

5.4.7 Corollary

$\sum_{j=1}^{n} a_{ij}\alpha_{kj} = 0$ if $i \neq k$, and $\sum_{i=1}^{n} a_{ij}\alpha_{ik} = 0$ if $j \neq k$.

Proof $\sum_{j=1}^{n} (-1)^{i+j} a_{ij} \det A_{kj}$ is the determinant of the matrix gotten from (a_{ij}) by replacing its kth row by its ith row. That matrix has the same ith and kth rows, and hence has determinant zero. The rest follows from 5.4.2.

Note that 5.4.6 and 5.4.7 can be combined. They assert that $\delta_{ik} \det(a_{ij}) = \sum_{i=1}^{n} a_{ij}\alpha_{ik} = \sum_{i=1}^{n} a_{ji}\alpha_{ki}$ for all j and k, where δ_{ik} is the Kronecker δ.

If 5.4.6 is looked at properly, it not only gives a method for calculating $\det(a_{ij})$, but it also gives a method for calculating the inverse of the matrix (a_{ij}), if it exists. Furthermore, it tells us exactly when (a_{ij}) does have an inverse.

Let $\operatorname{adj}(a_{ij})$, called the ***adjoint*** of (a_{ij}), be the matrix whose (i, j) entry is α_{ji}. Then 5.4.6 yields

$$\det(a_{ij}) I_n = (a_{ij}) \operatorname{adj}(a_{ij}) = (\operatorname{adj}(a_{ij}))(a_{ij}).$$

Thus, if $\det(a_{ij})$ is a unit in R, we get

$$I_n = (a_{ij})[\operatorname{adj}(a_{ij})/\det(a_{ij})] = [\operatorname{adj}(a_{ij})/\det(a_{ij})](a_{ij}).$$

In other words, $\text{adj}(a_{ij})/\det(a_{ij})$ is the inverse of (a_{ij}) in case $\det(a_{ij})$ is a unit in R. We noted earlier that $\det(a_{ij})$ is a unit if (a_{ij}) is a unit. Therefore, we have

5.4.8 Corollary

Let (a_{ij}) be in R_n. Then (a_{ij}) is a unit in R_n if and only if $\det(a_{ij})$ is a unit in R. Furthermore, if $\det(a_{ij})$ is a unit in R, then $(a_{ij})^{-1} = \text{adj}(a_{ij})/\det(a_{ij})$.

This corollary tells us a bit more. Suppose that a matrix (a_{ij}) in R_n has a right inverse. Then there is a matrix (b_{ij}) in R_n with $(a_{ij})(b_{ij}) = I_n$. Hence, $\det((a_{ij})(b_{ij})) = 1 = \det(a_{ij})\det(b_{ij})$. Since R is commutative, $\det(a_{ij})$ is a unit in R. Hence (a_{ij}) is a unit in R_n. Similarly, if (a_{ij}) has left inverse, it is a unit.

From 5.4.6 and 5.4.7, we can also get Cramer's Rule. Suppose that

$$\begin{aligned} a_{11}x_1 + a_{12}x_2 + \cdots + a_{1n}x_n &= b_1 \\ a_{21}x_1 + a_{22}x_2 + \cdots + a_{2n}x_n &= b_2 \\ &\vdots \\ a_{n1}x_1 + a_{n2}x_2 + \cdots + a_{nn}x_n &= b_n \end{aligned}$$

is a system of n equations in the n unknowns $x_1, x_2, \ldots, x_n$, with a_{ij} and b_i in R. The matrix (a_{ij}) is called the ***matrix of coefficients***, and $\det(a_{ij})$ is called the ***determinant*** of the system. Cramer's Rule gives a sufficient condition for such a system to have a unique solution.

5.4.9 Theorem (Cramer's Rule).

Let

$$\begin{aligned} a_{11}x_1 + a_{12}x_2 + \cdots + a_{1n}x_n &= b_1 \\ a_{21}x_1 + a_{22}x_2 + \cdots + a_{2n}x_n &= b_2 \\ &\vdots \\ a_{n1}x_1 + a_{n2}x_2 + \cdots + a_{nn}x_n &= b_n \end{aligned}$$

be a system of linear equations with coefficients a_{ij} and b_i in the commutative ring R with identity. If the matrix of coefficients (or equivalently, the determinant) of the system is a unit, then the system has a unique solution given by

$$x_k = \sum_{i=1}^{n} b_i \alpha_{ik}/\det(a_{ij}),$$

where α_{ij} is the cofactor of a_{ij}.

Proof If R were a field, then we know from vector space considerations that the system has a unique solution if (a_{ij}) is nonsingular. Just view (a_{ij})

as a linear transformation acting on the space of n-tuples of elements of the field. The question is whether there is a vector $(x_1, x_2, \ldots, x_n)$ whose image is $(b_1, b_2, \ldots, b_n)$. Since (a_{ij}) is nonsingular, there is exactly one such vector. However, our system here is over a commutative ring, but Cramer's Rule tells how to compute the desired vector. We compute. $\sum_{j=1}^{n} a_{ij}x_j = b_i$ implies that $\sum_{j=1}^{n} a_{ij}\alpha_{ik}x_j = b_i\alpha_{ik}$, and hence

$$\sum_{i=1}^{n} b_i\alpha_{ik} = \sum_{j=1}^{n}\sum_{i=1}^{n} a_{ij}\alpha_{ik}x_j = \sum_{i=1}^{n} a_{ik}\alpha_{ik}x_k = \det(a_{ij})x_k.$$

Thus if $\det(a_{ij})$ is a unit in R, we get $x_k = \sum_{i=1}^{n} b_i\alpha_{ik}/\det(a_{ij})$. Thus, there is at most one solution. But

$$\sum_{k=1}^{n} a_{jk} \sum_{i=1}^{n} b_i\alpha_{ik} = \sum_{i=1}^{n} b_i \sum_{k=1}^{n} a_{jk}\alpha_{ik} = b_j \det(a_{ij}),$$

and the theorem follows.

The numerator $\sum_{i=1}^{n} b_i\alpha_{ik}$ is the determinant of the matrix obtained by replacing the kth column of (a_{ij}) by $b_1, b_2, \ldots, b_n$. If R is a field, then $\det(a_{ij})$ is a unit if it is not 0. Therefore, such a system over a field has a unique solution if its determinant is not 0.

5.4.10 Lemma

Let $A_1, A_2, \ldots, A_n$ be square matrices with entries from R. Then

$$\det\begin{pmatrix} A_1 & & & \\ & A_2 & 0 & \\ & * & \ddots & \\ & & & A_n \end{pmatrix} = \det\begin{pmatrix} A_1 & & & \\ & A_2 & * & \\ & 0 & \ddots & \\ & & & A_n \end{pmatrix}$$

$$= \det(A_1)\det(A_2)\cdots\det(A_n).$$

Proof Let

$$(a_{ij}) = \begin{pmatrix} A_1 & * \\ 0 & A_2 \end{pmatrix}.$$

It suffices to show that $\det(a_{ij}) = \det(A_1)\det(A_2)$. Let A_1 be $r \times r$, A_2 be $s \times s$, and S be the symmetric group S_{r+s}. Then

$$\det(a_{ij}) = \sum_{\sigma\in S} (-1)^{\sigma} \prod_{i=1}^{r+s} a_{i\sigma(i)}.$$

If for some $k > r$, $\sigma(k) \le r$, then $\prod_{i=1}^{r+s} a_{i\sigma(i)} = 0$. Thus only those $\prod_{i=1}^{r+s} a_{i\sigma(i)}$ for which σ permutes $\{r+1, r+2, \ldots, r+s\}$, and hence also permutes

$\{1, 2, \ldots, r\}$, are nonzero. Every such permutation is uniquely a product $\sigma\tau$, where σ and τ are in S, σ fixes $\{r+1, r+2, \ldots, r+s\}$ elementwise, and τ fixes $\{1, 2, \ldots, s\}$ elementwise. It follows that

$$\det(a_{ij}) = \sum_{\sigma,\tau} (-1)^{\sigma\tau} \prod_{i=1}^{r} a_{i\sigma(i)} \prod_{i=r+1}^{r+s} a_{i\tau(i)}$$

$$= \sum_{\sigma} (-1)^{\sigma} \prod_{i=1}^{r} a_{i\sigma(i)} \sum_{\tau} (-1)^{\tau} \prod_{i=r+1}^{r+s} a_{i\tau(i)} = \det(A_1)\det(A_2),$$

where σ runs through the permutations with $\sigma(i) = i$ for $i > r$, and τ runs through those with $\tau(i) = i$ for $i \leq r$.

5.4.11 Corollary

The determinant of a triangular matrix is the product of the diagonal entries.

Let $f(x) = a_0 + a_1x + \cdots + a_{n-1}x^{n-1} + x^n$ be in $R[x]$. As in the case for fields, the ***companion matrix*** $C(f(x))$ is the matrix

$$\begin{pmatrix} 0 & 0 & 0 & \cdots & 0 & -a_0 \\ 1 & 0 & 0 & \cdots & 0 & -a_1 \\ 0 & 1 & 0 & \cdots & 0 & -a_2 \\ \vdots & & & & & \vdots \\ 0 & 0 & 0 & \cdots & 0 & -a_{n-2} \\ 0 & 0 & 0 & \cdots & 1 & -a_{n-1} \end{pmatrix}.$$

In case R is a field, we have proved (5.2.17) that the characteristic polynomial of $C(f(x))$ is $f(x)$.

5.4.12 Lemma

Let $f(x)$ be a monic polynomial in $R[x]$ of degree n. Then $\det(xI_n - C(f(x))) = f(x)$.

Proof Let $f(x) = a_0 + a_1x + \cdots + a_{n-1}x^{n-1} + x^n$. Then

$$xI_n - C(f(x)) = \begin{pmatrix} x & 0 & 0 & \cdots & 0 & a_0 \\ -1 & x & 0 & \cdots & 0 & a_1 \\ 0 & -1 & x & \cdots & 0 & a_2 \\ \vdots & & & & & \vdots \\ 0 & 0 & 0 & \cdots & x & a_{n-2} \\ 0 & 0 & 0 & \cdots & -1 & x + a_{n-1} \end{pmatrix}$$

Computing its determinant by expanding along the first row, inducting on n, and using 5.4.11, yields $\det(xI_n - C(f(x))) = x(a_1 + a_2x + \cdots + a_{n-1}x^{n-2}) + (-1)^{n-1}a_0(-1)^{n-1} = f(x)$.

Now we are in a position to prove the Cayley–Hamilton Theorem. It is a consequence of the following theorem which gives an explicit way to compute the characteristic polynomial of a matrix.

5.4.13 Theorem

Let (a_{ij}) be an $n \times n$ matrix over a field F. Then the characteristic polynomial of (a_{ij}) is $\det(xI_n - (a_{ij}))$.

Proof The matrix (a_{ij}) is similar to its rational canonical form

$$(c_{ij}) = \begin{pmatrix} C_1 & & & \\ & C_2 & & \\ & & \ddots & \\ & & & C_k \end{pmatrix},$$

where the C_i are the companion matrices of the invariant factors $f_i(x)$ of (a_{ij}). Thus for suitable (b_{ij}), $(c_{ij}) = (b_{ij})^{-1}(a_{ij})(b_{ij})$. We have

$$\begin{aligned}\det(xI_n - (a_{ij})) &= \det((b_{ij})^{-1})\det(xI_n - (a_{ij}))\det((b_{ij}) \\ &= \det((b_{ij})^{-1}xI_n(b_{ij}) - (b_{ij})^{-1}(a_{ij})(b_{ij})) \\ &= \det(xI_n - (c_{ij})) = \prod f_i(x),\end{aligned}$$

which is the characteristic polynomial of (a_{ij}).

An immediate corollary of this theorem and 5.4.11 is that if (a_{ij}) is any matrix of a linear transformation α which is triangular, then $c_\alpha(x) = \prod (x - a_{ii})$. In this connection, see 5.3.19.

We now have an effective way to compute the characteristic polynomial of a matrix (a_{ij}). Just compute $\det(xI_n - (a_{ij}))$. We have as an immediate consequence that every $n \times n$ matrix A over a field satisfies the polynomial $\det(xI_n - A)$. However, this fact holds for matrices over a commutative ring, and it follows from the identity $\det(A)I_n = A(\operatorname{adj}(A))$. Let A be in R_n,

$$c(x) = \det(xI_n - A) = x^n + a_{n-1}x^{n-1} + \cdots + a_1x + a_0,$$

and $Q(x) = \operatorname{adj}(xI_n - A)$. Then we have $c(x)I_n = (xI_n - A)Q(x)$, and $c(x)I_n - c(A) = \sum_{i=0}^{n} a_i(x^iI_n - A^i)$, where $a_n = 1$. Write $x^iI_n - A^i = (xI_n - A)Q_i(x)$. Then $c(x)I_n - c(A) = (xI_n - A)\sum_{i=0}^{n} a_iQ_i(x) = (xI_n - A)P(x) = (xI_n - A)Q(x) - c(A)$. Hence, $(xI_n - A)(Q(x) - P(x)) = c(A)$. If $Q(x) - P(x) \neq 0$, then the matrix $xI_n(Q(x) - P(x))$ has an entry which is a polynomial in x of higher degree than any entry in $A(Q(x) - P(x))$. It follows that $c(A) = 0$. We have proved the following theorem.

5.4.14 Cayley–Hamilton Theorem

Let A be an $n \times n$ matrix over a commutative ring R with identity. Then A satisfies the polynomial $\det(xI_n - A)$.

For any matrix A in R_n, we ***define*** its ***characteristic polynomial*** to be the polynomial $\det(xI_n - A)$.

There are two additional important points about determinants that we wish to make. We have seen that if α is a linear transformation on a vector space V, then we can define $\det(\alpha)$ to be $\det(a_{ij})$ for any matrix (a_{ij}) *of* α. There is a way to define $\det(\alpha)$ directly, that is, without going through the process of picking a basis of V and then taking the determinant of the matrix α relative to that basis. We will present this "coordinate-free" definition of $\det(\alpha)$. That is the first important point. The second is this. We want to *characterize* det as a function

$$\det: R_n \to R.$$

A number of its properties have been derived, and we will show that det is the *only* function $R_n \to R$ satisfying certain conditions. In fact, we take up this point first.

For our commutative ring R, let R^n be the module of n-tuples of elements of R, but view an element of R^n as a column

$$\begin{pmatrix} r_1 \\ r_2 \\ \vdots \\ r_n \end{pmatrix}$$

rather than as a row. Then an element of R_n, that is, an $n \times n$ matrix over R, can be thought of as an element of the Cartesian product $R^n \times R^n \cdots \times R^n$ of R^n with itself n times. Thus

$$\det: R^n \times R^n \times \cdots \times R^n \to R,$$

and we want to look at how det behaves with respect to the individual columns. We know already that if two columns of (a_{ij}) are equal then $\det(a_{ij}) = 0$, and that if any column of (a_{ij}) is 0, then $\det(a_{ij}) = 0$. We need two more facts that follow from the definition. We can add elements of R^n, and we can multiply them by scalars. This is done coordinatewise, by definition. Let

(a_{ij}) be in R_n, and let

$$c_i = \begin{pmatrix} a_{1i} \\ a_{2i} \\ \vdots \\ a_{ni} \end{pmatrix}.$$

Then $(a_{ij}) = (c_1, c_2, \ldots, c_n)$. Let c be in R. It should be clear from the definition of det that for any i,

$$\det(c_1, c_2, \ldots, c_{i-1}, cc_i, c_{i+1}, \ldots, c_n) = c \cdot \det(c_1, c_2, \ldots, c_n).$$

Further, if d_i is in R^n, then

$$\det(c_1, c_2, \ldots, c_{i-1}, c_i + d_i, c_{i+1}, \ldots, c_n)$$
$$= \det(c_1, \ldots, c_i, \ldots, c_n) + \det(c_1, \ldots, d_i, \ldots, c_n).$$

These two properties are expressed by saying that det is ***linear*** in each variable c_i. Here is a more general definition.

5.4.15 Definition

Let $C_1, C_2, \ldots, C_n$, and D be modules over R. A function

$$f: C_1 \times C_2 \times \cdots \times C_n \to D$$

is ***n*-multilinear** *if for $i = 1, 2, \ldots, n$, and for each r in R,*

$$f(c_1, \ldots, c_{i-1}, c_i + d_i, c_{i+1}, \ldots, c_n)$$
$$= f(c_1, \ldots, c_{i-1}, c_i, c_{i+1}, \ldots, c_n) + f(c_1, \ldots, c_{i-1}, d_i, c_{i+1}, \ldots, c_n),$$

and

$$f(c_1, \ldots, c_{i-1}, rc_i, c_{i+1}, \ldots, c_n) = r \cdot f(c_1, \ldots, c_{i-1}, c_i, c_{i+1}, \ldots, c_n).$$

*If $D = R$, then f is an n-**multilinear form**. If*

$$C = C_1 = C_2 = \cdots = C_n,$$

and $f(c_1, c_2, \ldots, c_n) = 0$ whenever $c_i = c_j$ for some $i \neq j$, then f is called **alternating**.

Thus, f is n-multilinear if f is linear as a function of each variable c_i when the others are held fixed. If $C_1 = C_2 = \cdots = C_n$ and f vanishes when two coordinates are equal, then f is alternating. Thus, det is an n-multilinear alternating form. We need one more fact. If C and D are R-modules, an

n-multilinear function $f: C \times C \times \cdots \times C \to D$ is ***skew-symmetric*** if for $i < j$,

$$f(c_1, c_2, \ldots, c_i, \ldots, c_j, \ldots, c_n) = \\ -f(c_1, \ldots, c_{i-1}, c_j, c_{i+1}, \ldots, c_{j-1}, c_i, c_{j+1}, \ldots, c_n).$$

Thus, switching two coordinates of $(c_1, c_2, \ldots, c_n)$ changes the sign of its value under f. An important fact is that if f is alternating, then it is skew-symmetric. Moreover, if f is skew-symmetric and σ is in S_n, then $f(c_{\sigma(1)}, \ldots, c_{\sigma(n)}) = (-1)^\sigma f(c_1, \ldots, c_n)$. Just note that if f is alternating, then (only writing the i and j coordinates) we get $f(c_i, c_j) = -f(c_j, c_i)$ from the equation

$$\begin{aligned} 0 &= f(c_i + c_j, c_i + c_j) \\ &= f(c_i, c_i) + f(c_i, c_j) + f(c_j, c_i) + f(c_j, c_j) \\ &= f(c_i, c_j) + f(c_j, c_i). \end{aligned}$$

This says that $f(c_{\sigma(1)}, \ldots, c_{\sigma(n)}) = (-1)^\sigma f(c_1, \ldots, c_n)$ for transpositions σ. Since every σ in S_n is a product of transpositions, the equality holds for arbitrary σ.

5.4.16 Theorem

Let r be in R. Then there is exactly one n-multilinear alternating form

$$\det_r: R^n \times R^n \times \cdots \times R^n \to R$$

such that $\det_r(I_n) = r$.

Proof There is at least one such form. Let $\det_r = r \cdot \det$. It is completely routine to check that $\det_r$ is an n-multilinear alternating form with $\det_r(I_n) = r$. Now let d_r be any such form. Let u_i be the ith column of I_n. Then $(u_1, u_2, \ldots, u_n) = I_n$, so $d_r(u_1, u_2, \ldots, u_n) = r$. Now we compute

$$d_r(c_1, c_2, \ldots, c_n).$$

Let $(a_{ij}) = (c_1, c_2, \ldots, c_n)$. Then $c_i = \sum_{j=1}^n a_{ji}u_j$, and we have $d_r(c_1, c_2, \ldots, c_n) = d_r(\sum_{j=1}^n a_{j1}u_j, \ldots, \sum_{j=1}^n a_{jn}u_j)$. We can expand this last expression by multilinearity, leaving only u_i's inside. But terms in this expansion will be nonzero only if the u_i's inside are distinct. In short, this last expression becomes

$$\begin{aligned} &\sum_\sigma \left(\prod_{i=1}^n a_{\sigma(i)i}\right) d_r(u_{\sigma(1)}, \ldots, u_{\sigma(n)}) \\ &\quad = \left(\sum_\sigma (-1)^\sigma \prod_{i=1}^n a_{\sigma(i)i}\right) d_r(u_1, \ldots, u_n) = r \cdot \det(a_{ij})' \\ &\quad = r \cdot \det(a_{ij}) = r \cdot \det(c_1, \ldots, c_n). \end{aligned}$$

There is a nice way to view 5.4.16. It is routine to check that the alternating forms on $R^n \times \cdots \times R^n$ form an R-module under the operation $(f+g)(X) = f(X) + g(X)$, and $(rf)(X) = r(f(X))$, where X is in $R^n \times \cdots \times R^n$ and r is in R. Our theorem then says that the module of n-multilinear alternating forms on $R^n \times \cdots \times R^n = R_n$ is free of rank 1, and in fact det is a generator. To generalize a bit, the R-module R^n is free of rank n. Let M be any free R-module of rank n. Then certainly the R-module of n-multilinear alternating forms on M^n, the Cartesian product of M with itself n times, is free of rank 1. The proof is the same as above. Just choose $\{u_1, u_2, \ldots, u_n\}$ to be a basis of M. Then for any n-multilinear alternating form f, $f(\sum a_{i1}u_1, \ldots, \sum a_{in}u_i) = \det(a_{ij}) \cdot f(u_1, \ldots, u_n)$. Hence, f is determined by $f(u_1, \ldots, u_n)$. Defining

$$d\left(\sum_i a_{i1}u_i, \ldots, \sum_i a_{in}u_i\right) = \det(a_{ij})$$

yields an n-multilinear alternating form with $d(u_1, \ldots, u_n) = 1$. Thus $f = f(u_1, \ldots, u_n) \cdot d$, or $\{d\}$ is a basis of the module of such forms.

Now suppose that α is any endomorphism of the free R-module M of rank n. Let $\mathrm{Alt}_n(M)$ denote the R-module of n-multilinear forms on M. For f in $\mathrm{Alt}_n(M)$, let

$$(\bar{\alpha}(f))(m_1, \ldots, m_n) = f(\alpha(m_1), \ldots, \alpha(m_n)).$$

Then $\bar{\alpha}$ is an endomorphism of the R-module $\mathrm{Alt}_n(M)$. But $\mathrm{Alt}_n(M)$ is free of rank 1. That is, $\bar{\alpha}$ is just multiplication by some scalar r_α in R. Thus, $\bar{\alpha}(f) = r_\alpha \cdot f$. With each endomorphism α of M we have associated a scalar r_α in R. We define $r_\alpha = \det(\alpha)$. When M is a vector space, is this consistent with defining $\det(\alpha)$ to be the determinant of any matrix of α? Let $\{u_1, \ldots, u_n\}$ be a basis of the free module M of rank n. Let d be the n-multilinear alternating form with $d(u_1, \ldots, u_n) = 1$. Then, as in the case for vector spaces, an endomorphism α of M is determined uniquely by the matrix (a_{ij}) given by

$$\alpha(u_j) = \sum_{i=1}^{n} a_{ij}u_i.$$

But

$$\begin{aligned}
\bar{\alpha}(d)(u_1, \ldots, u_n) &= d(\alpha(u_1), \ldots, \alpha(u_n)) \\
&= d\left(\sum_{i=1}^{n} a_{i1}u_i, \ldots, \sum_{i=1}^{n} a_{in}u_i\right) = \det(a_{ij})d(u_1, \ldots, u_n) \\
&= (\det(\alpha)) \cdot d(u_1, \ldots, u_n) = \det(a_{ij}) = \det(\alpha).
\end{aligned}$$

PROBLEMS

1 How many additions and multiplications are necessary in order to calculate the determinant of a 5×5 matrix? a 10×10?

2 Calculate the determinant of the following matrices.

$$\begin{pmatrix} 6 & -5 & 4 & 8 \\ 1 & 1 & 3 & 2 \\ 7 & 9 & -4 & 5 \\ 2 & 1 & 2 & 1 \end{pmatrix} \quad \begin{pmatrix} 3 & 7 & 2 & 4 \\ -3 & 6 & 0 & 5 \\ 0 & 5 & 4 & 3 \\ 3 & 1 & 0 & 8 \end{pmatrix}$$

$$\begin{pmatrix} 75 & 63 & -18 & 39 \\ -61 & 45/2 & -119 & 67 \\ 76 & 83 & 109/7 & 31 \\ 25 & 52 & -6 & 13 \end{pmatrix} \quad \begin{pmatrix} 1 & 4 & 9 & 16 \\ 4 & 9 & 16 & 25 \\ 9 & 16 & 25 & 36 \\ 16 & 25 & 25 & 49 \end{pmatrix}.$$

3 Are the vectors $(1, 0, 3, 2, 1)$, $(1, 1, 4, 3, 2)$, $(2, 0, 0, 1, 1)$, $(3, 4, 5, 0, 1)$, $(0, 1, 1, 1, 1)$ independent?

4 Prove that the determinant of

$$A = \begin{pmatrix} 1 & x_1 & x_1^2 & \cdots & x_1^{n-1} \\ 1 & x_2 & x_2^2 & \cdots & x_2^{n-1} \\ \vdots & & & & \vdots \\ 1 & x_n & x_n^2 & \cdots & x_n^{n-1} \end{pmatrix}$$

is $\prod_{i<j} (x_j - x_i)$. (The matrix A is ***Vandermonde's matrix***.)

5 Prove that every monic polynomial in $R[x]$ is the characteristic polynomial of some matrix over R.

6 Prove that AB and BA have the same characteristic polynomial.

7 Prove that the eigenvalues of a triangular matrix are the diagonal entries of that matrix.

8 Let K be a subfield of the field F, and let A be a matrix in K_n. Prove that A is invertible in K_n if and only if it is invertible in F_n.

9 Use Cramer's rule to solve the system

$$\begin{aligned} x - y + z &= 2 \\ 2x + y \phantom{{}+z} &= 5 \\ 3x + 7y - 2z &= 6. \end{aligned}$$

10 Solve the same system as in Problem 9 where the coefficients are taken modulo 11.

11 Let V be a finite-dimensional vector space over a field F, and let $A(V)$ be the vector space of all linear transformations on V. Let α be in $A(V)$, and let $f_\alpha: A(V) \to A(V)$ be defined by $f_\alpha(\beta) = \alpha\beta - \beta\alpha$. Prove that f_α is linear and that $\det(f_\alpha) = 0$.

12 Let $A(V)$ be the vector space of all linear transformations on the finite-dimensional vector space V. Let α be in $A(V)$. Let $f_\alpha(\beta) = \alpha\beta$ for β in $A(V)$. Prove that f_α is a linear transformation on the vector space $A(V)$, and find $\det(f_\alpha)$ in terms of $\det(\alpha)$.

13 Let S be a commutative subring of the ring of all $n \times n$ matrices over the commutative ring R. Let A be an $m \times m$ matrix over S. View A as an $(nm) \times (nm)$ matrix over R. As such a matrix, its determinant $\det_R(A)$ is an element of R. As a matrix over S, its determinant $\det_S(A)$ is an element of S. Prove that $\det_R(A) = \det(\det_S(A))$.

14 Prove that any 2×2 real matrix whose determinant is negative is similar over the real numbers to a diagonal matrix.

15 Let n be any positive integer. Prove that over the complex numbers the $n \times n$ matrix

$$\begin{bmatrix} 0 & 0 & 0 & \cdots & 0 & -1 \\ 1 & 0 & 0 & \cdots & 0 & 0 \\ 0 & 1 & 0 & \cdots & 0 & 0 \\ \vdots & & & & & \vdots \\ 0 & 0 & 0 & \cdots & 0 & 0 \\ 0 & 0 & 0 & \cdots & 0 & 0 \\ 0 & 0 & 0 & \cdots & 1 & 0 \end{bmatrix}$$

is similar to a diagonal matrix.

16 Let (a_{ij}) be an $n \times n$ matrix over the field F. Prove that there are at most n elements a in F such that $\det(aI_n - (a_{ij})) = 0$.

17 Let V be a vector space of dimension n over F, and let α and β be linear transformations on V. Suppose that α is nonsingular. Prove that there are at most n scalars a in F such that $a\alpha + \beta$ is singular.

18 Let A be an invertible matrix. Find the characteristic polynomial of A^{-1} in terms of that of A.

19 Prove that the constant term of the characteristic polynomial of an $n \times n$ matrix A is $(-1)^n\det(A)$.

20 Prove that A is a unit if and only if $\operatorname{adj}(A)$ is a unit.

21 Prove that $\operatorname{adj}(A') = (\operatorname{adj}(A))'$.

22 Prove that for an $n \times n$ matrix, $\det(\operatorname{adj}(A)) = (\det(A))^{n-1}$. What if $n = 1$?

23 A matrix A is ***skew*** if $A' = -A$. Prove that an odd $\times$ odd skew matrix over the complex numbers is singular.

24 Prove that the set of all real matrices of the form

$$\begin{pmatrix} a & -b & -c & -d \\ b & a & -d & c \\ c & d & a & -b \\ d & -c & b & a \end{pmatrix}$$

is a division ring. Prove that as a vector space over the real numbers, it is of dimension 4 and has a basis $\{1, i, j, k\}$ such that $i^2 = j^2 = k^2 = -1$, $ij = -ji = k$, $jk = -kj = i$, and $ki = -ik = j$.

25 If R is any commutative ring, the group of units of R_n is denoted $\mathrm{GL}_n(R)$. Let S be any subgroup of the group of units R^* of R. Prove that $N = \{A \in \mathrm{GL}_n(R)\colon \det(A) \in S\}$ is normal in $\mathrm{GL}_n(R)$, and that $\mathrm{GL}_n(R)/N \approx R^*/S$.

26 Prove that if the $n \times n$ matrix A over the field F has rank m, then A has an $m \times m$ nonsingular "submatrix" and has no larger nonsingular submatrix.

5.5 EQUIVALENCE OF MATRICES

We obtained similarity invariants for matrices over a field F by associating with a matrix a finitely generated module over $F[x]$, and showing that two matrices were similar if and only if the corresponding modules were isomorphic (5.2.9). Having a complete set of invariants for finitely generated $F[x]$-modules yielded complete sets of invariants for similarity classes of matrices over F (5.2.13). One way to find these invariants of a given matrix is to find (somehow) its rational canonical form (5.2.19). There is another quite different way. We have observed that the characteristic polynomial of a matrix (a_{ij}) in F_n can be gotten quite easily: Just pass to the matrix $xI_n - (a_{ij})$ and take its determinant. More generally, the invariant factors of (a_{ij}) may be obtained directly from the matrix $xI_n - (a_{ij})$. One purpose of this section is to see how this is done. To facilitate matters, we need to discuss matrices of homomorphisms between free modules over commutative rings. This will be a generalization of some of the material in 5.1.

Throughout, R will be a commutative ring with identity. If a free R-module has a basis of n elements, it is said to be ***free of rank n*** (4.6, page 168). Let X and Y be free R-modules with bases $\{x_1, x_2, \ldots, x_n\}$ and $\{y_1, y_2, \ldots, y_m\}$, respectively. A homomorphism $\alpha\colon X \to Y$ is determined by its action on the basis $\{x_1, x_2, \ldots, x_n\}$. Furthermore, if $w_1, w_2, \ldots, w_n$ are in Y, then there is exactly one homomorphism $\alpha\colon X \to Y$ such that $\alpha(x_j) = w_j$. Therefore, if $\alpha(x_j) = \sum_{i=1}^m a_{ij}y_i$, $j = 1, 2, \ldots, n$, then α is uniquely determined by the $m \times n$ matrix (a_{ij}). The association $\alpha \to (a_{ij})$ is a one-to-one correspondence between the set $\mathrm{Hom}(X, Y)$ of homomorphisms from X into Y and the set of $m \times n$ matrices over R. The matrix (a_{ij}) is called the ***matrix of*** α ***relative to the bases*** $\{x_1, x_2, \ldots, x_n\}$ and $\{y_1, y_2, \ldots, y_m\}$. If $X = Y$ and $x_i = y_i$ for all i, then the square matrix (a_{ij}) in R_n is the ***matrix of*** α ***relative to*** $\{x_1, x_2, \ldots, x_n\}$. This is the situation discussed in 5.1 in the case R is a field.

We need to find out what happens to (a_{ij}) if the bases are changed. So let $\{v_1, v_2, \ldots, v_n\}$ be a basis of X, and let $\{w_1, w_2, \ldots, w_m\}$ be a basis of Y. If $\alpha(v_j) = \sum_{i=1}^m b_{ij}w_i$, what is the relation between (a_{ij}) and (b_{ij})? Let $v_j = \sum_{i=1}^n c_{ij}x_i$, and let $w_j = \sum_{i=1}^m d_{ij}y_i$. Then

$$\begin{aligned}\alpha(v_j) &= \sum_i b_{ij}w_i = \sum_i b_{ij} \sum_k d_{ki}y_k = \sum_k \left(\sum_i b_{ij}d_{ki}\right)y_k = \alpha\left(\sum_i c_{ij}x_i\right) \\ &= \sum_i c_{ij}\alpha(x_i) = \sum_i c_{ij} \sum_k a_{ki}y_k = \sum_k \left(\sum_i c_{ij}a_{ki}\right)y_k.\end{aligned}$$

Therefore,

$$(d_{ij})(b_{ij}) = (a_{ij})(c_{ij}).$$

Note that (c_{ij}) and (d_{ij}) are invertible. Indeed, if $x_i = \sum_k e_{ki}v_k$, then $v_j = \sum_i c_{ij}x_i = \sum_i c_{ij} \sum_k e_{ki}v_k = \sum_k (\sum_i c_{ij}e_{ki})v_k$. Thus, $\sum_i e_{ki}c_{ij} = \delta_{ik}$. In other words, $(e_{ij})(c_{ij}) = I_n$. Similarly, $(c_{ij})(e_{ij}) = I_n$, and so $(c_{ij})^{-1} = (e_{ij})$. Similarly, (d_{ij}) is also invertible. We have

$$(d_{ij})(b_{ij}) = (a_{ij})(c_{ij})$$

with (c_{ij}) and (d_{ij}) invertible matrices. There is terminology for matrices so related.

5.5.1 Definition

Let A and B be $m \times n$ matrices over R. Then A is **equivalent** *to B if there are an invertible $m \times m$ matrix P over R and an invertible $n \times n$ matrix Q over R such that $B = PAQ$.*

Equivalent is an equivalence relation, and we have seen that matrices of the same homomorphism $\alpha: X \to Y$ are equivalent. The converse holds also.

5.5.2 Theorem

Let A and B be $m \times n$ matrices over R. Then A is equivalent to B if and only if they are matrices of the same homomorphism from a free R-module X of rank n into a free R-module Y of rank m relative to appropriate pairs of bases.

Proof Suppose that A is equivalent to B. Let X and Y be free of ranks n

and m, and with bases $\{x_1, x_2, \ldots, x_n\}$ and $\{y_1, y_2, \ldots, y_m\}$, respectively. There are invertible matrices P and Q such that $B = PAQ$. Let $P = (p_{ij})$, $Q = (q_{ij})$, and $v_j = \sum_i q_{ij}x_i$. We need $\{v_1, v_2, \ldots, v_n\}$ to be a basis. Let $(c_{ij}) = (q_{ij})^{-1}$. If $\sum_{i=1}^{n} a_i v_i = 0$, then

$$0 = \sum_i a_i \sum_j q_{ji}x_j = \sum_j \Big(\sum_i a_i q_{ji}\Big)x_j,$$

whence $\sum_i a_i q_{ji} = 0$ for all i. Thus,

$$0 = \sum_j \Big(\sum_i a_i q_{ji}\Big)c_{kj} = \sum_i a_i \sum_j q_{ji}c_{kj} = a_k.$$

Furthermore, $x_j = \sum_i c_{ij}v_i$, and it follows that $\{v_1, v_2, \ldots, v_n\}$ is a basis of X. Similarly, if $P^{-1} = (c_{ij})$ and $w_j = \sum_i c_{ij}y_i$, then $\{w_1, w_2, \ldots, w_m\}$ is a basis of Y. Define $\alpha: X \to Y$ by $\alpha(x_j) = \sum_i a_{ij}y_i$. We have

$$\begin{aligned}\alpha(v_j) = \alpha\Big(\sum_i q_{ij}x_i\Big) &= \sum_i q_{ij} \sum_k a_{ki}y_k = \sum_i q_{ij} \sum_k a_{ki} \sum_s p_{sk}w_s \\ &= \sum_s \Big(\sum_k \sum_i p_{sk}a_{ki}q_{ij}\Big)w_s = \sum_i b_{ij}w_i,\end{aligned}$$

since $B = PAQ$. Therefore A and B are both matrices of α, and the proof is complete.

Note that if $m = n$, $X = Y$, and $x_i = y_i$, then $Q = P^{-1}$, so that $B = PAP^{-1}$. Hence, as in the vector space case, two square matrices are similar if and only if they are matrices of the same endomorphism $\alpha: X \to X$ relative to appropriate (single) bases of X.

There are two fundamental ways to associate a module with a matrix. There is the one for square matrices over a field in 5.2. That procedure can be generalized. Let R^n be the free R-module of rank n, that is, the direct sum of n copies of the module R. As in 5.4, think of R^n as columns (r_i) of elements of R. If (a_{ij}) is in R^n, then $R^n \to R^n: (r_i) \to (\sum_j a_{ij}r_j)$ is an endomorphism of R^n. It is just the endomorphism whose matrix with respect to the canonical basis $\{e_1, e_2, \ldots, e_n\}$, where e_i is the column with 1 in the ith position and 0's elsewhere, is (a_{ij}). The action of (a_{ij}) on (r_i) is just matrix multiplication. Thus with a matrix A in R_n, we have an endomorphism of R^n, which we also denote by A. This enables us to make R^n into an $R[x]$-module by defining $f(x)v = f(A)(v)$ for v in R^n. Denote this $R[x]$-module by $M_1(A)$. Thus, with an $n \times n$ matrix A over R we have associated an $R[x]$-module $M_1(A)$. For fields, this is the construction carried out in 5.2. [$M_1(A)$ corresponds to V^α]. If A and B are in R_n, then $M_1(A) \approx M_1(B)$ if and only if A is similar to B. The proof is exactly the same as for the vector space case. Although $M_1(A)$ is a finitely generated $R[x]$-module, $R[x]$ is not necessarily a principal ideal domain, and so we have no nice complete set of invariants for $M_1(A)$ as we had in the case when R was a field.

The other module associated with A is $R^n/A(R^n)$. However, we want to be a little more general here. Let A be an $m \times n$ matrix over R. Then the homomorphism $R^n \to R^m$ whose matrix with respect to the canonical bases of R^n and R^m is A, is given by matrix multiplication, and is also denoted by A. Thus

$$(r_i) \to \left(\sum_{j=1}^{n} a_{ij} r_j \right) = A(r_i),$$

where $A = (a_{ij})$. Let $M_2(A)$ be the module $R^m/A(R^n)$. Now $M_2(A)$ is an R-module, while (in case A is a square matrix) $M_1(A)$ is an $R[x]$-module.

Here are the essential facts we are after. If R is a PID, and A and B are $m \times n$ matrices over R, then A is equivalent to B if and only if $M_2(A) \approx M_2(B)$. In this case, $M_2(A)$ is finitely generated over a PID, so is determined by its invariant factors. We would like to get the invariant factors of $M_2(A)$ directly from A. It turns out that there is a canonical form for A, that is, a matrix equivalent to A, which displays explicitly those invariant factors. Now let A be in R_n. Then both $M_1(A)$ and $M_2(xI_n - A) = R[x]^n/(xI_n - A)R[x]^n$ are modules over $R[x]$. These modules are isomorphic. If R is a field, then $R[x]$ is a PID. Therefore $xI_n - A$ is equivalent to $xI_n - B$ if and only if A is similar to B. The remainder of this section is devoted to establishing these facts.

5.5.3 Theorem

Let R be a commutative ring with identity. If A and B are equivalent $m \times n$ matrices over R, then $M_2(A) \approx M_2(B)$. That is,

$$R^m/A(R^n) \approx R^m/B(R^n).$$

Proof The proof is a consequence of a more general fact, which in turn is almost a formality. Suppose that $\alpha_i: M_i \to N_i$, $i = 1, 2$, are homomorphisms between the R-modules M_i and N_i. Suppose that $\beta_1: M_1 \to M_2$ and $\beta_2: N_1 \to N_2$ are isomorphisms such that $\alpha_1 = \beta_2^{-1}\alpha_2\beta_1$. Then

$$N_1/\alpha_1(M_1) \to N_2/\alpha_2(M_2): n_1 + \alpha_1(M_1) \to \beta_2(n_1) + \alpha_2(M_2)$$

is an isomorphism. The verification of this is routine in all aspects. Here is a picture.

$$\begin{array}{ccc} M_1 & \xrightarrow{\alpha_1} & N_1 \\ {\scriptstyle\beta_1}\downarrow & & \downarrow{\scriptstyle\beta_2} \\ M_2 & \xrightarrow{\alpha_2} & N_2 \end{array}$$

The relevant picture for our theorem is

$$\begin{array}{ccc} R^m & \xrightarrow{A} & R^n \\ {\scriptstyle Q}\downarrow & & \downarrow{\scriptstyle P^{-1}} \\ R^m & \xrightarrow{B} & R^n \end{array}$$

where $A = PBQ$ with P and Q appropriate nonsingular matrices.

Such pictures as above are called ***diagrams***, and the condition $\beta_2\alpha_1 = \alpha_2\beta_1$ is expressed by saying that the first diagram ***commutes***, or is a ***commutative diagram***.

The converse of 5.5.3 holds when R is a PID. That is, if R is a PID, if A and B are $m \times n$ matrices over R, and if $M_2(A) \approx M_2(B)$, then A and B are equivalent. In order to prove this, we need a theorem about submodules of free modules of finite rank. This theorem implies that finitely generated modules over a PID are direct sums of cyclic modules. We could actually derive it from our results in 4.7, but we prefer to give an alternate (direct) proof. This theorem will also give a nice canonical form for equivalence classes of equivalent matrices over PID's. Here is the theorem.

5.5.4 Theorem

Let M be a free module of finite rank n over a PID *R. Let N be a submodule of M. Then there are an integer $m \le n$, a basis $\{e_1, e_2, \ldots, e_n\}$ of M, and elements $r_1, r_2, \ldots, r_n$ of R such that $\{r_1e_1, r_2e_2, \ldots, r_me_m\}$ is a basis of N, $r_1 \mid r_2, r_2 \mid r_3, \ldots, r_{m-1} \mid r_m$, and $r_{n+1}, \ldots, r_n = 0$. The r_i are unique up to associates.*

Proof Induct on n. If $n = 1$, the theorem is clear. Suppose that the theorem is true for all free modules M of rank less than n. If $B = \{e_1, e_2, \ldots, e_n\}$ is a basis of M, then $M = Re_1 \oplus Re_2 \oplus \cdots \oplus Re_n$. Let B_i be the projection of M into Re_i followed by the natural map $Re_i \to R$. The set $B_i(N)$ is an ideal of R. It follows readily from 4.4.6 that this set of ideals, determined by all different bases of M, has a maximal element. Therefore, there is a basis $\{e_1, e_2, \ldots, e_n\} = B$ of M such that $B_1(N)$ is not properly contained in any $C_i(N)$ for any basis C of M. Let $B_1(N) = Ra_1$. Then $a_1e_1 + a_2e_2 + \cdots + a_ne_n$ is in N for some $a_2, \ldots, a_n$ in R. We want to show that a_1 divides a_2. Let d be a greatest common divisor of a_1 and a_2. Write $db_i = a_i$, and $s_1b_1 + s_2b_2 = 1$. Let $y_1 = b_1e_1 + b_2e_2$ and $y_2 = s_2e_1 - s_1e_2$. Then $s_1y_1 + b_2y_2 = e_1$ and $s_2y_1 - b_1y_2 = e_2$. It follows that $\{y_1, y_2, e_3, \ldots, e_n\}$ is a basis of M. Since

$$dy_1 + a_3e_3 + \cdots + a_ne_n = a_1e_1 + a_2e_2 + \cdots + a_ne_n$$

is in N, the ideal associated with y_1 in this basis contains Rd which contains Ra_1, hence is equal to Ra_1. Therefore d is in the ideal generated by a_1, and so a_1 divides a_2. Thus, a_1 divides a_i for all i.

Let $a_i = a_1 c_i$. Let $x_1 = e_1 + c_2 e_2 + \cdots + c_n e_n$. Then $\{x_1, e_2, e_3, \ldots, e_n\}$ is a basis of M and $a_1 x_1 = a_1 e_1 + a_2 e_2 + \cdots + a_n e_n$ is in N. Thus with respect to the basis $\{x_1, e_2, e_3, \ldots, e_n\}$, the image of the projection of N into Rx_1 is $Ra_1 x_1$, which is the intersection $N \cap Rx_1$. We have $N = (N \cap Rx_1) \oplus (N \cap (Re_2 \oplus \cdots \oplus Re_n))$. If this second intersection is 0, the basis $\{x_1, e_2, \ldots, e_n\}$ together with the elements $r_1 = a_1$, $r_2 = \cdots = r_n = 0$ satisfy our requirements. Otherwise, by the induction hypothesis, there are an integer m with $2 \leq m \leq n$, a basis $\{x_2, \ldots, x_n\}$ of $Re_2 \oplus \cdots \oplus Re_n$, and elements $r_2, \ldots, r_n$ in R such that $\{r_2 x_2, \ldots, r_m x_m\}$ is a basis of $N \cap (Re_2 \oplus \cdots \oplus Re_n)$, and such that

$$r_2 \mid r_3, r_3 \mid r_4, \ldots, r_{m-1} \mid r_m, r_{m+1} = \cdots = r_n = 0 \qquad (\text{if } m < n).$$

We need only that $a_1 \mid r_2$. But we already showed that if $\{e_1, e_2, \ldots, e_n\}$ is any basis and $a_1 e_1 + a_2 e_2 + \cdots + a_n e_n$ is in N, then $a_1 \mid a_2$. Now $a_1 x_1 + r_2 x_2$ is in N, whence $a_1 \mid r_2$. Setting $a_1 = r_1$ gets the required r_i's.

For the uniqueness of the r_i's, we appeal to 4.7.15; those not zero or a unit are just the invariant factors of M/N.

An immediate consequence is the fact that a finitely generated module X over a PID is a direct sum of cyclic modules. Let $\{x_1, x_2, \ldots, x_n\}$ be a set of generators of X. Let M be free of rank n with basis $\{e_1, e_2, \ldots, e_n\}$. Sending e_i to x_i induces an epimorphism $\alpha: M \to X$. Now there are a basis $\{y_1, y_2, \ldots, y_n\}$ and elements $r_1, r_2, \ldots, r_m$ $(m \leq n)$ of R such that $\{r_1 y_1, \ldots, r_m y_m\}$ is a basis of $\operatorname{Ker}(\alpha)$. Hence $X \approx M/\operatorname{Ker}(\alpha) \approx Ry_1/Rr_1 y_1 \oplus \cdots \oplus Ry_n/Rr_n y_n$ (where $r_{m+1} = \cdots = r_n = 0$), which is a direct sum of cyclic modules.

5.5.5 Corollary

Let R be a PID, *and let A and B be $m \times n$ matrices over R. Then A and B are equivalent if and only if the modules $R^m/A(R^n)$ and $R^m/B(R^n)$ are isomorphic.*

Proof By 5.5.3, we need only show that if $R^m/A(R^n) \approx R^m/B(R^n)$, then A and B are equivalent. By 5.5.4, there is a basis $\{x_1, x_2, \ldots, x_m\}$ of R^m and there are elements $r_1, r_2, \ldots, r_m$ of R such that $\{r_1 x_1, r_2 x_2, \ldots, r_k x_k\}$ is a basis of $A(R^n)$, and such that $r_1 \mid r_2, r_2 \mid r_3, \ldots, r_{k-1} \mid r_k$, $r_{k+1} = \cdots = r_m = 0$ (if $k < m$). The invariant factors of the module $R^m/A(R^n)$ are the nonunits among $r_1, r_2, \ldots, r_m$. Since $R^m/A(R^n) \approx R^m/B(R^n)$, there is a basis $\{y_1, y_2, \ldots, y_m\}$ of R^m

such that the nonzero elements of $\{r_1y_1, r_2y_2, \ldots, r_my_m\}$ are a basis of $B(R^n)$. Since $A(R^n)$ is free, $R^n = \mathrm{Ker}(A) \oplus S$, and A induces an isomorphism $S \to A(R^n)$. Similarly, $R^n = \mathrm{Ker}(B) \oplus T$, and B induces an isomorphism $T \to B(R^n)$. Thus, $\mathrm{Ker}(A) \approx \mathrm{Ker}(B)$ by 4.7.5. The map $R^m \to R^m: x_i \to y_i$ induces an isomorphism $\beta: A(R^n) \to B(R^n)$. This gives an isomorphism $\alpha: S \to T$ such that the diagram

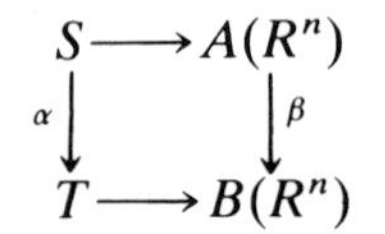

commutes. Extend α to an isomorphism $\mathrm{Ker}(A) \oplus S = R^n \to R^n = \mathrm{Ker}(B) \oplus T$ by sending $\mathrm{Ker}(A)$ isomorphically onto $\mathrm{Ker}(B)$. Then the diagram

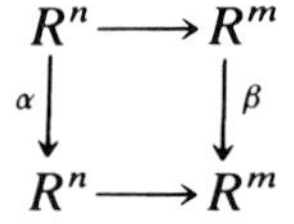

commutes. Let P and Q be the matrices of β and α, relative to the canonical bases of R^m and R^n, respectively. Then $A = QBP^{-1}$, whence A and B are equivalent.

Recall that the ***rank*** of a matrix A is the dimension of its row space, or equivalently, the dimension of its column space (3.4.16). The rank of an $m \times n$ matrix A over a field F is $\dim(A(F^n))$. Since $F^m/A(F^n) \approx F^m/B(F^n)$ if and only if $A(F^n)$ and $B(F^n)$ have the same dimension, we see that A and B are equivalent if and only if they have the same rank.

A matrix (a_{ij}), not necessarily square, is called ***diagonal*** if $a_{ij} = 0$ whenever $i \neq j$. Now we apply 5.5.5 to get a canonical form for equivalent matrices over PID's.

5.5.6 Theorem

Every $m \times n$ matrix A over a PID *is equivalent to a diagonal matrix with diagonal entries*

$$d_1, d_2, \ldots, d_r, 0, \ldots, 0,$$

where $d_1 | d_2, \ldots, d_{r-1} | d_r$. The d_i are unique up to associates.

Proof Let A be an $m \times n$ matrix. Pick a basis $\{x_1, x_2, \ldots, x_m\}$ of R^m and elements $d_1, d_2, \ldots, d_m$ of R such that $\{d_1x_1, d_2x_2, \ldots, d_mx_m\}$ generates $A(R^n)$, and such that $d_1 | d_2, d_2 | d_3, \ldots, d_{r-1} | d_r$, $d_{r+1} = \cdots =$

$d_m = 0$. No more than n of the d_i are nonzero. Let B be the $m \times n$ matrix $(\delta_{ij}d_j)$ with ij entry $\delta_{ij}d_j$. Then $R^m/B(R^n) \approx R^m/A(R^n)$. By 5.5.5, A and B are equivalent. The d_i are unique up to associates because those not zero or a unit are the invariant factors of $R^m/A(R^n)$.

5.5.7 Corollary

Every $m \times n$ matrix A over a field is equivalent to a diagonal matrix with diagonal entries

$$1, 1, \ldots, 1, 0, 0, \ldots, 0.$$

The number of 1's in the matrix is the rank of A.

5.5.8 Corollary

Every matrix of integers is equivalent to exactly one diagonal matrix with diagonal entries

$$d_1, d_2, \ldots, d_r, 0, 0, \ldots, 0$$

with the d_i positive and with d_i dividing d_{i+1}.

5.5.9 Corollary

Every matrix over $F[x]$, where F is a field, is equivalent to exactly one diagonal matrix with diagonal entries

$$f_1(x), f_2(x), \ldots, f_r(x), 0, \ldots, 0,$$

where the $f_i(x)$ are monic polynomials with $f_i(x)$ dividing $f_{i+1}(x)$.

Now consider an $n \times n$ matrix A over a field F. The matrix $xI_n - A$ over $F[x]$ is equivalent to a matrix of the form given in 5.5.9. Those polynomials $f_i(x)$ not 1 are the invariant factors of $F[x]^n/(xI_n - A)F[x]^n$. The invariant factors of A are the invariant factors of the $F[x]$-module F^n with scalar multiplication given by $f(x)\cdot v = f(A)(v)$. If we can show that this $F[x]$-module is isomorphic to the $F[x]$-module $F[x]^n/(xI_n - A)F[x]^n$, that is, that $M_1(A) \approx M_2(xI_n - A)$, then we will know that two matrices A and B in F_n are similar if and only if $xI_n - A$ and $xI_n - B$ are equivalent. Thus, one way to calculate the invariant factors of A is to put $xI_n - A$ in the equivalent form in 5.5.9 and read them off. Something more general is true.

5.5.10 Theorem

Let R be a commutative ring, and let A be an $n \times n$ matrix over R. Then $M_1(A) \approx M_2(xI_n - A)$.

Proof We need an $R[x]$-isomorphism

$$\alpha: R[x]^n/(xI_n - A)R[x]^n \to R^n.$$

Let $\{e_1, e_2, \ldots, e_n\}$ be the canonical basis of the free R-module R^n. Then $\{e_1, e_2, \ldots, e_n\}$ is also the canonical basis of the free $R[x]$-module $R[x]^n$, and it generates R^n as an $R[x]$-module. Thus, there is an $R[x]$-epimorphism

$$\alpha: R[x]^n \to R^n$$

such that $\alpha(e_i) = e_i$ for all i. We will show that $\mathrm{Ker}(\alpha) = (xI_n - A)R[x]^n$. First, $\alpha((xI_n - A)(e_i)) = \alpha(xe_i - Ae_i) = 0$. Thus, $\mathrm{Ker}(\alpha) \supset (xI_n - A)R[x]^n$. Now suppose that $\alpha(\sum_{i=1}^n f_i(x)e_i) = 0$. Write $f_i(x) = \sum_{j=0}^{n_i} b_{ij}x^j$. Since $\alpha(\sum_{i=1}^n f_i(x)e_i) = 0$, then $\sum_{i=1}^n f_i(A)e_i = 0$. Thus,

$$\begin{aligned}\sum_{i=1}^n f_i(x)e_i &= \sum_{i=1}^n f_i(x)e_i - \sum_{i=1}^n f_i(A)e_i \\ &= \sum_{i=1}^n \left(\sum_{j=1}^{n_i} (b_{ij}x^jI_n - b_{ij}A^j)\right)e_i = \sum_{i=1}^n \left(\sum_{j=1}^{n_i} x^jI_n - A^j\right)b_{ij}e_i\end{aligned}$$

is in $(xI_n - A)R[x]^n$ since $(xI_n - A)B = x^jI_n - A^j$ for the $n \times n$ matrix $B = \sum_{i=0}^{j-1} x^{j-i-1}A^i$ over $R[x]$. This completes the proof.

5.5.11 Corollary

Let A and B be $n \times n$ matrices over a field F. Then A and B are similar over F if and only if $xI_n - A$ and $xI_n - B$ are equivalent over $F[x]$. If $xI_n - A$ is equivalent to the diagonal matrix with diagonal entries

$$f_1(x), \ldots, f_r(x), 0, \ldots, 0,$$

where $f_i(x) \mid f_{i+1}(x)$, then the $f_i(x)$ not 1 are the invariant factors of A.

Proof $M_1(A) \approx F[x]/F[x]f_1(x) \oplus \cdots \oplus F[x]/F[x]f_r(x)$. If $f_i(x) = 1$, then $F[x]/F[x]f_i(x) = 0$. Thus those not 1 are the invariant factors of $M_1(A)$, and the corollary follows.

Actually, a new proof that a square matrix satisfies its characteristic polynomial is at hand. We need one additional fact.

5.5.12 Lemma

Let R be a commutative ring, and let A be an $n \times n$ matrix over R. Then $(\det(A))(R^n/A(R^n)) = 0$.

Proof From 5.4, page 225,

$$\det(A)R^n = \det(A)I_n(R^n) = A \operatorname{adj}(A)(R^n),$$

which is contained in $A(R^n)$. The lemma follows.

5.5.13 Corollary (Cayley–Hamilton Theorem)

An $n \times n$ matrix A over a commutative ring R satisfies its characteristic polynomial $\det(xI_n - A)$.

Proof By 5.5.12, $\det(xI_n - A)(R[x]^n/(xI_n - A)R[x]^n) = 0$. Thus by 5.5.10, $\det(xI_n - A)R^n = 0$. Let $\det(xI_n - A) = f(x)$. Then $f(x)R^n = f(A)R^n$. Thus, $f(A)$ is the 0 matrix since its product with any element of R^n is 0.

Let F be a field. In view of 5.5.11, the invariant factors of a matrix A in F_n are exhibited by the canonical form of $xI_n - A$. There is an algorithm for finding that canonical form, or equivalently, for computing these invariant factors. More generally, there is an algorithm for finding the d_i in 5.5.6. We proceed now to describe this algorithm.

Let R be any commutative ring, and let A be an $m \times n$ matrix over R. The ***elementary row operations*** on A are the following.

(1) Interchange two rows of A.
(2) Multiply a row of A by a unit of R.
(3) Add a multiple of a row of A to another row of A.

The ***elementary column operations*** on a matrix are defined similarly. Each of these operations on A can be effected by multiplying A by suitable nonsingular square matrices. In fact, if one of the row operations is applied to the $m \times m$ identity matrix I_m, yielding the matrix E, then the effect of applying that row operation to A is the matrix EA. This is easy to verify, and analogous statements hold for column operations. Notice that if an elementary (row or column) operation is applied to I_n, the resulting matrix is nonsingular.

A ***secondary matrix*** over R is a square matrix of the form

$$S = \begin{pmatrix} a & b & 0 & 0 & \cdots & 0 \\ c & d & 0 & 0 & \cdots & 0 \\ 0 & 0 & 1 & 0 & \cdots & 0 \\ \vdots & & & & & \vdots \\ 0 & 0 & & & 0 & 1 \end{pmatrix},$$

with $ad - bc = 1$.

A secondary matrix is of course invertible since its determinant is 1. Multiplying A on the left by S is a ***secondary row operation*** on A, and multiplying A on the right by S is a ***secondary column operation*** on A.

Recall that a matrix (a_{ij}) is in ***diagonal form*** if $a_{ij}=0$ whenever i is not j. The important fact is that any $m \times n$ matrix over a PID can be put into a canonical diagonal form with diagonal entries $d_1, d_2, \ldots, d_r, 0, 0, \ldots, 0$ with d_i dividing d_{i+1} by applying a succession of elementary and secondary operations to it. This gives an effective way to compute invariant factors. Here is how it is done. Let $A=(a_{ij})$ be an $m \times n$ matrix over a PID. We may interchange two rows or two columns by multiplying A by suitable elementary matrices. Hence, we can get $a_{11} \neq 0$ if $A \neq 0$. (If $A=0$, there is nothing to do.) Let d be a greatest common divisor of a_{11} and a_{12}. Write $a_{11}=a_1 d$, $a_{12}=a_2 d$, and $sa_{11}+ta_{12}=d$. Then $sa_1+ta_2=1$. Now

$$(a_{ij}) \cdot \begin{pmatrix} s & -a_2 & 0 & 0 & \cdots & 0 \\ t & a_1 & 0 & 0 & \cdots & 0 \\ 0 & 0 & 1 & 0 & \cdots & 0 \\ \vdots & & & & & \vdots \\ 0 & 0 & 0 & & & 1 \end{pmatrix} = \begin{pmatrix} d & 0 & a_{13} & \cdots & a_{1n} \\ a'_{21} & a'_{22} & a_{23} & \cdots & a_{2n} \\ a'_{31} & a'_{32} & a_{33} & \cdots & a_{3n} \\ \vdots & & & & \vdots \\ a'_{m1} & a'_{m2} & a_{m3} & \cdots & a_{mn} \end{pmatrix}.$$

Interchanging the second and third columns and repeating the process above yields an equivalent matrix with the $(1,2)$ and $(1,3)$ entries zero. Proceeding in this manner, we get a matrix with first row of the form $(d, 0, 0, \ldots, 0)$, and, in fact, with d a greatest common divisor of the original $a_{11}, a_{12}, \ldots, a_{1n}$. Now proceed in an analogous way to get a matrix with first column having only the $(1,1)$ entry nonzero. This process, however, disturbs the first row, and the matrix has the form

$$\begin{pmatrix} e & f_2 & f_3 & \cdots & f_n \\ 0 & & & & \\ \vdots & & & & \\ 0 & & & & \end{pmatrix}.$$

If e divides each f, then subtracting suitable multiplies of the first column from the other columns brings the matrix into the form

$$B = \begin{pmatrix} e & 0 & 0 & \cdots & 0 \\ 0 & & & & \\ \vdots & & & & \\ 0 & & & & \end{pmatrix}.$$

If e does not divide some f_i, then proceed as in the very first step. This yields a matrix of the form

$$\begin{pmatrix} e_1 & 0 & 0 & \cdots & 0 \\ g_2 & & & & \\ \vdots & & & & \\ g_m & & & & \end{pmatrix},$$

where e_1 is the greatest common divisor of e and f_i's. If e_1 divides each g_i, then subtracting suitable multiples of the first row from the other rows yields a matrix of the form B. If e_1 does not divide some g_i, then the matrix is equivalent to one of the form

$$\begin{pmatrix} e_2 & h_2 & h_3 & \cdots & h_n \\ 0 & & & & \\ \vdots & & & & \\ 0 & & & & \end{pmatrix},$$

where e_2 is a greatest common divisor of e_1 and the g_i's. The ideals generated by $e, e_1, e_2, \ldots$ are strictly increasing. Thus in a finite number of steps we reach an e_k which divides each element in the first row, and thus a matrix of the form B. At this point, our new $e = e_k$ may not divide every entry of the matrix B. If e does not divide a_{ij}, say, then add the ith row to the first and start over. We get a new e properly dividing the old e. The ascending chain of ideals generated by the element in the $(1, 1)$ position must stop. Thus we reach a stage where our matrix has the form B with e dividing every element of B. Setting $e = d_1$, we have the matrix in the form

$$\begin{pmatrix} d_1 & 0 & \cdots & 0 \\ 0 & a_{22} & \cdots & a_{2n} \\ \cdots & \cdots & \cdots & \cdots \\ 0 & a_{m2} & \cdots & a_{mn} \end{pmatrix}.$$

Now we can proceed as before without disturbing the first row or column. The rest should be clear. By induction on n, our procedure brings the matrix into the diagonal form

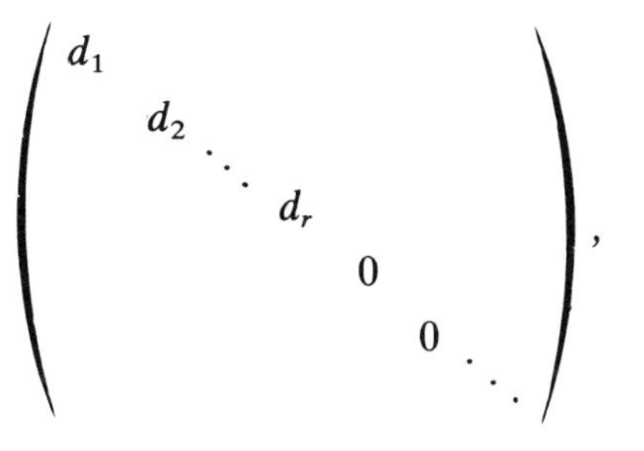

$$\begin{pmatrix} d_1 & & & & & \\ & d_2 & & & & \\ & & \ddots & & & \\ & & & d_r & & \\ & & & & 0 & \\ & & & & & 0 \\ & & & & & & \ddots \end{pmatrix},$$

with $d_i \mid d_{i+1}$.

Observe that the second elementary operation, namely, multiplying a row or column by a unit, has not been used. If our PID were $\mathbb{Z}$, or $F[x]$, we could make all the d_i positive, or monic, respectively, by so doing. In these cases, the d_i are unique. However, if the PID is $\mathbb{Z}$ or $F[x]$, we do not need the secondary operations in our algorithm. For instance, suppose our ring is $\mathbb{Z}$. By elementary operations, we can make $a_{11} > 0$. Write $a_{1i} = q_i a_{11} + r_i$, with $0 \leq r_i a_{11}$. Elementary operations then yield a matrix with first row $(a_{11}, r_2, r_3, \ldots, r_n)$. Start all over. Either all the r_i are 0 or there is an r_i with $0 < r_i < a_{11}$. Put r_i in the $(1, 1)$ position and repeat the process. This eventually brings us to a matrix with first row of the form $(e_1, 0, 0, \ldots, 0)$. Now work on the first column of this new matrix. We get a matrix with first column having only its $(1, 1)$ entry nonzero, but we know nothing about the first row any more. So start all over. The e_i are getting smaller, and are positive. Thus we eventually bring our matrix to the form with the only nonzero entry in the first row or first column the $(1, 1)$ entry. The rest should be clear. It is analogous to what we did in the general PID case. The division algorithm for polynomials (4.5.7) enables us to follow a similar procedure in the case that our ring is $F[x]$, where F is a field.

PROBLEMS

1 Prove that the matrix

$$\begin{pmatrix} x & -1 & 0 \\ 0 & x & -1 \\ 6 & 5 & x+2 \end{pmatrix}$$

is equivalent over $Q[x]$ to the diagonal matrix with diagonal entries $(1, 1, x^3 + 2x^2 + 5x + 6)$.

2 Reduce the following matrices of integers to canonical form.

$$(24 \quad 12 \quad 15), \qquad \begin{pmatrix} 2 & 8 & 6 \\ 4 & -2 & 10 \end{pmatrix}, \qquad \begin{pmatrix} 11 & 8 & 6 & 3 \\ 6 & 7 & 5 & 1 \\ 3 & -8 & 1 & -4 \\ 1 & -2 & -4 & 12 \\ 4 & 3 & 2 & 10 \end{pmatrix}.$$

3 Find the canonical form of the matrices of rational numbers

$$\begin{pmatrix} 0 & 1 & 1 & 1 \\ 0 & 0 & 1 & 1 \\ 0 & 0 & 0 & 1 \\ 0 & 0 & 0 & 0 \end{pmatrix}, \qquad \begin{pmatrix} a & b \\ c & d \end{pmatrix}, \qquad \begin{pmatrix} 5 & 6 & -6 \\ -1 & 4 & 2 \\ 3 & -6 & -4 \end{pmatrix}.$$

4 Reduce the following matrices over $Q[x]$ to the canonical diagonal form with diagonal entries

$$(f_1(x), f_2(x), \ldots, f_r(x), 0, \ldots, 0),$$

with $f_i(x)$ monic and with $f_i(x)$ dividing $f_{i+1}(x)$.

$$\begin{pmatrix} 1-x & x & 1+x^2 \\ x^2 & x & x^2 \\ x & -x & -x^2 \end{pmatrix},$$

$$\begin{pmatrix} x^4+1 & x^7-4x^3+1 & x^4-4x^3+4x-5 \\ 2x^4+3 & 2x^7-2x^4+4x^3-2 & 3x^4+4x^3+x^2+10x-14 \\ x^4+2 & x^7-x^4+2x^3-2 & 3x^4-6x^3+x^2+6x-9 \end{pmatrix}.$$

5 Prove that if the matrix $A \neq 0$ over a PID has canonical form with diagonal entries $(d_1, d_2, \ldots, d_r, 0, \ldots, 0)$, then d_1 is the greatest common divisor of all the entries of A.

6 Prove that any invertible matrix over a PID is a product of elementary and secondary matrices.

7 Prove that an $n \times n$ matrix over a PID is invertible if and only if it is equivalent to I_n.

8 Let R be a commutative ring, and let M and N be R-modules. Prove that the set $\mathrm{Hom}_R(M, N)$ of R-homomorphisms from M into N is an R-module via $(\alpha+\beta)(m) = \alpha(m) + \beta(m)$ and $(r\alpha)(m) = r(\alpha(m))$. Prove that $E_R(M) = \mathrm{Hom}_R(M, M)$ is an R-algebra, where multiplication is composition of maps.

9 Let M and N be free R-modules of finite ranks m and n, respectively. Pick bases of M and N, and associate with each α in $\mathrm{Hom}_R(M, N)$ its matrix with respect to this pair of bases. Prove that this association is an R-module isomorphism. If $M = N$ and the two bases are equal, show that it is an R-algebra isomorphism.

10 Let M be a module over the commutative ring R, and let α be an endomorphism of M. For m in M and $f(x)$ in $R[x]$, let $f(x)\cdot m = f(\alpha)(m)$. Prove that this makes M into an $R[x]$-module. Denote this module by M^α. Prove that $M^\alpha \approx M^\beta$ if and only if there is an automorphism γ of the R-module M such that $\alpha = \gamma\beta\gamma^{-1}$.

11 Let M be a module over the commutative ring R. Let $M[x]$ be the set of all "polynomials" $m_0 + m_1x + \cdots + m_kx^k$ in x with coefficients in M. Prove that $M[x]$ is, in a natural way, an $R[x]$-module. Let α be an endomorphism of the R-module M, and let

$$(x-\alpha)\left(\sum m_ix^i\right) = \sum m_ix^{i+1} - \sum \alpha(m_i)x^i.$$

Prove that $x - \alpha$ is an $R[x]$-endomorphism of $M[x]$ and that $M[x]/(x - \alpha)M[x] \approx M^{\alpha}$.

5.6 EUCLIDEAN SPACES

Let $\mathbb{R}$ be the field of real numbers. In the vector space $\mathbb{R}^2$, that is, in the plane, there are notions of distance, length, angles, and so on. In vector spaces over arbitrary fields F, these geometric notions are not available. To define appropriately such concepts in F^2 (or in F^n), F must have special properties. In our previous discussions of vector spaces, the field itself played no special role. Our vector spaces were over just any field F. Two fields of special interest are the field $\mathbb{R}$ of real numbers and the field $\mathbb{C}$ of complex numbers. In this section, we will study vector spaces over $\mathbb{R}$ that come equipped with an added structure that will enable us to define such geometric notions as length and angle. A parallel development for $\mathbb{C}$ will be indicated in the exercises.

Various special properties of the real numbers will be used. For example, $\mathbb{R}$ is ***ordered***. That is, in $\mathbb{R}$ there is a notion of an element a being ***less than*** an element b, written $a < b$, or equivalently, element b being ***greater than*** element a, written $b > a$, and this relation satisfies certain properties. We write $a \leq b$ if $a < b$ or $a = b$. Numbers a such that $0 < a$ are ***positive***, and positive numbers have unique square roots. If $a \leq b$, and if c is positive, then $ac \leq bc$. Such familiar properties of the real numbers will be used throughout our discussion, and we assume that the reader is acquainted with them.

How does one define length, and so on, in $\mathbb{R}^2$? First, if $x = (x_1, y_1)$ and $y = (x_2, y_2)$ are in $\mathbb{R}^2$, then the ***inner product***, or ***dot product***, of x and y is $(x, y) = x_1x_2 + y_1y_2$. The product (x, y) is also denoted by $x \cdot y$. The other geometric notions are defined in terms of this product. For example, the ***length*** of x is $(x, x)^{1/2}$, and the ***angle*** θ between x and y is given by $\cos(\theta) = (x, y)/((x, x)(y, y))^{1/2}$. Analogous definitions are made for $\mathbb{R}^n$. The inner product above is a map $\mathbb{R}^2 \times \mathbb{R}^2 \rightarrow \mathbb{R}$. What we will consider will be vector spaces V over $\mathbb{R}$ that come equipped with a map $V \times V \rightarrow \mathbb{R}$ in terms of which the requisite definitions can be made.

5.6.1 Definition

A **Euclidean space** *is a vector space V over the field $\mathbb{R}$ of real numbers, together with a map $V \times V \rightarrow \mathbb{R}$, called the* **inner product**, *such that if (v, w) denotes the image of the pair (v, w) in $V \times V$, then for all u, v, w in V and a and b in $\mathbb{R}$,*

(a) $(v, w) = (w, v)$,

(b) $(v, v) \geq 0$, *and* $(v, v) = 0$ *if and only if* $v = 0$, *and*

(c) $(a \cdot u + b \cdot v, w) = a(u, w) + b(v, w)$.

There are two classical examples. For the vector space $\mathbb{R}^n$ of n-tuples of real numbers, define $\mathbb{R}^n \times \mathbb{R}^n \to \mathbb{R}$ by $(v, w) \to \sum_{i=1}^n v_i w_i$, where $v = (v_1, v_2, \ldots, v_n)$ and $w = (w_1, w_2, \ldots, w_n)$. This makes $\mathbb{R}^n$ into an n-dimensional Euclidean space. When we speak of the Euclidean space $\mathbb{R}^n$, we mean $\mathbb{R}^n$ with this particular inner product. The second example is the vector space V of all continuous real functions on the closed interval $[0, 1]$ with an inner product given by $(f(x), g(x)) = \int_0^1 f(x)g(x)\,dx$. This makes V into an infinite-dimensional Euclidean space. Our concern will be almost wholly with finite-dimensional spaces.

5.6.2 Definition

The **length** *of a vector v in a Euclidean space V is the real number $\|v\| = (v, v)^{1/2}$. If v and w are in V, then the* **distance** *from v to w is $\|v - w\|$.*

The basic properties of length are given in the following theorem.

5.6.3 Theorem

Let V be a Euclidean space. Then for all r in $\mathbb{R}$ and v and w in V

(a) $\|rv\| = |r|\,\|v\|$,

(b) $\|v\| \geq 0$, *and* $\|v\| = 0$ *if and only if* $v = 0$,

(c) $|(v, w)| \leq \|v\|\,\|w\|$,

(d) $\|v + w\| \leq \|v\| + \|w\|$.

Proof The equalities $\|rv\| = (rv, rv)^{1/2} = (r^2(v, v))^{1/2} = |r|\,\|v\|$ establish (a).

(b) follows immediately from 5.6.1(b).

Since

$$0 \leq (\|w\|\, v \pm \|v\|\, w, \|w\|\, v \pm \|v\|\, w) = 2\,\|v\|^2\,\|w\|^2 \pm 2\,\|v\|\,\|w\|(v, w),$$

we have $|(v, w)| \leq \|v\|\,\|w\|$, and so (c) holds.

$$\begin{aligned}\|v + w\|^2 &= (v + w, v + w) = (v, v) + 2(v, w) + (w, w)\\ &\leq \|v\|^2 + 2\,\|v\|\,\|w\| + \|w\|^2 = (\|v\| + \|w\|)^2,\end{aligned}$$

whence

$$\|v + w\| \leq \|v\| + \|w\|,$$

and we have (d).

Part (c) is ***Schwarz's inequality***, and part (d) is the ***triangle inequality***. From 5.6.2, it follows easily that our notion of distance satisfies the usual properties of ordinary distance, namely that

(1) the distance from v to w is ≥ 0, and is 0 if and only if $v = w$,
(2) the distance from v to w is the distance from w to v, and
(3) the distance from u to v plus the distance from v to w is not less than the distance from u to w.

If M is a set with a function $d: M \times M \rightarrow \mathbb{R}$, and if we call $d(x, y)$ the ***distance*** from x to y, then M is a ***metric space*** if (1), (2), and (3) above are satisfied. Property (2) expresses the symmetry of distance, and property (3) is the ***triangle inequality***. Our notion of distance in a Euclidean space V makes V into a metric space.

Part (c) of 5.6.3 asserts that for nonzero vectors v and w, $-1 \leq (v, w)/(\|v\| \, \|w\|) \leq 1$. This makes it possible to define the ***angle between*** two nonzero vectors in a Euclidean space as that angle θ between 0° and 180° such that

$$\cos(\theta) = (v, w)/(\|v\| \, \|w\|).$$

5.6.4 Definition

Two vectors v and w in a Euclidean space V are **orthogonal** *if $(v, w) = 0$. A subset S of V is an* **orthogonal set** *if any two distinct vectors of S are orthogonal. The set S is* **orthonormal** *if it is orthogonal and $\|s\| = 1$ for all s in S.*

Note that the Euclidean space $\mathbb{R}^n$ has an orthonormal basis, namely

$$\{(1, 0, \ldots, 0), (0, 1, 0, \ldots, 0), \ldots, (0, 0, \ldots, 0, 1)\}.$$

If $\{v_1, v_2, \ldots, v_n\}$ is an orthonormal basis of a Euclidean space V, and if $v = \sum_{i=1}^{n} a_i v_i$ and $w = \sum_{i=1}^{n} b_i v_i$, then

$$(v, w) = \left(\sum_i a_i v_i, \sum_i b_i v_i\right) = \sum_{i,j} a_i b_j (v_i, v_j) = \sum_i a_i b_i.$$

Furthermore, if $\{v_1, v_2, \ldots, v_n\}$ is a basis of any n-dimensional real vector space V, then defining $(\sum_{i=1}^{n} a_i v_i, \sum_{i=1}^{n} b_i v_i) = \sum_i a_i b_i$ makes V into a Euclidean space, and $\{v_1, v_2, \ldots, v_n\}$ is an orthonormal basis of that space. Our immediate goal is to show that every finite-dimensional Euclidean space so arises. That is, we want to show that every finite-dimensional Euclidean space has an orthonormal basis.

5.6.5 Theorem

Let $\{v_1, v_2, \ldots, v_n\}$ be an independent subset of the Euclidean space V. Then there is an orthogonal subset $\{w_1, w_2, \ldots, w_n\}$ of V such that for

$i \leq n$,

$$\{v_1, v_2, \ldots, v_i\} \quad \text{and} \quad \{w_1, w_2, \ldots, w_i\}$$

generate the same subspace of V.

Proof Let $w_1 = v_1$, and proceeding inductively, let $w_j = v_j - \sum_{i<j} ((v_j, w_i)/(w_i, w_i))w_i$. Then for $k < j$,

$$\begin{aligned}(w_j, w_k) &= (v_j, w_k) - \sum_{i<j} ((v_j, w_i)/(w_i, w_i))(w_i, w_k) \\ &= (v_j, w_k) - ((v_j, w_k)/(w_k, w_k))(w_k, w_k) = 0.\end{aligned}$$

Therefore, the set $\{w_1, w_2, \ldots, w_j\}$ is orthogonal. By the construction, w_j is a linear combination of $\{v_1, v_2, \ldots, v_j\}$, and not a linear combination of $\{v_1, v_2, \ldots, v_{j-1}\}$. Thus $w_j \neq 0$, and so the set $\{w_1, w_2, \ldots, w_j\}$ is independent. The theorem follows.

The construction of the orthogonal set $\{w_1, w_2, \ldots, w_n\}$ is called the ***Gram–Schmidt orthogonalization process***. Note that if $\{v_1, v_2, \ldots, v_i\}$ is orthogonal, then $v_j = w_j$ for $j \leq i$.

5.6.6 Theorem

Every finite-dimensional Euclidean space has an orthonormal basis.

Proof By 5.6.5, a finite-dimensional Euclidean space has an orthogonal basis $\{w_1, w_2, \ldots, w_n\}$. The basis

$$\{w_1/\|w_1\|, \ldots, w_n/\|w_n\|\}$$

is orthonormal.

An immediate consequence is the following.

5.6.7 Theorem

Let V be a finite-dimensional Euclidean space. Then V has a basis $\{v_1, v_2, \ldots, v_n\}$ such that if $v = \sum_i a_i v_i$ and $w = \sum_i b_i v_i$, then $(v, w) = \sum_i a_i b_i$.

Two Euclidean spaces V and W are ***isomorphic*** if there is a vector space isomorphism $\alpha\colon V \to W$ which also preserves inner products; that is, that also satisfies $(x, y) = (\alpha(x), \alpha(y))$ for all x and y in V. Now 5.6.6 asserts that every Euclidean space of dimension n is isomorphic to $\mathbb{R}^n$ with the usual inner product. Consequently, any two Euclidean spaces of the same dimension are isomorphic.

If S is a subset of a Euclidean space V, let $S^\perp = \{v \in V: (s, v) = 0 \text{ for all } s \in S\}$. That is, $S^\perp$ is the set of all vectors in V which are orthogonal to every vector in S. It is easy to verify that $S^\perp$ is a subspace of V.

5.6.8 Theorem

Let V be a finite-dimensional Euclidean space, and let S be a subspace. Then $V = S \oplus S^\perp$, and $S^{\perp\perp} = S$.

Proof The subspace S of V is a Euclidean space in its own right, so it has an orthogonal basis $\{w_1, w_2, \ldots, w_m\}$. Let $\{w_1, \ldots, w_m, w_{m+1}, \ldots, w_n\}$ be a basis of V. Applying the Gram–Schmidt process to this basis yields an orthonormal basis $\{v_1, \ldots, v_m, v_{m+1}, \ldots, v_n\}$ of V. Clearly, $v_{m+1}, \ldots, v_n$ are in $S^\perp$. If w is in $S^\perp$, then $w = \sum_{i=1}^n a_i v_i$, and for $j \le m$, $0 = (v_j, w) = a_j$, whence w is in the subspace generated by $\{v_{m+1}, \ldots, v_n\}$. Therefore, $V = S \oplus S^\perp$. That $S = S^{\perp\perp}$ follows readily.

The subspace $S^\perp$ of V is called the ***orthogonal complement*** of S. Associating $S^\perp$ with S is reminiscent of the duality between the subspaces of V and those of its dual, or adjoint, V^*. In fact, the inner product on a Euclidean space V provides a natural way to identify V with its dual V^*. The mapping $f: V \to V^*$ given by $f(v)(w) = (v, w)$ is an isomorphism, and for a subspace W of V, $f(W^\perp) = K(W)$. This readily yields that $S \to S^\perp$ is a duality on the subspaces of a Euclidean space, that is, is a one-to-one correspondence that reverses inclusions (Problem 8).

Since the notion of length is fundamental in Euclidean spaces, those linear transformations on a Euclidean space which preserve length are of special interest.

5.6.9 Definition

Let V be a Euclidean space. A linear transformation a on V is **orthogonal** *if $\|\alpha(v)\| = \|v\|$ for every v in V.*

Equivalently, α ***is orthogonal if and only if it preserves inner products.*** Indeed, if α is orthogonal, then $\|\alpha(v)\| = \|v\|$, and so $(\alpha(v), \alpha(v))^{1/2} = (v, v)^{1/2}$, whence $(\alpha(v), \alpha(v)) = (v, v)$. For any v and w in V,

$$\begin{aligned}(v + w, v + w) = (v, v) + 2(v, w) + (w, w) &= ((\alpha(v + w), \alpha(v + w)) \\ &= (\alpha(v), \alpha(v)) + 2(\alpha(v), \alpha(w)) + (\alpha(w), \alpha(w)).\end{aligned}$$

It follows that $(v, w) = (\alpha(v), \alpha(w))$, and thus that α preserves inner products. If α preserves inner products, then $(\alpha(v), \alpha(v)) = (v, v)$, and so $\|\alpha(v)\| = \|v\|$ and α is orthogonal.

In particular, if α is orthogonal then α preserves orthogonality, or $(v, w) = 0$ if and only if $(\alpha(v), \alpha(w)) = 0$. Also it is useful to observe that if $\{v_1, v_2, \ldots, v_n\}$ is a basis, then α is orthogonal if $(v_i, v_j) = (\alpha(v_i), \alpha(v_j))$ for all i and j.

We now examine the matrix of an orthogonal transformation relative to an orthonormal basis. Let α be such a linear transformation on V, and let $\{v_1, v_2, \ldots, v_n\}$ be such a basis of V. If (a_{ij}) is the matrix of α relative to this basis, then $\alpha(v_j) = \sum_i a_{ij}v_i$, and

$$(v_i, v_j) = (\alpha(v_i), \alpha(v_j)) = \delta_{ij} = \left(\sum_k a_{ki}v_k, \sum_k a_{kj}v_k\right) = \sum_k a_{ki}a_{kj}.$$

This means that $(a_{ij})'(a_{ij}) = I_n$, the identity $n \times n$ real matrix, where $(a_{ij})'$ denotes the transpose of the matrix (a_{ij}). Thus, $(a_{ij})^{-1} = (a_{ij})'$. Another way to look at it is that the rows of (a_{ij}) form an orthonormal basis of the Euclidean space $\mathbb{R}^n$. On the other hand, let (a_{ij}) be an $n \times n$ matrix such that $(a_{ij})' = (a_{ij})^{-1}$, and let $\{v_1, v_2, \ldots, v_n\}$ be an orthonormal basis of V. Then the linear transformation α given by $\alpha(v_j) = \sum_i a_{ij}v_i$ is orthogonal. Indeed, $(\alpha(v_i), \alpha(v_j)) = (\sum_k a_{ki}v_k, \sum_k a_{kj}v_k) = \sum_k a_{ki}a_{kj} = \delta_{ij} = (v_i, v_j)$.

5.6.10 Definition

An $n \times n$ real matrix (a_{ij}) is **orthogonal** *if its transpose is its inverse. Thus, (a_{ij}) is orthogonal if $(a_{ij})' = (a_{ij})^{-1}$.*

We have proved the following theorem.

5.6.11 Theorem

A linear transformation on a Euclidean space is orthogonal if its matrix relative to some orthonormal basis is an orthogonal matrix. If a linear transformation is orthogonal, then its matrix relative to every orthonormal basis is orthogonal.

Note that orthogonal linear transformations and matrices are nonsingular, and that the inverse of an orthogonal transformation or matrix is orthogonal.

5.6.12 Theorem

If $\{v_1, v_2, \ldots, v_n\}$ is an orthonormal basis, then the matrix (a_{ij}) is orthogonal if and only if $\{w_1, w_2, \ldots, w_n\}$ given by $w_j = \sum_i a_{ij}v_i$ is an orthonormal basis.

Proof 5.6.12 is essentially a restatement of 5.6.11. The linear transformation α given by $\alpha(v_j) = \sum_i a_{ij}v_i = w_j$ is orthogonal if $\{w_1, w_2, \ldots, w_n\}$ is an orthonormal basis. By 5.6.11, (a_{ij}) is orthogonal. If (a_{ij}) is orthogonal, then by 5.6.11, α given by $\alpha(v_j) = \sum_i a_{ij}v_i = w_j$ is orthogonal, whence $\{w_1, w_2, \ldots, w_n\}$ is an orthonormal basis.

By 5.5.12, a change of orthonormal bases is effected only by orthogonal matrices.

We would like to find canonical forms for orthogonal matrices, but we are equally interested in ***symmetric matrices***. A matrix (a_{ij}) is ***symmetric*** if $a_{ij} = a_{ji}$ for all i and j. Thus, a symmetric matrix is one that is equal to its transpose. Suppose that α is a linear transformation on a Euclidean space V and that $(\alpha(v), w) = (v, \alpha(w))$ for all v and w in V. Then α is called ***self-adjoint***. In fact, for any linear transformation α on V, there is a unique linear transformation α^* on V such that $(\alpha(v), w) = (v, \alpha^*(w))$ for all v and w in V. Thus α^* is called the ***adjoint*** of α. If $\{v_1, v_2, \ldots, v_n\}$ is an orthonormal basis of V, α^* defined by $\alpha^*(w) = \sum_i (\alpha(v_i), w)v_i$ is the unique such map. Thus α is self-adjoint if $\alpha = \alpha^*$. If α is self-adjoint and if $\{v_1, v_2, \ldots, v_n\}$ is an orthonormal basis, then the matrix (a_{ij}) of α relative to this basis is symmetric. Indeed,

$$(\alpha(v_i), v_j) = \left(\sum_k a_{ki}v_k, v_j\right) = a_{ji} = (v_i, \alpha(v_j)) = \left(v_i, \sum_k a_{kj}v_k\right) = a_{ij}.$$

Conversely, if (a_{ij}) is symmetric, then α defined by $\alpha(v_j) = \sum_i a_{ij}v_i$ is easily seen to be self-adjoint. Thus, we have the following theorem.

5.6.13 Theorem

A linear transformation α on a Euclidean space is self-adjoint if its matrix relative to some orthonormal basis is symmetric. If α is self-adjoint, then its matrix relative to every orthonormal basis is symmetric.

We will now prove that if α is a self-adjoint linear transformation, then there is an orthonormal basis such that the matrix of α relative to that basis is diagonal. The following lemma paves the way.

5.6.14 Lemma

Let α be a self-adjoint linear transformation on a Euclidean space V. Then the minimum polynomial of α is a product of linear factors.

Proof We are going to have to use some special facts about the real numbers. The one we need here is that any polynomial of degree at least 1 over the real numbers factors into a product of linear and quadratic factors, and we assume this fact. Let $f(x)$ be the minimum polynomial of α. Let $q(x)$ be a quadratic factor of $f(x)$. There is a nonzero vector v in V such that $q(\alpha)v = 0$. Then v and $\alpha(v)$ generate a two-dimensional subspace W of V that is invariant under α. Now W is a Euclidean space and so has an orthonormal basis $\{v_1, v_2\}$, and α is a self-adjoint linear transformation on W. Let the matrix of α relative to $\{v_1, v_2\}$ be

$$A = \begin{pmatrix} a & b \\ c & d \end{pmatrix}.$$

Then A is symmetric, so $b = c$, and the characteristic polynomial of α on W is $(x - a)(x - d) - b^2 = x^2 + (-a - d)x - b^2 + ad$. Since

$$\begin{aligned}(-a - d)^2 - 4(ad - b^2) &= a^2 + 2ad + d^2 - 4ad + 4b^2 \\ &= (a - d)^2 + 4b^2 \geq 0,\end{aligned}$$

this polynomial has real roots. Thus, $f(x)$ is a product of linear factors.

The name of the following theorem derives from its role in determining the principal axis of an ellipse. In plane analytic geometry, the standard form of the equation of an ellipse is $x^2/a^2 + y^2/b^2 = 1$. In this form, the axes of the ellipse are parallel to the coordinate axes. For certain values of a, b, c, and d, $ax^2 + bxy + cy^2 = d$ is the equation of an ellipse, but unless $b = 0$, the axes are not parallel to the coordinate axes. The ***Principal Axis Theorem*** implies that by a rotation of the axes, that is, by an orthogonal transformation, the equation can be put in standard form, from which the lengths of the axes and other information about the ellipse can be read off. It can also be used to deduce results about the higher dimensional analogues of conic sections.

5.6.15 The Principal Axis Theorem

Let α be a self-adjoint linear transformation on the Euclidean space V. Then V has an orthonormal basis such that the matrix of α relative to that basis is diagonal. The diagonal elements are the eigenvalues of α, and each appears with its algebraic multiplicity.

Proof By 5.6.14, α has an eigenvector v, and $v/\|v\| = v_1$ is an eigenvector of unit length. Let $W = (\mathbb{R}v_1)^\perp = \{v \in V: (v_1, v) = 0\}$. Then since $(\alpha(v_1), v) = (av_1, v) = 0 = (v_1, \alpha(v))$ for all v in W, the subspace W is invariant under α, so α is a self-adjoint linear transformation on W. By induction on dimension, W has an orthonormal basis $\{v_2, v_3, \ldots, v_n\}$ such that the matrix of α relative to it is of the desired form. The orthonormal basis $\{v_1, v_2, \ldots, v_n\}$ of V is the desired basis.

5.6.16 Corollary

Let A be a real symmetric matrix. Then there is an orthogonal matrix B such that $B^{-1}AB$ is diagonal. The diagonal elements are the eigenvalues of A, each appearing with its algebraic multiplicity.

Proof The matrix A is the matrix relative to an orthonormal basis of a self-adjoint linear transformation α on a Euclidean space. There is an orthonormal basis such that the matrix C of α relative to it has the required form. Such a change of basis is effected only by orthogonal matrices (5.6.12).

5.6.17 Corollary

The characteristic polynomial of a symmetric matrix or of a self-adjoint linear transformation is a product of linear factors.

5.6.18 Corollary

Two real symmetric matrices are similar if and only if they have the same characteristic polynomial.

Now we will get a canonical form for orthogonal matrices. The theorem is this.

5.6.19 Theorem

Let α be an orthogonal linear transformation on a Euclidean space V. Then V has an orthonormal basis such that the matrix of α relative to this *basis is of the form*

$$\begin{pmatrix} 1 &&&&&&&&&& \\ & 1 &&&&&&&&& \\ && \ddots &&&&&&&& \\ &&& 1 &&&&&&& \\ &&&& -1 &&&&&& \\ &&&&& -1 &&&&& \\ &&&&&& \ddots &&&& \\ &&&&&&& \cos(\theta_1) & -\sin(\theta_1) && \\ &&&&&&& \sin(\theta_1) & \cos(\theta_1) & \ddots && \\ &&&&&&&&& \cos(\theta_k) & -\sin(\theta_k) \\ &&&&&&&&& \sin(\theta_k) & \cos(\theta_k) \end{pmatrix}$$

Proof Note first that if W is invariant under α, then so is $W^{\perp}$. To see this, we need that for any w in W and v in $W^{\perp}$, $(\alpha(v), w) = 0$. Since W is invariant under α and α is nonsingular, then α maps W onto W. Thus $w = \alpha(w_1)$ for some w_1 in W, and $(\alpha(v), w) = (\alpha(v), \alpha(w_1)) = (v, w_1) = 0$.

If α has an eigenvector v, then $\alpha(v) = av$, and since α is orthogonal, $(v, v) = (av, av)$, so $a = \pm 1$. Now $V = \mathbb{R}v \oplus (\mathbb{R}v)^{\perp}$. The restriction of α to $(\mathbb{R}v)^{\perp}$ is an orthogonal transformation, and by induction on dimension, $(\mathbb{R}v)^{\perp}$ has an orthonormal basis of the required form. That basis together with $v/\|v\|$ is the basis we seek. Now suppose that α has no eigenvector. Then the characteristic polynomial is the product of irreducible quadratic factors. Let $q(x)$ be one of them. There is a nonzero v in V such that $q(\alpha)v = 0$. The vectors v and $\alpha(v)$ generate a two-dimensional subspace X which is invariant under α. The restriction of α to X is an orthogonal transformation on X, and $q(x)$ is the characteristic polynomial. Since α is orthogonal, any matrix A of α relative to an orthonormal basis is orthogonal. Thus $\det(A) = \det(A') = \det(A^{-1})$, so $\det(A) = \pm 1$. But $\det(A)$ is the constant term. So $q(x) = x^2 + bx \pm 1$. If $q(x) = x^2 + bx - 1$, then $q(x)$ has real roots, so α has an eigenvector. Hence $q(x) = x^2 + bx + 1$. Let the matrix of α (on X) relative to an orthonormal basis of X be

$$\begin{pmatrix} c & d \\ e & f \end{pmatrix}.$$

This matrix is orthogonal and satisfies $q(x)$. Hence

(1) $cf - de = 1$,
(2) $c^2 + d^2 = 1$,
(3) $e^2 + f^2 = 1$, and
(4) $ce + df = 0$.

These equations imply that $c = f$ and $d = -e$. Since $e^2 + f^2 = 1$, we can write $e = \cos(\theta)$ and $f = \sin(\theta)$ for some θ. Hence the matrix becomes

$$\begin{pmatrix} \cos(\theta) & -\sin(\theta) \\ \sin(\theta) & \cos(\theta) \end{pmatrix}.$$

By induction on dimension, $X^\perp$ has an orthonormal basis of the required kind, and the rest follows easily.

5.6.16 Corollary

Let A be an orthogonal matrix. Then there exists an orthogonal matrix B such that BAB^{-1} has the form in 5.6.15.

Proof A is the matrix relative to an orthonormal basis of an orthogonal transformation α. There is an orthonormal basis such that the matrix of α is of the form in 5.6.15. Such a change of orthonormal bases is effected only by orthogonal matrices.

PROBLEMS

1 Let V be a finite-dimensional real vector space over $\mathbb{R}$, and let $\{v_1, v_2, \ldots, v_n\}$ be a basis of V. For $v = \Sigma_i a_i v_i$ and $w = \Sigma_i b_i v_i$, define $(v, w) = \Sigma_i a_i b_i$. Prove that this makes V into a Euclidean space.

2 Prove that if V is a Euclidean space with inner product (v, w), then for any positive real number r, $r(v, w)$ is an inner product on V.

3 Prove that $\mathbb{R}^2$ is a Euclidean space if inner products are given by $((a, b), (c, d)) = (a - b)(c - d) + bd$.

4 Prove that a Euclidean space V is a metric space via $V \times V \to \mathbb{R}$: $(v, w) \to \|v - w\|$.

5 Prove that a linear transformation α on a Euclidean space is orthogonal if for some basis $\{v_1, v_2, \ldots, v_n\}$, $(v_i, v_j) = (\alpha(v_i), \alpha(v_j))$.

6 Let $\{v_1, v_2, \ldots, v_n\}$ be an orthonormal basis of the Euclidean space V, and let v be in V.

(a) Prove that $v = \Sigma_i (v_i, v)v_i$.

(b) Prove that if α is a linear transformation on V, then the matrix (a_{ij}) of α relative to $\{v_1, v_2, \ldots, v_n\}$ is given by $a_{ij} = (v_i, \alpha(v_j))$.

(c) Prove that $\alpha^*(v) = \Sigma_i (\alpha(v_i), v)v_i$ defines the unique linear transformation on V such that $(\alpha(v), w) = (v, \alpha^*(w))$ for all v and w in V. Prove that $\alpha^{**} = \alpha$.

7 If α is a linear transformation on a Euclidean space, prove that $\alpha + \alpha^*$ is self-adjoint.

8 Let V be a finite-dimensional Euclidean space. Let V^* be the dual space of V. (See 3.4.1) Define $f: V \to V^*$ by $f(v)(w) = (v, w)$ for v in V and w in V. Prove that f is an isomorphism. Prove that for any subspace W of V, $f(W^\perp) = K(W)$.

9 Prove that the mapping of the set of all subspaces of a finite-dimensional Euclidean space to itself given by $S \to S^\perp$ is a duality. That is, prove that it is a one-to-one correspondence that reverses inclusions.

10 Let α be a linear transformation on the Euclidean space V. Let V^* be the dual space of V, let $\beta: V^* \to V^*$ be the conjugate, or adjoint of α in the sense of 3.4.11. Let $\alpha^*: V \to V$ be the adjoint of α in the sense of this section. Prove that $\alpha^* = \gamma^{-1}\beta\gamma$, where $\gamma: V \to V^*$ is the map described in Problem 8.

11 Prove that if α is self-adjoint and $\alpha^k = 1$ for some $k \geq 1$, then $\alpha^2 = 1$.

12 For each of the matrices below, find an orthogonal matrix B such that BAB^{-1} is diagonal.

$$\begin{pmatrix} 5 & -3 \\ -3 & 5 \end{pmatrix}, \quad \begin{pmatrix} 1 & 1 & 1 \\ 1 & 0 & 1 \\ 1 & 1 & 1 \end{pmatrix}, \quad \begin{pmatrix} 4 & 2(5^{1/2}) \\ 2(5^{1/2}) & -4 \end{pmatrix}.$$

13 Let V be the vector space of n-tuples of complex numbers. For $v = (c_1, c_2, \ldots, c_n)$ and $w = (d_1, d_2, \ldots, d_n)$ in V, define $(v, w) = \sum_i c_i\bar{d}_i$, where $\bar{c}$ denotes the conjugate of the complex number c. Prove that for all u, v, and w in V, and a and b in $\mathbb{C}$,

(a) $(v, w) = \overline{(w, v)}$,

(b) $(v, v) \geq 0$, and $(v, v) = 0$ if and only if $v = 0$, and

(c) $(au + bv, w) = a(u, w) + b(v, w)$.

A vector space V over $\mathbb{C}$ with a map $V \times V \to \mathbb{C}$ satisfying (a), (b), and (c) above is called a ***unitary space***. ***Length*** and ***distance*** are defined exactly as for Euclidean spaces.

14 Prove 5.6.3 for unitary spaces, and show that the notion of distance makes a unitary space a metric space.

15 Let V be the vector space of all continuous complex-valued functions on $[0, 1]$. Prove that V together with the map $V \times V \to \mathbb{C}$ given by $(f(x), g(x)) = \int_0^1 f(x)\overline{g(x)}\,dx$ is an infinite-dimensional unitary space.

16 Prove that a finite-dimensional unitary space has an orthonormal basis.

17 A linear transformation α on a unitary space V is called ***unitary*** if $(\alpha(v), \alpha(w)) = (v, w)$ for all v and w in V. Prove that if $(\alpha(v), \alpha(v)) = (v, v)$ for all v in V, then α is unitary.

18 Prove that a linear transformation on a unitary space is unitary if and only if it takes orthonormal bases to orthonormal bases.

19 If $C = (c_{ij})$ is an $n \times n$ complex matrix, then C^* is the matrix whose (i, j)

entry is $\overline{c_{ji}}$, and it is called the ***adjoint*** *of* C. The matrix C is called ***unitary*** if $CC^* = 1$. Prove that a linear transformation on a unitary space is unitary if and only if its matrix relative to any orthonormal basis is unitary. Prove that if $\{v_1, v_2, \ldots, v_n\}$ is an orthonormal basis of a unitary space V, then $\{w_1, w_2, \ldots, w_n\}$ given by $w_j = \sum_i c_{ij} v_i$ is an orthonormal basis of V if and only if (c_{ij}) is unitary.

20 Let α be a linear transformation on the finite-dimensional unitary space V. Let $\{v_1, v_2, \ldots, v_n\}$ be an orthonormal basis of V, and let the matrix of α relative to this basis be (a_{ij}). Define $a^*: V \to V$ by $a^*(v_i) = \sum_j \bar{a}_{ij} v_j$. Prove that for all v and w in V, $(\alpha(v), w) = (v, \alpha^*(w))$. Prove that if β is any linear transformation on V satisfying $(\alpha(v), w) = (v, \beta(w))$ for all v and w in V, then $\beta = \alpha^*$. As in the case of Euclidean spaces, α^* is called the ***adjoint*** of α.

21 A linear transformation α on a unitary space is called ***normal*** if $\alpha\alpha^* = \alpha^*\alpha$. Let a be an eigenvalue of a normal α, and let v be an eigenvector belonging to a; that is, let $\alpha(v) = av$. Prove that $\alpha^*(v) = \bar{a}v$. Thus, prove that the conjugate of a is an eigenvalue of α^*, and prove that v is an eigenvector of α^* belonging to $\bar{a}$.

22 Let v be an eigenvector of the linear transformation α defined on a unitary space V. Prove that

$$(\mathbb{C}v)^{\perp} = \{w \in V: (v, w) = 0\}$$

is invariant under α.

23 Let α be a normal linear transformation on the finite-dimensional unitary space V. Prove that there exists an orthonormal basis of V such that the matrix of α relative to it is diagonal.

24 Let C be an $n \times n$ complex matrix such that $CC^* = C^*C$; that is, C is a ***normal*** matrix. Prove that there exists a unitary matrix U such that $U^{-1}CU$ is diagonal. (Note that unitary matrices A satisfy $AA^* = A^*A$.)

25 A linear transformation α on a unitary space V is called ***self-adjoint***, or ***Hermitian***, if $\alpha = \alpha^*$. Prove that a self-adjoint α on a finite-dimensional space V has a real diagonal matrix relative to some orthonormal basis of V. Prove that the eigenvalues of such a Hermitian linear transformation are real.

26 Let C be a complex $n \times n$ matrix such that $C = C^*$. That is, $c_{ij} = \overline{c_{ji}}$ for all entries c_{ij} of C. Prove that there is a unitary matrix U such that $U^{-1}CU$ is diagonal, and that all the eigenvalues of C are real.

27 Prove that a linear transformation α on a unitary space is Hermitian if and only if $(\alpha(v), v)$ is real for all v in V.

28 Prove that a normal linear transformation is unitary if and only if its eigenvalues are all of absolute value 1.

29 Prove that if α is a normal linear transformation, then α^* is a polynomial in α; that is, $\alpha^* = p(\alpha)$, where $p(x)$ is in $\mathbb{C}[x]$.

30 Prove that if α and β are linear transformations on a unitary space, and if $\alpha\beta = 0$, then $\alpha\beta^* = 0$.

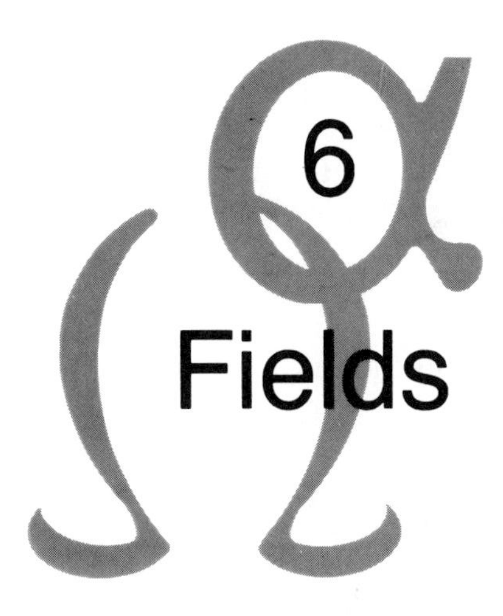

6 Fields

6.1 SUBFIELDS AND EXTENSION FIELDS

A knowledge of fields is essential for almost any algebraic endeavor. We will concentrate on the very basic notions but will proceed far enough to show the impossibility of "duplicating cubes," of "trisecting angles," and, more deeply, of "solving by radicals" polynomials of degree greater than 4. We thus have two purposes—to present some of the notions of field theory and to show how these notions lead to the solution of some classical problems.

We begin by listing some examples of fields.

6.1.1 Examples of fields

(a) The field $\mathbb{Q}$ of rational numbers;
(b) the field $\mathbb{R}$ of real numbers;
(c) the field $\mathbb{C}$ of complex numbers;
(d) the field $\mathbb{Z}/p\mathbb{Z}$ of integers modulo the prime p;
(e) $\{a + bi: a, b \in \mathbb{Q}\}$;
(f) $\{a + b(2)^{1/2}: a, b \in \mathbb{Q}\}$;
(g) $\{a + b(q)^{1/2}: a, b \in \mathbb{Q}, q$ a fixed positive rational number$\}$;
(h) the quotient field $F(x)$ of the integral domain $F[x]$ of all polynomials in x with coefficients from a field F;
(i) $\mathbb{Q}[x]/(x^2 + x + 1)$;
(j) $(\mathbb{Z}/2\mathbb{Z})[x]/(x^3 + x + 1)$;
(k) $F[x]/I$, where I is any prime ideal in $F[x]$.

Any field F has a unique smallest subfield P, called the ***prime subfield*** of F. It may be obtained by taking the intersection of all the subfields of F, or alternately by taking the subfield generated by 1. If the integral multiples $n \cdot 1$ of 1 are distinct in F, then the map

$$\mathbb{Z} \to \mathbb{Z} \cdot 1 : n \to n \cdot 1$$

is an isomorphism, and thus F contains a copy of the ring $\mathbb{Z}$ of integers and hence of the field $\mathbb{Q}$ of rationals. It should be clear that in this case the prime subfield of F is isomorphic to $\mathbb{Q}$. Furthermore, if $n \cdot a = 0$ for any $a \in F$, then $0 = n \cdot a = (n \cdot 1)a$, so either a or n is zero. Therefore, the nonzero elements of F all have infinite additive order. This is expressed by saying that F has ***characteristic zero***. If $m \cdot 1 = n \cdot 1$ with $m \neq n$, then $(m - n) \cdot 1 = 0$, and thus there is a smallest positive integer p such that $p \cdot 1 = 0$. If $p = qr$ with $1 \leq q < p$, then $0 = p \cdot 1 = (qr) \cdot 1 = (q \cdot 1)(r \cdot 1)$, whence $r \cdot 1 = 0$, and $r = p$. It follows that p is a prime. In this case, the map

$$\mathbb{Z} \to \mathbb{Z} \cdot 1 : n \to n \cdot 1$$

has kernel $p\mathbb{Z}$. This is a ring homomorphism and is clearly onto. Thus $\mathbb{Z} \cdot 1 \approx \mathbb{Z}/p\mathbb{Z}$, which is a field. Therefore, $\mathbb{Z} \cdot 1$ is the smallest subfield of F. For a nonzero element a of F, $p \cdot a = (p \cdot 1)a = 0$. Therefore, the additive order of any nonzero element of F is p. This is expressed by saying that F has ***characteristic*** p.

6.1.2 Theorem

The prime subfield of a field of characteristic 0 *is isomorphic to the field* $\mathbb{Q}$ *of rational numbers. The prime subfield of a field of prime characteristic* p *is isomorphic to the field* $\mathbb{Z}/p\mathbb{Z}$ *of integers modulo* p.

If F is a subfield of a field K, then K is called an ***extension*** of F. In particular, every field is an extension of its prime subfield. If K is an extension of F, then K is, in particular, a vector space over F.

6.1.3 Definition

Let K be an extension of F. The dimension of K as a vector space over F is denoted by $[K{:}F]$ *and called the* **degree** *of the extension.*

The vector space K over F may not have a finite basis, in which case we have not really defined its dimension, and we will just say that it is ***infinite***. Our real concern with degrees will be with finite ones. An extension of finite degree is called a ***finite extension***.

The field of complex numbers $\mathbb{C}$ is an extension of degree 2 of the field of real numbers $\mathbb{R}$. The set $\{1, i\}$ is a basis of $\mathbb{C}$ as a vector space over $\mathbb{R}$. For any field F, the quotient field $F(x)$ of $F[x]$ is an extension of infinite degree of F. The set $\{1, x, x^2, x^3, \ldots\}$ is linearly independent over F.

Let K be an extension of F. If $a \in K$, let $F(a)$ denote the smallest subfield of K containing F and a, and let $F[a]$ denote the smallest subring of K containing F and a. Note that $F[a]$ is just the set of all polynomials $c_0 + c_1 a + \cdots + c_n a^n$ in a with coefficients in F, and that $F(a)$ is the field of quotients of $F[a]$. We distinguish two principal kinds of elements a in the extension K of F—those with $[F(a):F]$ finite, and those with $[F(a):F]$ not finite.

6.1.4 Theorem

Let K be an extension of F and let $a \in K$. Then $[F(a):F]$ is finite if and only if a is a root of a nonzero polynomial in $F[x]$.

Proof Suppose that $[F(a):F] = n$. Then the family $\{1, a, a^2, \ldots, a^n\}$ is dependent. Thus there are c_i in F, not all zero, such that $\sum_{i=0}^{n} c_i a^i = 0$. Hence, a is a root of the nonzero polynomial $\sum_{i=0}^{n} c_i x^i$ in $F[x]$.

Now suppose that a is a root of a nonzero polynomial in $F[x]$. The set of all polynomials in $F[x]$ of which a is a root is an ideal in $F[x]$, and this ideal has a unique monic generator $p(x)$. If $p(x) = q(x)r(x)$, then $0 = p(a) = q(a)r(a)$, whence either $q(a)$ or $r(a)$ is zero. Therefore, $p(x)$ is prime in $F[x]$. The map

$$\varphi: F[x] \to F[a]: f(x) \to f(a)$$

is a ring homomorphism with kernel $(p(x))$. Hence $F[x]/(p(x)) \approx \mathrm{Im}(\varphi)$. Since $p(x)$ is prime, $F[x]/(p(x))$ is a field. But $\mathrm{Im}(\varphi) = F[a]$. Thus,

$$F[x]/(p(x)) \approx F[a] = F(a).$$

Let $x + (p(x)) = y$, and let $\deg(p(x)) = n$. It is trivial to check that $\{1, y, y^2, \ldots, y^{n-1}\}$ is a basis of $F[x]/(p(x))$ over F. The map φ induces an isomorphism between the vector spaces $F[x]/(p(x))$ and $F(a)$ over F. Therefore, $[F(a):F]$ is finite. In fact, $\{1, a, a^2, \ldots, a^{n-1}\}$ is a basis of $F(a)$ over F.

6.1.5 Definition

Let K be an extension of F. The element a in K is **algebraic** *over F it is a root of a nonzero polynomial in $F[x]$. The (unique) monic polynomial in $F[x]$ of smallest degree that a satisfies is the* **minimum polynomial** *of a. Its* *degree is the* **degree** *of a over F. The extension K of F is an* **algebraic extension** *if every element of K is algebraic over F.*

Therefore, the elements a of K that are algebraic over F are those that generate finite extensions $F(a)$ of F. In the proof of 6.1.4, we saw that if $[F(a):F]=n$, then $\{1, a, \ldots, a^{n-1}\}$ is a basis of $F(a)$ over F, n is the degree of a over F, and $F[a]=F(a)$.

Suppose that two elements a and b in K are algebraic over F. The elements $a+b$, ab, and a^{-1} if $a \neq 0$ are also algebraic over F, but this is not exactly obvious. We will show that the set of all elements of K which are algebraic over F is a subfield of K. We need some important preliminaries.

6.1.6 Theorem

Let K be a finite extension of F, and let L be a finite extension of K. Then $[L:F]=[L:K][K:F]$. In particular, L is a finite extension of F.

Proof The idea is simple. Let $\{a_1, a_2, \ldots, a_m\}$ be a basis of K over F, and let $\{b_1, b_2, \ldots, b_n\}$ be a basis of L over K. We will show that

$$\{a_i \cdot b_j : i=1, 2, \ldots, m; j=1, 2, \ldots, n\}$$

is a basis of L over F. Let $c \in L$. Then $c = \sum_{j=1}^{n} e_j b_j$ with $e_i \in K$, and $e_j = \sum_{i=1}^{m} f_{ij} a_i$ with $f_{ij} \in F$. Thus,

$$c = \sum_{j=1}^{n} \left(\sum_{i=1}^{m} f_{ij} a_i \right) b_j = \sum_{i,j} f_{ij} a_i b_j.$$

Hence, the set $\{a_i b_j\}$ generates L over F. We need it to be independent over F. Suppose that $\sum_{i,j} f_{ij} a_i b_j = 0$ with $f_{ij} \in F$. Then $0 = \sum_{j=1}^{n} (\sum_{i=1}^{m} f_{ij} a_i) b_j$, with $\sum_{i=1}^{m} f_{ij} a_i \in K$. Since $\{b_1, \ldots, b_n\}$ is independent over K, $\sum_{i=1}^{m} f_{ij} a_i = 0$ for all j. Since $\{a_1, \ldots, a_m\}$ is independent over F, $f_{ij} = 0$ for all j. This completes the proof.

Note that if $F \subset K \subset L$ and $[L:F]=n$ is finite, then the degrees $[K:F]$ and $[L:K]$ divide n. In particular, if $[L:K]$ is prime, then there are no fields between L and K.

6.1.7 Theorem

Let K be an extension of F. Then the set of all elements of K that are algebraic over F is a subfield of K.

Proof Let a and b be in K and algebraic over F. Let $b \neq 0$. We need

$a+b$, ab, and $1/b$ algebraic over F. This follows immediately from 6.1.6. We have

$$[F(a)(b):F] = [F(a)(b):F(a)][F(a):F].$$

Both $[F(a):F]$ and $[F(a)(b):F(a)]$ are finite, the latter because b is algebraic over F and thus certainly over $F(a)$. Thus, $[F(a)(b):F]$ is finite. Therefore, every element in $F(a)(b)$ is algebraic over F. The elements $a+b$, ab, and $1/b$ are in $F(a)(b)$.

It should be clear that $F(a)(b) = F(b)(a)$. This field is denoted $F(a, b)$. More generally, if $\{a_1, a_2, \ldots, a_n\}$ is any finite set of elements in K, where K is an extension of F, then $F(a_1, a_2, \ldots, a_n)$ denotes the subfield of K generated by F and the elements $a_1, a_2, \ldots, a_n$.

6.1.8 Corollary

If K is an algebraic extension of F, and if L is an algebraic extension of K, then L is an algebraic extension of F.

Proof Let a be an element of L. We need $[F(a):F]$ finite. The element a satisfies a nonzero polynomial $\sum_{i=0}^{n} a_i x^i$ in $K[x]$. The field $F(a_1, a_2, \ldots, a_n) = G$ is a finite extension of F, and $G(a)$ is a finite extension of G. Therefore, $[G(a):F]$ is finite. Since $F(a)$ is a subfield of $G(a)$, it follows that $[F(a):F]$ is finite.

At this point we are able to answer some classical questions concerning constructions with ruler and compass. The key result is 6.1.6.

The real number a is called ***constructible*** if it belongs to a subfield K of the field of real numbers such that K can be obtained from $\mathbb{Q}$ by a succession of extensions of degree 2. Thus for K to be constructible, there must be fields $K_0, K_1, \ldots, K_n$ such that $K = K_n \supset K_{n-1} \supset \cdots \supset K_1 \supset K_0 = \mathbb{Q}$, with $[K_{i+1}:K_i] = 2$, and $a \in K$. The degree of the extension K of $\mathbb{Q}$ is a power of 2, but it is a special kind of extension. We will see the reason for the terminology "constructible" in a moment.

6.1.9 Theorem

The set of constructible real numbers is a field.

Proof Let a and b be constructible real numbers. Then there are fields F and K with $a \in F$, $b \in K$, and appropriate fields F_i between F and $\mathbb{Q}$ and K_i between K and $\mathbb{Q}$. Let L be the subfield generated by F and K. Let

$$F_i = \mathbb{Q}(a_1, a_2, \ldots, a_i)$$

with $[F_{i+1}:F_i]=2$ and $F_n=F$. Let $K_i=\mathbb{Q}(b_1, b_2, \ldots, b_i)$ with $[K_{i+1}:K_i]=2$ and $K_m=K$. Let $F_{n+i}=F(b_1, b_2, \ldots, b_i)$. Then $[F_{n+i+1}:F_{n+i}]$ is either 1 or 2. Thus, the sequence

$$Q \subset F_1 \subset F_2 \subset \cdots \subset F_n = F \subset F_{n+1} \subset \cdots \subset F_{n+m} = L$$

is one of successive extensions each of degree 1 or 2. Since $a \pm b$, ab, and a/b (if $b \neq 0$) are in L, the theorem follows.

Note that a constructible number has degree a power of 2 over $\mathbb{Q}$.

Constructible numbers are numbers that can be constructed with ruler and compass. What does this mean? Ruler and compass constructions are generally done in elementary geometry courses. These are constructions done in a finite number of steps, starting with only the unit segment (unless specifically stated otherwise). To use the ruler means to regard as drawn the line determined by two given points. To use the compass means to regard as drawn the circle having a given point as center and the length of a given line segment as radius. Ruler and compass constructions are carried out in the plane. For example, such constructions as those of a perpendicular bisector of a given line segment, and of a line through a given point parallel to a given line are no doubt familiar. Other possible ruler (more appropriately, straightedge) and compass constructions are those of points partitioning a given segment into n equal segments, of line segments of length equal to the sum, difference, product, or quotient of the lengths of two given line segments. To say that a real number x is constructible means that a line segment of length $|x|$ is constructible. In particular, all rational numbers are constructible. One can construct the square root of a given positive real number. Thus, square roots of positive rational numbers are constructible with ruler and compass. Therefore, starting with the unit segment, we can construct $\mathbb{Q}$, and for any positive rational number x, we can construct $x^{1/2}$. Hence, we can construct $\mathbb{Q}(x^{1/2})$. Similarly, we can construct $\mathbb{Q}(x^{1/2}, y^{1/2})$ for any other positive number $y \in \mathbb{Q}(x^{1/2})$. Thus, we see that a real number r is constructible (with ruler and compass) if r is in a field of real numbers obtainable from $\mathbb{Q}$ by successive extensions of degree 2.

Suppose that we are *given* a field F of real numbers. What new numbers are constructible from it with ruler and compass? We can draw all circles whose centers have coordinates in F and whose radii are in F, and we can draw all lines determined by points whose coordinates are in F. The intersections of these circles and lines give us new points, but each such new point has coordinates in $F(r^{1/2})$ for some positive real number r (depending on the point). This is not particularly difficult to show. Now from the $F(r^{1/2})$ we can get additional fields of the form $F(r^{1/2}, s^{1/2})$, and so on. The reader should be convinced at this point that a real number is constructible according to our earlier definition if and only if it is constructible with ruler and compass.

Now suppose that an angle θ is constructible with ruler and compass. This means that starting with the unit segment and using ruler and compass constructions, two lines can be obtained which intersect at an angle θ. Then the numbers $\sin\theta$ and $\cos\theta$ are constructible. If the angle θ can be trisected, then the length $\cos\theta/3$ is constructible. The angle 60° is constructible. If it could be trisected, then $\cos 20°$ would be constructible. But the trigonometric identity $\cos 3\theta = 4\cos^3\theta - 3\cos\theta$ yields $1/2 = 4x^3 - 3x$, where $x = \cos 20°$. Hence $8x^3 - 6x - 1 = 0$. But the polynomial $8x^3 - 6x - 1$ is irreducible over $\mathbb{Q}$. Hence x is of degree 3 over $\mathbb{Q}$ and so is not constructible. Therefore, the angle 60° cannot be trisected with ruler and compass. It is easy to find other angles that cannot be trisected.

To duplicate the (unit) cube boils down to constructing a number x such that $x^3 = 2$, that is, constructing a cube whose volume is twice that of the unit cube. But clearly $2^{1/3}$ is not constructible. It is of degree 3 over $\mathbb{Q}$.

To "square the circle" means to construct a square whose area is that of a circle whose radius is 1, thus to construct a number x such that $x^2 = \pi$. But π is not algebraic over $\mathbb{Q}$, whence one cannot "square the circle." (To prove that π is not algebraic over $\mathbb{Q}$ is a difficult matter, and we will not do it.)

The gist of the whole matter is that one cannot make ruler and compass constructions that result in constructing real numbers that are not algebraic of degree a power of 2 over $\mathbb{Q}$.

PROBLEMS

1 Let K be an extension of F, and let $a \in K$. Prove that $F[a] = F(a)$ if and only if a is algebraic over F.

2 Let K be an extension of F, and let a and b be in K. Prove that $F(a)(b) = F(b)(a)$.

3 Let K be a finite extension of F, and let a and b be in K. Prove that $[F(a, b):F] \leq [F(b):F][F(a):F]$.

4 Prove that if $f(x)$ is irreducible over $F[x]$ and $[K:F]$ is relatively prime to $\deg(f(x))$, then $f(x)$ is irreducible over $K[x]$.

5 Let F be a subfield of an integral domain K. Prove that if K has finite dimension as a vector space over F, then K is a field.

6 Prove that the usual formula for the roots of a quadratic equation holds for any field of characteristic not 2.

7 Suppose that $[K:\mathbb{Q}] = 2$. Prove that $K = \mathbb{Q}(\sqrt{n})$, where n is a square-free integer.

8 Let F be the quotient field of a principal ideal domain D of characteristic not 2. Suppose that $[K:F] = 2$. Prove that $K = F(d^{1/2})$, where d is a square-free element of D.

9 Suppose that the minimum polynomial of the number a over $\mathbb{Q}$ is x^3+2x^2-5x-1. Write $(a^5+2a-7)/(a^2+4)$ in the form $r_0+r_1a+r_2a^2$ with $r_i \in \mathbb{Q}$.

10 Let K be a finite extension of F, and let $a \in K$ with $a \neq 0$. Prove that the map $K \to K: b \to ab$ is a nonsingular linear transformation on the vector space K over F, and that the minimum polynomial of this linear transformation is the same as the minimum polynomial of a as an element algebraic over F.

11 Find the minimum polynomial over $\mathbb{Q}$ of each of the following.
(a) $1+2^{1/2}$
(b) $2^{1/2}+3^{1/2}$
(c) $i+3^{1/2}$
(d) $3^{1/3}+9^{1/3}$
(e) $2^{1/2}+3^{1/3}$

12 Prove that the order of a finite field is a power of a prime.

13 Prove that any two fields of order 4 are isomorphic.

14 Prove that any two fields of order 9 are isomorphic.

15 Let $a_1, a_2, \ldots, a_{n+1}$ be distinct elements of the field F, and let $b_1, b_2, \ldots, b_{n+1}$ be any elements of F. Prove that there is exactly one polynomial $f(x)$ in $F[x]$ of degree n such that $f(a_i)=b_i$ for all i.

16 Prove that the angles of 72° and of 60° are constructible, whereas the angle of $(360/7)°$ is not constructible.

17 Prove that a regular pentagon and a regular hexagon are constructible, whereas a regular septagon is not constructible.

6.2 SPLITTING FIELDS

If K is an algebraic extension of F, then an element a in K satisfies a polynomial $f(x)$ in $F[x]$ of degree at least 1. Conversely, given such a polynomial, is there an algebraic extension K of F in which that polynomial has a root? The existence of such a field is a fundamental fact of field theory.

6.2.1 Theorem

Let F be a field, and let $f(x)$ be in $F[x]$ and have degree at least 1. *Then there is a finite extension K of F containing a root of $f(x)$.*

Proof Let $p(x)$ be a prime factor of $f(x)$. Then $F[x]/(p(x))$ is a field, and it contains F via the embedding

$$F \to F[x]/(p(x)): a \to a+(p(x)).$$

The desired field is $K = F[x]/(p(x))$. In the proof of 6.1.4, we observed that $[K:F] = \deg(p(x))$. We need only that K contains a root of $f(x)$. We will show that it contains a root of $p(x)$. Let $p(x) = a_0 + a_1x + \cdots + a_nx^n$. For K to contain a root of $p(x)$ means that there is an element $k \in K$ such that

$$a_0 + a_1k + \cdots + a_nk^n = 0.$$

The element $x + (p(x))$ works. Indeed

$$\begin{aligned} a_0 + a_1(x + (p(x))) + \cdots + a_n(x + (p(x)))^n \\ = a_0 + a_1x + \cdots + a_nx^n + (p(x)) = 0. \end{aligned}$$

Therefore, we may "adjoin a root" of a polynomial $f(x) \in F[x]$ to F simply by forming the field $K = F[x]/(p(x))$, where $p(x)$ is any nonlinear prime factor of $f(x)$. The field K, as a vector space over F, has basis

$$\{1 + (p(x)), x + (p(x)), \ldots, x^{n-1} + (p(x))\},$$

and as an extension of F, $K = F(x + (p(x)))$. In $K[x]$, $p(x)$ has a linear factor. Now adjoin a root of an irreducible nonlinear factor of $f(x)$ over K. Using 4.5.10, we arrive by iteration at a field L such that $f(x)$ factors into linear factors in $L[x]$. In other words, L "contains all the roots" of $f(x)$, and in fact is generated by F and those roots. By our construction, $[L:F] \leq m!$, where $m = \deg(f(x))$. Strictly speaking, the field L is not unique. However, it is as unique as one can expect.

6.2.2 Definition

Let $f(x)$ be a polynomial in $F[x]$. An extension K of F is a **splitting field** *of $f(x)$ over F if $f(x)$ factors into linear factors in $K[x]$ and K is generated by F and the roots of $f(x)$.*

Splitting fields are also called ***root fields***. If K is a root field of $f(x) \in F[x]$, then $f(x) = a(x - a_1)(x - a_2) \cdots (x - a_n)$ in $K[x]$, and $K = F(a_1, a_2, \ldots, a_n)$. We know that every polynomial $f(x)$ in $F[x]$ has a splitting field. We want to show that any two splitting fields of $f(x)$ are essentially the same.

Let $\alpha: F \to G$ be an isomorphism between the fields F and G. Then α induces an isomorphism $\alpha: F[x] \to G[x]$ by

$$\alpha(a_0 + a_1x + \cdots + a_nx^n) = \alpha(a_0) + \alpha(a_1)x + \cdots + \alpha(a_n)x^n.$$

If $p(x)$ is a prime polynomial in $F[x]$, then $\alpha(p(x))$ is prime in $G[x]$, and α induces an isomorphism

$$\alpha: F[x]/(p(x)) \to G[x]/(\alpha(p(x)))$$

by

$$\alpha(f(x)+(p(x)))=\alpha(f(x))+(\alpha(p(x))).$$

In particular, $\alpha(x+(p(x)))=x+(\alpha(p(x)))$. We saw in the proof of 6.1.4 that if K is an extension of F, and if a is an element of K which is a root of $p(x)$, then $F(a)$ is isomorphic to $F[x]/(p(x))$ via the mapping given by $f(x)+(p(x))\to f(a)$. We can make the analogous statement about G and $\alpha(p(x))$. We have proved the following theorem.

6.2.3 Theorem

Let $\alpha\colon F\to G$ be an isomorphism between the fields F and G. Let $p(x)$ be a prime polynomial in $F[x]$. If a is a root of $p(x)$ in an extension of F, and if b is a root of $\alpha(p(x))$ in an extension of G, then there is an isomorphism $\beta\colon F(a)\to G(b)$ such that $\alpha=\beta$ on F and $\beta(a)=b$.

6.2.4 Corollary

Let $p(x)$ be prime in $F[x]$. If a and b are roots of $p(x)$ in an extension of F, then there is an isomorphism $\alpha\colon F(a)\to F(b)$ such that α is the identity on F and such that $\alpha(a)=b$.

Proof Let $F=G$ and let α be the identity map in 6.2.3.

6.2.5 Theorem

Let $\alpha\colon F\to G$ be an isomorphism, and let $f(x)$ be in $F[x]$. If K is a root field for $f(x)$ and if L is a root field of $\alpha(f(x))$, then there is an isomorphism $\beta\colon K\to L$ which agrees with α on F.

Proof We will induct on the degree of $f(x)$. If $\deg(f(x))=1$, then take $\beta=\alpha$. Let $p(x)$ be a prime factor of $f(x)$. Let a be a root of $p(x)$ in K, and let b be a root of $\alpha(p(x))$ in L. By 6.2.3, there is an isomorphism $\gamma\colon F(a)\to G(b)$ such that $\gamma=\alpha$ on F, and such that $\gamma(a)=b$. Write $f(x)=(x-a)g(x)$ in $F(a)[x]$. Now K is a splitting field for $g(x)$ over $F(a)$, and L is a splitting field for $\alpha(g(x))$ over $G(b)$. Since $\deg(g(x))<\deg(f(x))$, there is an isomorphism $\beta\colon K\to L$ such that β agrees with γ on $F(a)$. This completes the proof.

The following corollary is the result we are really after. In order to proceed by induction, we were forced to prove the stronger result 6.2.5.

6.2.6 Corollary

Let $f(x) \in F[x]$, and let K and L be splitting fields for $f(x)$. Then there is an isomorphism $\alpha: K \to L$ such that α is the identity on F.

Let F be a field, and let $f(x)$ be an element of $F[x]$ of degree >1. There is an algebraic extension K of F such that in $K[x]$, $f(x)$ factors into a product of linear factors. It is natural to wonder if there is an algebraic extension K of F such that every such $f(x) \in F[x]$ factors completely in $K[x]$. This is indeed the case. In fact, there exists an algebraic extension K of F such that every $f(x)$ in $K[x]$ of degree ≥ 1 factors into linear factors in $K[x]$, and any two such extensions of F are ***F*-isomorphic**, that is, if K and L are two such extensions of F, then there is an isomorphism $\varphi: K \to L$ such that $\varphi(a) = a$ for all $a \in F$. A field K such that every $f(x) \in K[x]$ of degree ≥ 1 factors into a product of linear factors is said to be ***algebraically closed***. This is equivalent to every polynomial in $K[x]$ of degree ≥ 1 having a root in K, and is equivalent to K having no algebraic extensions. If K is an algebraic extension of F and if K is algebraically closed, then K is called an ***algebraic closure*** of F. The classic example of an algebraically closed field is the field of complex numbers. A proof that that field is algebraically closed in given in 7.6. A proof of the existence of algebraic closures for arbitrary fields is somewhat beyond us at this point. It involves the Axiom of Choice, which we will not meet until the Appendix. However, we will now sketch a proof that a countable field has an algebraic closure.

Recall that a set S is ***countably infinite*** if there exists a one-to-one map f from the set of positive integers onto S, and is ***countable*** if it is finite or countably infinite (1.3). For example, the field $\mathbb{Q}$ is countably infinite. Now suppose that F is any countable field. Then $F = \{a_1, a_2, \ldots, a_n\}$ for some n, or $F = \{a_1, a_2, \ldots\}$. In either case, let $S_m = \{a_i: i \leq m\}$. There are finitely many nonzero elements in $F[x]$ of degree $\leq m$ with coefficients in S_m. Let $f_m(x)$ be their product. Let $F = F_0$, and for $m \geq 0$, let F_{m+1} be the root field of $f_{m+1}(x)$ over F_m. We have the chain $F_0 \subset F_1 \subset F_2 \subset \ldots$. Let $K = \bigcup_m F_m$. We assert that K is an algebraic closure of F. First, K is algebraic over F since each F_m is algebraic over F. If $k(x) \in K[x]$, then $k(x) \in F_m[x]$ for some m. Thus the roots of $k(x)$ are algebraic over F, whence they satisfy some polynomial $f(x)$ in $F[x]$. But $f(x)$ divides some $f_p(x)$, whence the roots of $k(x)$ are in F_{p+1}. Therefore, K is algebraically closed.

How unique is an algebraic closure of a countable field? Let K and L be algebraic closures of the countable field F. Let $f_1(x)$, $f_2(x), \ldots$ be the polynomials in $F[x]$ constructed earlier. Then $f_1(x)$ splits completely in $K[x]$ and in $L[x]$, so that K and L contain root fields K_1 and L_1, respectively, of $f_1(x)$ over F. By 6.2.5, there is an isomorphism $\varphi_1: K_1 \to L_1$ fixing F elementwise. Now $f_2(x)$ is in $K_1[x]$ and $L_1[x]$, and splits completely in $K[x]$ and $L[x]$, so that K and L contain root fields K_2 and L_2 of $f_2(x)$ over $K_1[x]$ and

$L_1[x]$, respectively. By 6.2.5, there is an isomorphism $\varphi_2: K_2 \to L_2$ extending $\varphi_1: K_1 \to L_1$. Continue this process. We get an isomorphism $\varphi: \bigcup_i K_i \to \bigcup_i L_i$. Since K is algebraic over F, $K = \bigcup_i K_i$, and similarly for L. Thus, we have an isomorphism $\varphi: K \to L$ which is the identity on F. We have proved the following theorem.

6.2.7 Theorem

A countable field has an algebraic closure. If K and L are algebraic closures of a countable field F, then there is an isomorphism $\varphi: K \to L$ fixing F elementwise.

The theorem is true with the countability hypothesis removed, as we shall see in the Appendix.

The algebraic closure of $\mathbb{Q}$ is the field of ***algebraic numbers***. Finite extensions of $\mathbb{Q}$ are called ***algebraic number fields***, or simply ***number fields***. They have an elaborate theory. Of particular interest is the set of all algebraic numbers that satisfy a monic polynomial with integer coefficients. This turns out to be a ring, the ring of ***algebraic integers***. The algebraic integers in a given algebraic number field are called the ***integers*** of that field.

A more detailed study of splitting fields is complicated by the existence of irreducible polynomials of degree n that have fewer than n roots in their splitting fields. We need some information about them at this point. If K and L are splitting fields of $f(x)$, then from 6.2.6 it follows that $f(x)$ factors into distinct linear factors over K if and only if it so factors over L.

6.2.8 Definition

A polynomial of degree n is **separable** *if it has n distinct roots in its splitting field. A finite extension K of F is* **separable** *if every element of K is a root of a separable polynomial in $F[x]$.* **Inseparable** *means not separable.*

There is an easy test for separability of polynomials. If $f(x)$ is any polynomial in $F[x]$, we can take its (formal) derivative. For instance, if $f(x) = a_0 + a_1x + \cdots + a_nx^n$, then

$$f'(x) = a_1 + a_2x + 2a_3x^2 + \cdots + na_nx^{n-1}.$$

The usual rules for differentiation hold, and they are easily verified. Here is the test for separability.

6.2.9 Theorem

A polynomial $f(x)$ over the field F is separable if and only if $f(x)$ and $f'(x)$ are relatively prime.

Proof Since $f(x)$ and $f'(x)$ are both in $F[x]$, their greatest common divisor in $K[x]$ for any extension K of F is the same as their greatest common divisor in $F[x]$. Let K be the root field of $f(x)$, and write $f(x) = c\prod_{i=1}^{n}(x-a_i)^{k_i}$ in $K[x]$, with the a_i distinct. We get

$$f'(x) = c\sum_{i=1}^{n}\left[k_i(x-a_i)^{k_i-1}\prod_{i\neq j}(x-a_j)^{k_j}\right].$$

If for some i, $k_i > 1$, then $x - a_i$ is a common factor of $f(x)$ and $f'(x)$. If $k_i = 1$ for all i, then

$$f'(a_i) = c\left[\prod_{i\neq j}(a_i - a_j)\right] \neq 0,$$

so that $f(x)$ and $f'(x)$ have no root, and hence no factor, in common.

6.2.10 Corollary

A prime polynomial $f(x)$ is inseparable if and only if $f'(x) = 0$. A prime polynomial over a field F of characteristic 0 is separable. A prime polynomial over a field F of characteristic p is inseparable if and only if it is a polynomial in x^p.

Proof If $f'(x) = 0$, then $f(x)$ and $f'(x)$ are not relatively prime, whence $f(x)$ is inseparable. If $f(x)$ is inseparable, then $f(x)$ and $f'(x)$ have a factor of degree at least 1 in common. Since $f(x)$ is prime, $f(x)$ divides $f'(x)$. But $\deg(f'(x)) < \deg(f(x))$. Thus, $f'(x) = 0$. If F has characteristic 0 and $f(x)$ is prime, then $f(x)$ has degree at least 1, and hence $f'(x) \neq 0$. If F has characteristic p and $f(x)$ is prime, then it is apparent that $f'(x) = 0$ if and only if $f(x) = g(x^p)$.

It may not be clear that there exist prime inseparable polynomials. Here is an example. Let $P = \mathbb{Z}/p\mathbb{Z}$ be the field with p elements. Form the quotient field $K = P(y)$ of the polynomial ring $P[y]$. Let F be the subfield $P(y^p)$ of K. The polynomial $x^p - y^p \in F[x]$ is inseparable since its derivative is 0. In $K[x]$, $x^p - y^p = (x-y)^p$. If $x^p - y^p = f(x)g(x)$ is a nontrivial factorization in $F[x]$, then $f(x) = (x-y)^n$ for $0 < n < p$. Therefore, $y^n \in F$. Now $F = P(y^p)$ consists of quotients of polynomials in y^p with coefficients in P. Hence $y^n = h(y^p)/k(y^p)$ with $h(y^p)$ and $k(y^p)$ in $P(y^p)$, and so $k(y^p)y^n = h(y^p)$ is in $P[y^p]$. But

$0<n<p$ clearly makes this impossible. Therefore $x^p - y^p$ is prime in $F[x]$, is separable, and has y as a root of multiplicity p in K.

We are in a position to use splitting fields to describe completely all finite fields. We begin by noticing a few elementary properties that a finite field F must have. The characteristic of F is a prime p. Hence, F is a vector space of finite dimension n over its prime subfield P of p elements. Therefore, F has p^n elements. Let $q = p^n$. Every element of F satisfies the polynomial $x^q - x$ in $P[x]$. To see this, note that the multiplicative group F^* of nonzero elements of F has $q-1$ elements. Thus for $a \in F^*$, $a^{q-1} = 1$, whence $a^q = a$. Now 0 certainly satisfies $x^q - x$. The polynomial $x^q - x$ has q roots in F, hence splits completely in F. Therefore, F is the root field of $x^q - x$ over the field P of p elements. Any two such root fields are isomorphic by 6.2.5. We have shown that there is at most one field of order p^n (up to isomorphism).

To show that there is a finite field with $p^n = q$ elements, the natural thing to do is to take the splitting field of $x^q - x$ over the field P of p elements. This gets a field F, but conceivably it is too small. If a and b are roots of $x^q - x$ in F, then

$$(a+b)^q - (a+b) = a^q + b^q - a - b = 0,$$

and

$$(ab)^q - ab = a^q b^q - ab = 0,$$

using the facts that the characteristic of P is p, so that $py = 0$ for any $y \in F$, that $a^q = a$, and that $b^q = b$. Similarly, if $a \neq 0$, then $1/a$ is a root of $x^q - x$. Therefore, the root field F of $x^q - x$ consists entirely of roots of $x^q - x$. If this polynomial has no multiple roots, then F has $q = p^n$ elements. By 6.2.9, this is the case. Thus, we have the following theorem about finite fields.

6.2.11 Theorem

Let p be a prime, and let n be a nonnegative integer. There is, up to isomorphism, exactly one field with $p^n = q$ elements. That field is the splitting field of the polynomial $x^q - x$ over the field of integers modulo p.

The multiplicative group of a finite field is cyclic. This is a special case of a more general result that is just as easy to prove.

6.2.12 Theorem

A finite multiplicative subgroup of a field is cyclic.

Proof Let G be a finite multiplicative subgroup of a field. If q is the largest invariant factor of G, then $x^q = 1$ for all x in G, so that $x^q - 1$ has o(G) roots in G. Since $q \leq \text{o}(G)$ and $x^q - 1$ has no more than q roots, it follows that $q = \text{o}(G)$, and that G is cyclic.

If F is any field of characteristic p, then

$$F \to F: a \to a^p$$

is an isomorphism of F onto its subfield of pth powers. If F is finite, this map must be onto as well. Thus, the following theorem is true.

6.2.13 Theorem

Let F be a finite field of characteristic p. Then the map $F \to F: a \to a^p$ is an automorphism of F. In particular, every element of F has a pth root.

Let F be a finite field, and let K be any finite extension of F. If K has $p^n = q$ elements, then it is a root field of $x^q - x$ over $F[x]$, and hence is a separable extension. If a is any generator of K^*, then $K = F(a)$. Therefore any extension of finite degree over a finite field F is a ***simple*** extension, that is, it is of the form $F(a)$ for some a in the extension field. The key to this is the separability of the extension.

6.2.14 Theorem

Let K be a finite separable extension of any field F. Then $K = F(a)$ for some $a \in K$.

Proof Since K is a finite extension of F,

$$K = F(a_1, a_2, \ldots, a_n)$$

for some $a_i \in K$. By induction, we may assume that $K = F(a, b)$ for a and b in K. Let $f(x)$ be the minimum polynomial of a over F, and let $g(x)$ be the minimum polynomial of b over F. We will work in the root field L of $f(x)g(x)$ over K. Let $a_1, a_2, \ldots, a_m$ be the roots of $f(x)$, and let $b_1, b_2, \ldots, b_n$ be the roots of $g(x)$, with $a = a_1$ and $b = b_1$. Since K is separable, the b_i are distinct. We already observed that finite fields are simple extensions of any subfield. Thus, we may assume that F is infinite. Choose an element c in F different from $(a_i - a_1)/(b_1 - b_k)$, $1 \leq i \leq m$, $1 < k \leq n$. Let $d = a + bc$. We assert that $K = F(d)$. Now $f(d - cx)$ and $g(x)$ have the root b in common. Since $d - cb_k = a + bc - cb_k = a_i$ implies that $c = (a_i - a_1)/(b_1 - b_k)$, it is the only root they have in common. Thus, the greatest common factor of $f(c - dx)$ and $g(x)$ is $x - b$. But $f(d - cx)$ and $g(x)$ are in $F(d)[x]$. Therefore, so is their greatest common divisor. Hence $b \in F(d)$. Since b, c, and d are in $F(d)$, so is $a = d - bc$. Thus, $K \subset F(d)$. But $d \in K$ implies that $K = F(d)$, and our proof is complete.

Notice that we really proved something stronger, namely that if $K = F(a_1, a_2, \ldots, a_n)$ and if $a_2, a_3, \ldots, a_n$ are roots of separable polynomials, then K is a simple extension of F. Just note that our proof shows that $F(a_1, a_2) = F(d_1)$ for some $d_1 \in K$. Hence $K = F(d_1, a_3, \ldots, a_n)$ with $a_3, \ldots, a_n$ roots of separable polynomials, and induction finishes the job. Note that even if a_1 were also separable, we do not know yet that $F(a_1, a_2, \ldots, a_n)$ is a separable extension. That fact will not be available until 6.3.9.

This theorem simplifies the study of finite separable extensions K of a field F, since K then has a basis of the form $\{1, a, a^2, \ldots, a^{n-1}\}$ over F. We will make good use of it in the next section.

6.2.15 Corollary

Any finite extension of a field of characteristic 0 *is a simple extension.*

PROBLEMS

1 Find the degrees of the splitting fields of the following polynomials over $\mathbb{Q}$.
(a) $x^2 - 1$
(b) $x^2 + 1$
(c) $x^2 - 2$
(d) $x^3 - 2$
(e) $x^3 - 3$
(f) $x^4 - 5$
(g) $x^4 - 4$
(h) $x^4 + x^2 + 1$
(i) $x^p - 1$, p a prime.

2 Let K be an algebraic extension of F, and let $f(x) \in K[x]$. Prove that there is a finite extension L of K such that in $L[x]$, $f(x)$ divides some nonzero element of $F[x]$.

3 Find the group of automorphisms of the fields $\mathbb{Q}(2^{1/3})$, $\mathbb{Q}(2^{1/2})$, and $\mathbb{Q}(i)$.

4 Let C be the algebraic closure of a field F, and let K be a field between F and C. Prove that C is the algebraic closure of K.

5 Prove that the algebraic closure of a countable field is countable.

6 Prove that if F is countable, then so is $F[x]$.

7 Prove that the usual rules for differentiation of polynomials hold. Thus, show that
(a) $(f(x)+g(x))' = f'(x)+g'(x)$,
(b) $(f(x)g(x))' = f'(x)g(x)+f(x)g'(x)$, and
(c) $(f(x)^n)' = nf(x)^{n-1}f'(x)$.

8 Let $f(x)$ be a polynomial over a field of characteristic 0. Let $d(x)$ be the greatest common divisor of $f(x)$ and $f'(x)$. Prove that $f(x)/d(x)$ has the same roots as $f(x)$, and prove that $f(x)/d(x)$ is separable.

9 Prove that if F is a field with p^n elements, then F contains a subfield with p^m elements if and only if m divides n.

10 Prove that if F and G are subfields of a finite field, and $F \approx G$, then $F = G$.

11 Determine the order of the automorphism $F \to F: a \to a^p$ of a finite field F of characteristic p.

12 Prove that the derivative of a polynomial $f(x)$ over a finite field F of characteristic p is 0 if and only if $f(x) = (g(x))^p$ for some $g(x) \in F[x]$.

13 Let F be a field of characteristic $p \neq 0$. Prove that if $F \to F: a \to a^p$ is not onto, then F has a finite inseparable extension.

14 Let K be an extension of the field F of characteristic $p \neq 0$. Let $L = \{a \in K: a^q \in F \text{ for } q \text{ some power of } p\}$. Prove that L is a subfield of K. Prove that $[L:F]$ is a power of p, if finite.

15 Prove that if K is a separable extension of F, and L is any field between K and F, then K is a separable extension of L.

16 Let K be a finite extension of F, and let

$$\{b_1, b_2, \ldots, b_n\}$$

be a basis of K over F. For $a \in K$, let $ab_i = \sum_{j=1}^{n} a_{ij}b_j$. Prove that the polynomial $\det(xI_n - (a_{ij}))$ is a multiple of the minimum polynomial of a over F. Prove that it is a power of the minimum polynomial of a over F.

6.3 GALOIS THEORY

Let K be an extension of F. Galois theory relates, in certain special cases, extensions of F contained in K with subgroups of the group of automorphisms of K which fix F elementwise. It provides a way of reducing questions about fields to questions about groups. The insolvability of quintic equations is proved via Galois theory, for example. The subject is difficult, but it is important.

Let K be an extension of F. The set of automorphisms of K is a group under composition of maps. Those automorphisms of K which fix F elementwise form a subgroup of that group, and is denoted $G(K/F)$. It is called the ***group of automorphisms of*** K ***over*** F, or ***the group of*** F-

automorphisms of K. We will also speak of F-automorphisms from an extension K of F into another extension L of F. It is an isomorphism from K into L that fixes F elementwise.

If $\mathbb{C}$ is the field of complex numbers, and $\mathbb{R}$ is the field of real numbers, then $G(\mathbb{C}/\mathbb{R})$ is the two element group, and $G(\mathbb{C}/\mathbb{Q})$ consists of all automorphisms of $\mathbb{C}$. If F is any field, and $K = F(x_1, x_2, \ldots, x_n)$ is the field of quotients of the ring $F[x_1, x_2, \ldots, x_n]$ of all polynomials in the indeterminants x_i with coefficients in F, then any permutation of

$$\{x_1, x_2, \ldots, x_n\}$$

induces an automorphism of K which fixes F elementwise. On the other hand, $G(\mathbb{Q}(2^{1/3})/\mathbb{Q})$ has just one element. To see this, note that any automorphism of $\mathbb{Q}(2^{1/3})$ must fix $2^{1/3}$ since 2 has only one real cube root. But every element of $\mathbb{Q}(2^{1/3})$ is of the form $a + b(2^{1/3}) + c(2^{1/3})^2$, with a, b, and c in $\mathbb{Q}$. Therefore any automorphism that fixes $\mathbb{Q}$ elementwise, and fixes $2^{1/3}$, must be the identity automorphism.

Let S be a subgroup of the group of all automorphisms of a field K. The subfield $K(S) = \{a \in K\colon \alpha(a) = a \text{ for all } \alpha \in S\}$ is called the ***fixed field*** of S. Note that indeed it is a subfield of K. If S is a subgroup of $G(K/F)$, then $K(S)$ contains F. In this case, we have associated with each subgroup S of $G(K/F)$ a field between K and F. The subgroup $\{e\}$ corresponds to K, and $G(K/F)$ corresponds to some subfield of K containing F, but not necessarily to F, as the extension $\mathbb{Q}(2^{1/3})$ of $\mathbb{Q}$ testifies. The Fundamental Theorem of Galois Theory asserts that in certain special cases, this association is a one-to-one correspondence between the subgroups of $G(K/F)$ and the extensions of F contained in K. Our goal is this fundamental theorem. The crux of the matter is determining when the fixed field of $G(K/F)$ is F itself.

The following definition is a fundamental one.

6.3.1 Definition

An extension K of F is **normal** *if it is algebraic and if every irreducible polynomial in $F[x]$ which has a root in K factors into linear factors in $K[x]$.*

Thus if an irreducible polynomial $f(x) \in F[x]$ has a root in the normal extension K of F, then it has all its roots in K. Another way to put it is that K contains a root field of any irreducible polynomial in $F[x]$ of which it contains a root. The precise connection between root fields and normal extensions is given in the following theorem.

6.3.2 Theorem

A field K is a finite normal extension of a field F if and only if it is the root field of some polynomial in $F[x]$.

Proof Let K be the root field of $f(x) \in F[x]$. Then K is a finite, and thus an algebraic, extension of F. Suppose that an irreducible polynomial $g(x) \in F[x]$ has a root a in K. Let L be a root field of $g(x)$ over K, and let b be a root of $g(x)$ in L. Then there is an F-isomorphism $\alpha: F(a) \to F(b)$ which takes a to b. Further, $K(a)$ and $K(b)$ are root fields of $f(x)$ and $\alpha(f(x)) = f(x)$ over $F(a)[x]$ and $F(b)[x]$, respectively. Therefore, by 6.2.5, there is an isomorphism $\beta: K(a) \to K(b)$ extending α. But $K(a) = K$ since a is in K. Since β fixes F elementwise, β must permute the roots of $f(x)$, which all lie in K and generate K over F. Thus, $\beta(K) = K$. Hence $b \in K$, and K is a normal extension of F.

Now suppose that K is a finite normal extension of F. Let $K = F(a_1, a_2, \ldots, a_n)$, and let $f_i(x)$ be the minimum polynomial of a_i over F. Then K is the root field of $\prod_{i=1}^n f_i(x)$ over F.

The next lemma and its corollaries will dispose of many of the technical details in proving the Fundamental Theorem of Galois Theory. One additional notion is needed. Let $a_1, a_2, \ldots, a_n$ be elements of a field K. The ***elementary symmetric polynomials*** in $a_1, a_2, \ldots, a_n$ are the coefficients of the polynomial $\prod_{i=1}^n (x + a_i) \in K[x]$. For example, if $n = 4$, they are

$$1,$$

$$a_1 + a_2 + a_3 + a_4,$$

$$a_1a_2 + a_1a_3 + a_1a_4 + a_2a_3 + a_2a_4 + a_3a_4,$$

$$a_1a_2a_3 + a_1a_2a_4 + a_1a_3a_4 + a_2a_3a_4,$$

and

$$a_1a_2a_3a_4.$$

For our purposes at the moment, the pertinent fact is this. If σ is any automorphism of K which induces a permutation of $\{a_1, a_2, \ldots, a_n\}$, then σ fixes each elementary symmetric polynomial in the a_i's. This should be obvious.

6.3.3 Lemma

Let G be a subgroup of the group of automorphisms of a field K, and let F be the fixed field of G. Suppose that each element in K has only finitely many images under G; that is, for each $a \in K$, the set $\{\alpha(a): \alpha \in G\}$ is finite. Then K is a normal separable extension of F.

Proof Let $a \in K$, and let $a_1, a_2, \ldots, a_n$ be the distinct images of a under the automorphisms in G. One of them is a, of course. Each $\alpha \in G$ permutes the set $\{a_1, a_2, \ldots, a_n\}$. Therefore each $\alpha \in G$ fixes each coefficient of $f(x) = \prod_{i=1}^{n} (x - a_i)$, whence $f(x)$ must be in $F[x]$. Since the roots of $f(x)$ are the distinct elements $a_1, a_2, \ldots, a_n$, $f(x)$ is a separable polynomial, and so K is a separable extension of F. To get K normal over F, let $g(x)$ be an irreducible polynomial in $F[x]$ with a root $a \in K$. Using the notation above, we see that $g(x)$ divides $f(x)$ since $g(x)$ is the minimum polynomial of a over F, and a is a root of $f(x) \in F[x]$. The polynomial $f(x)$ splits completely in $K[x]$, and thus so must $g(x)$. Therefore, K is normal over F.

It is useful to observe that the polynomial $f(x) = \prod_{i=1}^{n} (x - a_i)$ constructed in the proof above is irreducible over F. Indeed, if $f(x) = g(x)h(x)$ with $g(x)$ and $h(x)$ nonunits in $F[x]$, and if a_i is a root of $g(x)$, then for some $\alpha \in G$, $\alpha(a_i)$ is a root of $h(x)$. But each automorphism of K which fixes F elementwise must take roots of a polynomial over F into roots of that same polynomial. But $g(x)$ and $h(x)$ have no roots in common since $f(x)$ has no multiple roots. Hence, $f(x)$ is prime in $F[x]$.

There are two important cases when each element of K has only finitely many images under G. If K is algebraic over F, each element a in K satisfies its minimum polynomial $f(x) \in F[x]$, and each element of G must carry a to a root of $f(x)$. Thus, each element of K has only finitely many images under G. The other case is when G itself is finite.

6.3.4 Corollary

Let G be a subgroup of the group of automorphisms of a field K, and let F be the fixed field of G. If K is algebraic over f, then K is a normal separable extension of F.

6.3.5 Corollary

Let G be a finite subgroup of the group of automorphisms of a field K, and let F be the fixed field of G. Then

(a) *K is a finite normal separable extension of F,*
(b) $[K:F] = |G|$, and
(c) $G(K/F) = G$.

Proof By 6.3.3, K is a normal separable extension of F. Therefore any finite extension L of F contained in K is separable and hence simple. If $L = F(a)$, then the degree of the minimum polynomial of a over F is $[L:F]$. But the minimum polynomial of a is the polynomial $f(x)$ constructed for a in 6.3.3, which in our present case is of degree at most

$|G|$. It follows that $[K:F] \le |G|$. Let $K = F(a)$. Only the identity element of G fixes a. Therefore the set of images of a under G is $\{\alpha(a): \alpha \in G\}$, and it has $|G|$ elements. Thus the polynomial $f(x) = \prod_{\alpha \in G}(x - \alpha(a))$ is the polynomial constructed for a in the lemma, and it is the minimum polynomial for a. Its degree on the one hand is $|G|$, and on the other hand is $[K:F]$. This proves (b). To prove (c), we note first that $|G(K/F)| \le [K:F]$. Since $K = F(a)$, an F-automorphism of K is determined by its effect on a, and there are only $[K:F]$ possible images for a, namely the roots of the minimum polynomial of a over K, which has degree $[K:F]$. Now (c) follows from the inclusion $G \subset G(K/F)$.

Notice that we do not know yet that if K is a finite extension of F, then $G(K/F)$ is a finite group. We need this fact.

6.3.6 Lemma

Let K be a finite extension of F. Then $|G(K/F)| \le [K:F]$. In fact, $|G(K/F)| = [K:H]$, where H is the fixed field of $G(K/F)$.

Proof By 6.3.4, K is a separable extension of the fixed field H of $G(K/F)$. Thus, $K = H(a)$ for some a in K. Each element of $G(K/F) = G(K/H)$ is uniquely determined by its action on a, and a can only go to other roots of its minimum polynomial over H. Therefore, $G(K/F)$ is finite. By 6.3.5, $|G(K/F)| \le [K:F]$.

At this point we need a variant of 6.2.5.

6.3.7 Lemma

Let $f(x)$ be a separable polynomial in $F[x]$, and let $\alpha: F \to K$ be an isomorphism. If M is a root field of $f(x)$ and if N is a root field of $\alpha(f(x))$, then there are $[M:F]$ extensions of α to an isomorphism $\beta: M \to N$.

Proof We induct on $[M:F]$. If $[M:F] = 1$, then the lemma is clear. Let $p(x)$ be a prime factor of $f(x)$, and let a be a root of $p(x)$ in M. There are $\deg(p(x))$ isomorphisms from $F(a)$ into N extending α, one for each root b of $\alpha(p(x))$. Consider such a map $\beta: F(a) \to K(b)$. Now M is a root field of the separable polynomial $f(x) \in F(a)[x]$, and N is a root field of $\beta(f(x)) \in K(b)[x]$. By the induction hypothesis, there are $[M:F(a)]$ extensions of β to an isomorphism $M \to N$. The result follows.

Now we are in a position to prove a basic theorem, which sums up much of what we have done so far.

6.3.8 Theorem

Let K be a finite extension of F. Then the following are equivalent.
(a) *K is a normal separable extension of F.*
(b) *K is the root field of a separable polynomial in $F[x]$.*
(c) $|G(K/F)| = [K:F]$.
(d) *The fixed field of $G(K/F)$ is F.*

Proof Assume (a). By 6.2.14, $K = F(a)$ for some $a \in K$. The minimum polynomial of a has all its roots in K since K is normal over F. Thus, (a) implies (b). By 6.3.7 applied to the identity map $F \to F$, (b) implies (c). By 6.3.6, (c) implies (d), and by 6.3.4, (d) implies (a).

The Fundamental Theorem of Galois Theory is now at hand. However, 6.3.8 enables us to establish two important properties of separability, and we digress to do that. Suppose that K is any extension of a field F. It is a quite nonobvious fact that the elements of K which are separable over F form a subfield of K. To prove this fact, let a and b be elements of K which are separable over F. Let $f(x)$ and $g(x)$ be the minimum polynomials of a and b, respectively. If $f(x) = g(x)$, then the root field N of $f(x)$ over F is separable. If $f(x)$ and $g(x)$ are distinct, then they are relatively prime, so the root field N of $f(x)g(x)$ over F is a separable extension of F by 6.3.8. In either case, N contains a copy of $F(a, b)$. Hence $F(a, b)$ is separable over F. Therefore the elements of K separable over F form a subfield K_s of K, called the ***separable closure*** of F in K.

Now we will show that separability is transitive. Let K be a separable extension of F, and let L be a separable extension of K. Let L_s be the separable closure of F in L. It clearly contains K, so L is separable over L_s. We need $L = L_s$. Let a be an element of L, and let $f(x)$ be its minimum polynomial over F. If $f(x)$ is not separable, then $f(x) = f_1(x^p)$, and $f_1(x)$ is the minimum polynomial of a^p over F. If $f_1(x)$ is not separable, then $f_1(x) = f_2(x^p)$, and $f_2(x)$ is the minimum polynomial of a^q over F, where $q = p^2$. Continuing, there is a smallest integer e such that if $q = p^e$, then a^q is separable over F. Now $a^q \in L_s$, and a satisfies $(x - a)^q = x^q - a^q \in L_s[x]$. Therefore, its minimum polynomial over L_s must divide $x^q - a^q$. But a is separable over L_s, so its minimum polynomial over L_s has distinct roots. It follows that its minimum polynomial is $x - a$, so that a is in L_s, and $L = L_s$, as desired. We have the following theorem.

6.3.9 Theorem

(a) *If K is an extension of F, then the set of elements of K which are separable over F is a subfield of K.*

(b) *If K is a separable extension of F, and if L is a separable extension of K, then L is a separable extension of F.*

6.3.10 Definition

A **Galois extension** *of a field F is a finite normal separable extension K of F. In this case, $G(K/F)$ is called the* **Galois group of K over** *F. If $f(x) \in F[x]$ is separable, then the root field K of $f(x)$ over F is a Galois extension, and $G(K/F)$ is called the* **Galois group of** *of $f(x)$* **over** *F.*

Note that if K is a Galois extension of F, and if L is any field between K and F, then K is a Galois extension of L.

6.3.11 The Fundamental Theorem of Galois Theory

Let K be a Galois extension of F. Then

$$L \to G(K/L)$$

is a one-to-one correspondence between the subfields L between K and F and the subgroups of $G(K/F)$. The extension L of F is normal if and only if $G(K/L)$ is a normal subgroup of $G(K/F)$. In this case, $G(L/F) \approx G(K/F)/G(K/L)$.

Proof Since K is a Galois extension of any L between K and F, L is the fixed field of $G(K/L)$ by 6.3.8. Thus, $L \to G(K/L)$ is one-to-one. If S is a subgroup of $G(K/L)$, then by 6.3.5, $S = G(K/F(S))$, where $F(S)$ is the fixed field of S. Thus, our correspondence $L \to G(K/F)$ is also onto.

Suppose that L is a normal extension of F. Since L is a simple extension $F(a)$ of F, for $\alpha \in G(K/F)$, $\alpha(a)$ must be a root of the minimum polynomial of a. All those roots lie in L since L is normal. Therefore, L is mapped into itself by every element of $G(K/F)$. Since L is a finite-dimensional vector space over F, every element of $G(K/F)$ induces an automorphism of L. We then have a homomorphism $G(K/F) \to G(L/F)$, namely the restriction to L. The kernel of this mapping is clearly $G(K/L)$, so $G(K/L)$ is a normal subgroup of $G(K/F)$.

If $G(K/L)$ is a normal subgroup of $G(K/F)$, then $|G(K/F)/G(K/L)| = [K{:}F]/[K{:}L] = [L{:}F]$, and it follows that $G(K/F) \to G(L/F)$ is an epimorphism since $|G(L/F)| \leq [L{:}F]$ by 6.3.6. Thus $G(K/F)/G(K/L) \approx G(L/F)$, and $|G(L/F)| = [L{:}F]$. By 6.3.8, L is a normal extension of F, and this completes the proof.

To illustrate the theory, we will compute the Galois group of the polynomial x^3-5 over the field $\mathbb{Q}$ of rational numbers. So let K be the root field over $\mathbb{Q}$ of x^3-5. The roots are $5^{1/3}$, $\omega 5^{1/3}$, and $\omega^2 5^{1/3}$, where ω is the cube root $(1+i3^{1/2})/2$ of 1. Since $[\mathbb{Q}(5^{1/3}):\mathbb{Q}]=3$, and $\mathbb{Q}(5^{1/3})$ does not contain ω or ω^2, and since $[K:\mathbb{Q}]\leq 6$, it follows that $[K:\mathbb{Q}]=6$. Since $G(K/\mathbb{Q})$ is a subgroup of permutations of $\{5^{1/3}, \omega 5^{1/3}, \omega^2 5^{1/3}\}$, and since $G(K/\mathbb{Q})$ has 6 elements, it must be the group of all permutations of these roots of x^3-5. Thus $G(K/\mathbb{Q})$ is S_3, the symmetric group on 3 objects. This group has one nontrivial normal subgroup, the alternating group A_3. This normal subgroup corresponds to the normal extension $\mathbb{Q}(\omega)$, which is the root field over $\mathbb{Q}$ of the polynomial x^2+x+1. There are three subgroups of $G(K/\mathbb{Q})$ of order 2. These correspond to the extensions $\mathbb{Q}(5^{1/3})$, $\mathbb{Q}(\omega 5^{1/3})$, and $\mathbb{Q}(\omega^2 5^{1/3})$. It is easy to check that these three extensions are distinct.

In any field F, an nth root of 1 is ***primitive*** if it is not an mth root of 1 for $0<m<n$. For example, in the field of complex numbers, -1 is a primitive second root of 1, $(1+i3^{1/2})/2$ is a primitive cube root of 1, i is a primitive 4th root of 1, -1 is not a primitive 4th root of 1, and $i\omega$ is a primitive 12th root of 1. The primitive nth roots of 1 are precisely the generators of the multiplicative group of all nth roots of 1, and are in one-to-one correspondence with the positive integers less than and prime to n, or equivalently, to the multiplicative units in $\mathbb{Z}/n\mathbb{Z}$. We will find the Galois group of the root field K of x^8-1 over $\mathbb{Q}$. First, note that $K=\mathbb{Q}(a)$, where a is any primitive 8th root of 1. Thus, any element of $G(\mathbb{Q}(a)/\mathbb{Q})$ is determined by its effect on a. The element a can be taken precisely to the roots of the minimum polynomial of a over $\mathbb{Q}$. What is this polynomial? It must divide $x^8-1=(x-1)(x+1)(x^2+1)(x^4+1)$. Clearly all the primitive 8th roots of 1 are roots of x^4+1, and x^4+1 is irreducible over $\mathbb{Q}$ (8.2, Problem 9.). Hence the minimum polynomial of a is x^4+1, and so $G(\mathbb{Q}(a)/\mathbb{Q})$ has four elements. But there are two groups with four elements. Which is it? The four primitive 8th roots of 1 are a, a^3, a^5, and a^7, and the four elements of $G(\mathbb{Q}(a)/\mathbb{Q})$ are 1, α, β, and γ characterized by $1(a)=a$, $\alpha(a)=a^3$, $\beta(a)=a^5$, and $\gamma(a)=a^7$. Thus, $G(\mathbb{Q}(a)/\mathbb{Q})$ is isomorphic to the group $\{1,3,5,7\}$ with multiplication being ordinary multiplication modulo 8. In other words, it is isomorphic to the group $(\mathbb{Z}/8\mathbb{Z})^*$ of units of $\mathbb{Z}/8\mathbb{Z}$. Every element of this group except the identity has order 2, whence the group is the direct product of two groups of order 2. Since it is Abelian, every field between $\mathbb{Q}$ and $\mathbb{Q}(a)$ is a normal extension of $\mathbb{Q}$. Our group $G(\mathbb{Q}(a)/\mathbb{Q})$ has three nontrivial subgroups, so that there are three fields between $\mathbb{Q}(a)$ and $\mathbb{Q}$. What are they? Above, a was any primitive 8th root of 1. The primitive 8th roots of 1 are the four numbers $\pm(2^{1/2}\pm i2^{1/2})/2$. It is easy to see that the three intermediate fields are $\mathbb{Q}(i)$, $\mathbb{Q}(2^{1/2})$, and $\mathbb{Q}(i2^{1/2})$. These are distinct fields, and there are only three such. To which subgroups of $G(\mathbb{Q}(a)/\mathbb{Q})$ do they correspond? Letting $a=(2^{1/2}+i2^{1/2})/2$, we have $a^3=(-2^{1/2}+i2^{1/2})/2$, $a^5=(-2^{1/2}-i2^{1/2})/2=-a$, and $a^7=(2^{1/2}-i2^{1/2})/2=-a^3$. Again, letting α, β, and γ be characterized by $\alpha(a)=a^3$, $\beta(a)=-a$, and $\gamma(a)=-a^3$, we see that

$\alpha(a+a^3)=a+a^3$, $\beta(a^2)=(-a)^2=a^2$, and $\gamma(a-a^3)=a-a^3$. Thus the subgroup $\{1, \alpha\}$ fixes $a+a^3=i2^{1/2}$ and hence corresponds to the extension $\mathbb{Q}(i2^{1/2})$, the subgroup $\{1, \beta\}$ fixes $a^2=i$ and hence corresponds to the extension $\mathbb{Q}(i)$, and the subgroup $\{1, \gamma\}$ fixes $a-a^3=2^{1/2}$ and hence corresponds to the extension $\mathbb{Q}(2^{1/2})$.

There are some features of the example above which merit attention. All the primitive 8th roots of 1 have the same minimum polynomial over $\mathbb{Q}$. That minimum polynomial was $\prod_{i=1}^{4}(x-a_i)$, where a_i ranged over the primitive 8th roots of 1. In particular, this product has rational coefficients and is irreducible over $\mathbb{Q}$. It even has integer coefficients. The Galois group of the splitting field of x^8-1 was isomorphic to the group of units of $\mathbb{Z}/8\mathbb{Z}$, which is an ***Abelian*** group. These facts prevail for any n, as we will now see.

Let F be a field of characteristic 0 containing the nth roots of 1. Since F has characteristic 0, the polynomial x^n-1 is separable, and so there are n nth roots of 1. Let $\varphi_n(x)$ be the polynomial in $F[x]$ which has as its roots the primitive nth roots of 1. Thus $\varphi_n(x)=\prod_{i=1}^{\varphi(n)}(x-\omega_i)$, where the ω_i range over the primitive nth roots of 1. The function $\varphi(n)$ is ***Euler's φ-function***. It is the number of positive integers less than n and prime to n; equivalently, $\varphi(n)$ is the number of primitive nth roots of 1. Our first goal is to show that $\varphi_n(x)$ is irreducible over $\mathbb{Q}[x]$. The polynomial $\varphi_n(x)$ is called the nth ***cyclotomic polynomial***. We need a lemma.

6.3.12 Lemma

Let $f(x)$ be a monic polynomial with integer coefficients. Let $f(x)=g(x)h(x)$ in $\mathbb{Q}[x]$ with $g(x)$ and $h(x)$ monic. Then $g(x)$ and $h(x)$ are in $\mathbb{Z}[x]$.

Proof Let a and b be the smallest positive integers such that $ag(x)=m(x)$ and $bh(x)=n(x)$ are in $\mathbb{Z}[x]$. We have $abf(x)=m(x)n(x)$. Let $m(x)=\sum_{i=0}^{r}a_ix^i$, and let $n(x)=\sum_{i=0}^{s}b_ix^i$. Note that $\gcd\{a_0, a_1, \ldots, a_r\}=\gcd\{b_0, b_1, \ldots, b_s\}=1$. If $ab=1$, there is nothing to do. Otherwise, let p be a prime dividing ab. Let i and j be the smallest indices such that p does not divide a_i and b_j, respectively. Now p must divide the coefficient $\cdots a_{i-2}b_{j+2}+a_{i-1}b_{j+1}+a_ib_j+a_{i+1}b_{j-1}+\cdots$ of x^{i+j}, and it divides every term except a_ib_j. Therefore $ab=1$, and the lemma is proved.

This lemma is a classic. We will treat factorizations of polynomials over certain integral domains in Chapter 8.

6.3.13 Theorem

The nth cyclotomic polynomial $\varphi_n(x)$ is in $\mathbb{Z}[x]$ and is irreducible over $\mathbb{Q}[x]$.

Proof Let ω be a primitive nth root of 1, and let $f(x)$ be its minimum polynomial over $\mathbb{Q}$. Let p be a prime not dividing n, and let $g(x)$ be the minimum polynomial over $\mathbb{Q}$ of the primitive nth root ω^p. By the lemma above, $f(x)$ and $g(x)$ are in $\mathbb{Z}[x]$. If $f(x) \neq g(x)$, then they are relatively prime in $\mathbb{Q}[x]$, and hence have no roots in common. They both divide $x^n - 1 = f(x)g(x)h(x)$, with $h(x) \in \mathbb{Z}[x]$ also. The polynomial $g(x^p)$ has ω as a root, whence $g(x^p) = f(x)k(x)$, again with $k(x) \in \mathbb{Z}[x]$. Now view the equalities

$$x^n - 1 = f(x)g(x)h(x), \quad \text{and} \quad g(x^p) = f(x)k(x)$$

modulo p. That is, regard the polynomials in $(\mathbb{Z}/p\mathbb{Z})[x]$. Operating modulo p, we have $g(x^p) = g(x)^p = f(x)k(x)$, and any prime factor $q(x)$ of $f(x)$ must divide $g(x)$, whence $q(x)^2$ must divide $x^n - 1 = f(x)g(x)h(x)$. Hence $x^n - 1$ has multiple roots. But $x^n - 1$ and nx^{n-1} are relatively prime since p does not divide n. Therefore, back in $\mathbb{Q}[x]$ we have $f(x) = g(x)$.

Now let ω^m be any primitive nth root of 1. Then $m = p_1 p_2 \cdots p_t$, with the p_i prime and prime to n. By what we just did, the minimum polynomial of ω^{p_1} is $f(x)$. But then for the same reason, the minimum polynomial of $\omega^{p_1 p_2}$ is $f(x)$. It follows that $f(x)$ is the minimum polynomial over $\mathbb{Q}$ of all the primitive nth roots of 1. Thus, $f(x)$ divides $\varphi_n(x)$ in $\mathbb{Z}[x]$. But the degree of $f(x)$ is at least as great as that of $\varphi_n(x)$, and hence $\varphi_n(x) = f(x)$. Our theorem is proved.

The previous theorem enables us to compute the Galois group over $\mathbb{Q}$ of $\varphi_n(x)$, or equivalently of $x^n - 1$.

6.3.14 Theorem

The Galois group over $\mathbb{Q}$ of the nth cyclotomic polynomial $\varphi_n(x)$ is isomorphic to the group of units in the ring $\mathbb{Z}/n\mathbb{Z}$. In particular, it is Abelian and has order $\varphi(n)$.

Proof Let K be a root field of $\varphi_n(x)$ over $\mathbb{Q}$, and let ω be a primitive nth root of 1. Then $K = \mathbb{Q}(\omega)$, and if $\sigma \in G(K/\mathbb{Q})$, then $\sigma(\omega) = \omega^i$, where $0 < i < n$ and $(i, n) = 1$. Since $\varphi_n(x)$ is the minimum polynomial of ω over $\mathbb{Q}$, there is a $\sigma \in G(K/\mathbb{Q})$ for each such i. It is now completely routine to check that $\sigma \to i$ gives the desired isomorphism.

The proof above works with $\mathbb{Q}$ replaced by any field F of characteristic 0, except that $\varphi_n(x)$ may not be irreducible over F and hence the map $\sigma \to i$ may not be onto. Thus, we have

6.3.15 Theorem

Let F be any field of characteristic 0. The Galois group of $x^n - 1$ over F is isomorphic to a subgroup of the group of units of $\mathbb{Z}/n\mathbb{Z}$. In particular, it is Abelian.

An extension K of F is called ***Abelian*** if $G(K/F)$ is Abelian. Similarly one speaks of ***cyclic*** extensions. Abelian extensions and cyclic extensions are classical topics of study in algebraic number theory. The previous theorems and the following one barely scratch the surface of these important subjects.

6.3.16 Theorem

Let F be a field of characteristic 0 which contains all the nth roots of 1, and let $a \in F$. Then the Galois group of $x^n - a$ over F is isomorphic to a subgroup of the additive group of $\mathbb{Z}/n\mathbb{Z}$. In particular, it is cyclic, and if $x^n - a$ is irreducible, it is cyclic of order n.

Proof Let K be a root field of $x^n - a$ over F. If $K \neq F$, then let $b^n = a$ with $b \in K$ and $b \notin F$. Let ω be a primitive nth root of 1. Then the roots of $x^n - a$ are the distinct elements $b, b\omega, b\omega^2, \ldots, b\omega^{n-1}$. Therefore $K = F(b)$, and the elements of $G(K/F)$ are determined by their effect on b. Let σ and τ be in $G(K/F)$. Then $\sigma(b) = b\omega^i$ and $\tau(b) = b\omega^j$ for unique integers satisfying $0 \le i < n$ and $0 \le j < n$. Hence

$$\sigma\tau(b) = \sigma(b\omega^j) = \sigma(b)\sigma(\omega^j) = b\omega^i\omega^j = b\omega^{i+j}.$$

It follows that associating σ with i is a monomorphism from $G(K/F)$ into the additive group of integers modulo n, as we needed to show.

6.3.17 Corollary

Let F be a field of characteristic 0, let p be a prime, let F contain the pth roots of 1, and let $a \in F$. Then $x^p - a$ is either irreducible or splits completely over F. If $x^p - a$ is irreducible over F, then its Galois group over F is cyclic of order p.

Proof The Galois group of $x^p - a$ is either cyclic of order p or 1 by 6.3.16. The corollary follows.

PROBLEMS

1 Prove that the root field of a separable polynomial is the root field of an irreducible separable polynomial.

2 Prove that the Galois group of a polynomial of degree n has order dividing $n!$.

3 Prove that K is a finite normal extension of F if and only if for any extension L of K, every element of $G(L/F)$ induces an automorphism of K.

4 Prove that if K is normal over F, if $F \subset L \subset K$, and if α is an F-isomorphism of L into K, then α is induced by an F-automorphism of K.

5 Prove that if K is a normal extension of F and if α is an F-isomorphism of K into itself, then α is an F-automorphism of K.

6 Let K be a finite extension of F. Prove that there is a (finite) normal extension N of F containing K such that if M is any normal extension of F containing K, then there is a K-isomorphism of N into M. The extension N of F is called a ***normal closure*** of the extension K of F. Prove that any two normal closures of K over F are K-isomorphic.

7 Let K be a Galois extension of F, let $\alpha \in G(K/F)$, and let L and M be fields between K and F. Prove that $\alpha(L) = M$ if and only if $\alpha G(K/L)\alpha^{-1} = G(K/M)$. [That is, prove that L and M are "conjugate" subfields if and only if $G(K/L)$ and $G(K/M)$ are conjugate subgroups.]

8 Prove that the Galois group $G(K/F)$ of a separable polynomial $f(x) \in F[x]$ is transitive on the roots of $f(x)$ if and only if $f(x)$ is irreducible in $F[x]$.

9 Prove that if K is a finite separable extension of F, then the number of fields between K and F is finite.

10 Prove that $K(5^{1/3})$ is the root field of $x^3 - 5$ over the field $K = \mathbb{Q}(i3^{1/2})$.

11 Find the Galois group of the polynomials
(a) $x^3 - 3$
(b) $x^4 - 2$
(c) $x^3 + 2x + 1$
(d) $x^4 + x^2 + 1$
(e) $x^6 - 1$
(f) $x^{12} - 1$
(g) $x^5 - 1$
over the field $\mathbb{Q}$. In each case, make explicit the correspondence between subgroups and intermediate extensions, and determine which are normal.

12 Prove that $x^n - 1 = \prod_{d|n} \varphi_d(x)$. Prove from this fact that for all n, $\varphi_n(x) \in \mathbb{Z}[x]$.

13 Calculate $\varphi_n(x)$ for $n \leq 12$.

14 Let F be a field of prime characteristic p, and let n be a positive integer prime to p. Prove that there are primitive nth roots of 1 in some extension of F.

15 Prove that if p is prime, then the Galois group of $x^p - 1$ over $\mathbb{Q}$ is cyclic of order $p - 1$. Prove that the Galois group of $x^p - a$ over $\mathbb{Q}$ is not necessarily Abelian.

16 Let p be a prime. Let $F \subset L \subset K$ with $G(K/F)$ and $G(L/F)$ cyclic of orders p^n and p^{n-1}, respectively. Prove that if $K = L(a)$, then $K = F(a)$.

17 Let F be a field of characteristic 0, and let $x^n - a \in F[x]$. Prove that if $x^n - a$ has a factor in $F[x]$ of degree prime to n, then $x^n - a$ has a root in F.

18 Let x be an indeterminate, and let $y \in F(x)$. Prove that $F(x) = F(y)$ if and only if there exist elements a, b, c, $d \in F$ such that $ad - bc \neq 0$ and such that $y = (ax + b)/(cx + d)$. Find $G(F(x)/F)$.

6.4 SOLVABILITY BY RADICALS

To simplify our discussion somewhat, we will assume throughout this section that *all our fields have characteristic* 0. First, we make precise the notion of "solving by radicals." Let $f(x) \in F[x]$. If $f(x)$ is linear, its roots are in F. If $f(x)$ is quadratic, then its roots lie in a radical extension K of F, that is, in a field K such that $K = F(a)$ where $a^n \in F$ for some positive integer n. In fact, if $f(x) = ax^2 + bx + c$ with $a \neq 0$, then the roots are given by the familiar expression $(-b \pm (b^2 - 4ac)^{1/2})/2a$, whence the roots lie in $K = F((b^2 - 4ac)^{1/2})$, and $(b^2 - 4ac) \in F$. If $f(x)$ is cubic, then it can be shown that its roots lie in a field K that can be obtained from F by two successive radical extensions. This means that K is a radical extension of a radical extension of F. The standard formulas for the roots of cubics bear this out. A similar statement is true if $f(x)$ is a quartic. The roots of $f(x)$ lie in a field that can be gotten from F by a succession of radical extensions. A classic theorem proved by Abel around 1821 asserts that this is not necessarily true if $f(x)$ is of degree 5 or higher. Galois theory, introduced by Galois around 1830, elucidates this phenomenon by providing a condition on the Galois group of $f(x)$ that is decisive for the solution of $f(x)$ by radicals.

6.4.1 Definition

Let $f(x) \in F[x]$. Then $f(x)$ is **solvable by radicals over** *F if there is a finite sequence*

$$F = F_0 \subset F_1 \subset \cdots \subset F_n$$

of fields with each F_i a radical extension of F_{i-1}, and with F_n containing the root field of $f(x)$ over F.

If $f(x)$ is not solvable by radicals, then in particular there is no formula expressing the roots of $f(x)$ in terms of various roots of rational functions of the coefficients of $f(x)$. For any polynomials of degree less than 5, such formulas exist, but not for arbitrary polynomials of degree 5 or greater.

The condition on the Galois group of $f(x)$ that is decisive for the solvability of $f(x)$ by radicals follows.

6.4.2 Definition

A group G is **solvable** *if it has a sequence of subgroups $1 = G_0 \subset G_1 \subset G_2 \subset \cdots \subset G_n = G$ such that each G_i is normal in G_{i+1} and such that G_{i+1}/G_i is Abelian.*

In particular, every Abelian group is solvable. A deep theorem proved in 1963 by Walter Feit and John Thompson asserts that every group of odd order is solvable. However, the alternating groups A_n with $n \geq 5$ are simple, and since such A_n are non-Abelian, they are not solvable.

One should note that if G is finite and solvable, then there is a sequence of subgroups $1 = G_0 \subset G_1 \subset \cdots \subset G_n = G$ such that G_i is normal in G_{i+1}, and such that G_{i+1}/G_i is cyclic of prime order. Indeed, if G_{i+1}/G_i is Abelian and has more than one element, then it has a subgroup H/G_i of prime order. The subgroup H is normal in G_{i+1}, and by induction, there is a finite chain $H = H_0 \subset H_1 \subset \cdots \subset H_r = G_{i+1}$ with H_i normal in H_{i+1} and with H_{i+1}/H_i cyclic of prime order. The rest should be clear. Therefore, in the case G is finite, the condition in 6.4.2 that G_{i+1}/G_i be Abelian could be replaced by the condition that G_{i+1}/G_i be cyclic of prime order (or simply cyclic, if one preferred).

Our goal is to show that $f(x) \in F[x]$ is solvable by radicals over F if and only if the Galois group of $f(x)$ over F is solvable. Keep in mind that our fields all have characteristic 0. First we record some basic properties of solvable groups.

6.4.3 Theorem

Subgroups and quotient groups of solvable groups are solvable. If N is a normal subgroup of G and if N and G/N are solvable, then G is solvable.

Proof Let S be a subgroup of the solvable group G. There is a chain $1 = G_0 \subset G_1 \subset \cdots \subset G_n = G$ of subgroups G_i of G such that G_i is normal in G_{i+1} and G_{i+1}/G_i is Abelian. Let $S_i = S \cap G_i$. Then $1 = S_0 \subset S_1 \subset \cdots \subset S_n = S$ is such a chain for S. Indeed, if $x \in S_{i+1}$ and $y \in S_i$, then $xyx^{-1} \in S$ since x and y are both in S, and $xyx^{-1} \in G_i$ since $x \in G_{i+1}$, $y \in G_i$, and G_i is normal in G_{i+1}. Hence S_i is normal in S_{i+1}. The map $S_{i+1}/S_i \to G_{i+1}/G_i$: $xS_i \to xG_i$ is a monomorphism, whence S_{i+1}/S_i is Abelian and S is solvable.

Now let G/N be a factor group of the solvable group G. In a similar vein, one can verify that $G_0N/N \subset G_1N/N \subset \cdots \subset G_nN/N$ satisfies the conditions necessary to make G/N solvable.

Suppose that N is a normal subgroup of G and that N and G/N are solvable. Then there are the usual chains of subgroups

$$1 = N_0 \subset N_1 \subset \cdots \subset N_r = N,$$

and

$$1 = N/N \subset G_1/N \subset \cdots \subset G_s/N = G/N$$

of N and G/N respectively. It is completely routine to check that the chain of subgroups

$$1 = N_0 \subset N_1 \subset \cdots \subset N_r \subset G_1 \subset G_2 \subset \cdots \subset G_s = G$$

has the requisite properties.

The following lemma eases our work a bit.

6.4.4 Lemma

Let $f(x) \in F[x]$ be solvable by radicals over F. Then there is a finite sequence $F = F_0 \subset F_1 \subset \cdots \subset F_n$ of fields with each F_i a radical extension of F_{i-1} and a normal extension of F, and with F_n containing the root field of $f(x)$ over F.

Proof By the definition of $f(x)$ being solvable by radicals over F, there is a chain $F = F_1 \subset F_2 \subset \cdots \subset F_n$ with $F_i = F_{i-1}(a_i)$, with $a_i^{r_i} \in F_{i-1}$, and with F_n containing a root field of $f(x)$ over F. If $r_i = rs$, then $F_i = F_{i-1}(a_i^r)(a_i)$, $a_i^r \in F_{i-1}(a_i^r)$, and $(a_i^r)^s = a^{r_i} \in F_{i-1}$. Hence, we may suppose that the r_i are prime numbers. Let $m = \prod_{i=1}^n r_i$, and let K_0 be the root field of $x^m - 1$ over F. For $i \geq 1$, let $K_i = K_{i-1}(a_i)$. By 6.3.17, each extension in the chain $F \subset K_0 \subset K_1 \subset \cdots \subset K_n$ is a normal radical extension, and the lemma is proved.

6.4.5 Theorem

Let F be a field of characteristic 0, and let $f(x) \in F[x]$. Then $f(x)$ is solvable by radicals over F if and only if the Galois group of $f(x)$ over F is solvable.

Proof Let $f(x)$ be solvable by radicals over F. Let $F \subset K_0 \subset K_1 \subset \cdots \subset K_n$ be the chain constructed in the proof of 6.4.4 Consider the chain of groups

$$1 = G(K_n/K_n) \subset G(K_n/K_{n-1}) \subset \cdots \subset G(K_n/K_0) \subset G(K_n/F).$$

Each subgroup is normal in the next since each extension K_{i+1} of K_i and K_0 of F is normal. For $i \geq 0$, $G(K_n/K_i)/G(K_n/K_{i+1})$ is isomorphic to $G(K_{i+1}/K_i)$ by part of the Fundamental Theorem of Galois Theory (6.3.11), and $G(K_{i+1}/K_i)$ is Abelian by 6.3.17. Similarly, $G(K_n/F)/G(K_n/K_0)$ is isomorphic to $G(K_0/F)$. By 6.3.15, $G(K_0/F)$ is Abelian. It follows from 6.4.3 that $G(K_n/F)$ is solvable. The root field K of $f(x)$ over F lies between F and K_n, and is normal over F. By the Fundamental Theorem (6.3.11), the Galois group $G(K/F)$ of $f(x)$ over F is isomorphic to $G(K_n/F)G(K_n/K)$, and by 6.4.3, this quotient is solvable. Hence the Galois group of $f(x)$ over F is solvable.

The proof of the converse is a little complicated. In order not to lose sight of it, we put the essence of the matter in the following lemma.

6.4.6 Lemma

Let K be a normal extension of F of prime degree p, and let F contain all the pth roots of 1. Then K is a radical extension of F.

Proof Let ω be a pth root of 1 with $\omega \neq 1$. Let $K = F(a)$, let σ generate $G(K/F)$, and for $0 \leq i < p$, consider the quantities

$$a_i = a + \omega^i\sigma(a) + (\omega^2)^i\sigma^2(a) + \cdots + (\omega^{p-1})^i\sigma^{p-1}(a).$$

We have $\sigma(a_i) = \omega^{-i}a_i$. Therefore $\sigma(a_i^p) = (\omega^{-i})^p a_i^p = a_i^p$, whence a_i^p is in F. Noting that $\sum_{j=0}^{p-1}(\omega^i)^j = 0$ for $0 < i \leq p-1$, because ω satisfies $1 + x + x^2 + \cdots + x^{p-1} = 0$, we get $(1/p)\sum_{i=0}^{p-1} a_i = a$. Since a is not in F, some a_i is not in F. For that i, $K = F(a_i)$, and $a_i^p \in F$. Our lemma is proved.

Now we will reduce the remaining half of the proof of 6.4.5 to the situation in Lemma 6.4.6. So suppose that the Galois group of $f(x)$ over F is solvable. Let K be the root field of $f(x)$ over F, and let $[K:F] = n$. Let N be the root field of $x^n - 1$ over K. We have $F \subset K \subset N$, $G(N/K)$ Abelian, and hence solvable, $G(N/K)$ normal in $G(N/F)$, and $G(K/F)$ solvable by hypothesis. By 6.4.3, $G(N/F)$ is solvable. Since $G(K/F)$ is solvable and finite, there is a chain of fields $F = F_0 \subset F_1 \subset \cdots \subset F_r = K$ such that each F_{i+1} is a normal extension of F_i, and such that $G(F_{i+1}/F_i)$ is cyclic of prime order p_i. Each F_{i+1} is the root field of some polynomial $f_i(x) \in F_i[x]$. The root field of $x^n - 1$ over F is contained in N. Let K_0 be this root field, and for $i \geq 0$, let K_{i+1} be the root field in N of $f_i(x)$ over K_i. If an element of $G(K_{i+1}/K_i)$ fixes elementwise the roots of $f_i(x)$, then it is the identity. Therefore, the restriction of an element of $G(K_{i+1}/K_i)$ to F_{i+1} is an element of $G(F_{i+1}/F_i)$. This gives rise to a homomorphism $G(K_{i+1}/K_i) \to G(F_{i+1}/F_i)$. An element in the kernel of this homomorphism fixes the roots of $f_i(x)$, and hence fixes K_{i+1}. Therefore, we have a monomorphism $G(K_{i+1}/K_i) \to G(F_{i+1}/F_i)$. Since $|G(F_{i+1}/F_i)|$ is prime, it follows that $[K_{i+1}:K_i] = p_i$ or 1. The extension K_0 of F is a radical extension, and K_0 contains all the p_ith roots of 1. We need the extension K_{i+1} of K_i to be radical. That was the purpose of 6.4.6.

In order to show that not every polynomial is solvable by radicals, it suffices to find one whose Galois group is not solvable. For $n \geq 5$, the symmetric groups S_n are not solvable since the normal subgroups A_n are simple and non-Abelian.

6.4.7 Theorem

For every positive integer n, there is a field F and a polynomial $f(x) \in F[x]$ of degree n whose Galois group over F is the symmetric group S_n.

This theorem will be a consequence of the following discussion which involves symmetric polynomials, and which is of interest in its own right.

Let F be a field, and adjoin n indeterminates $x_1, x_2, \ldots, x_n$ to F. Thus, the resulting field $F(x_1, x_2, \ldots, x_n) = K$ consists of quotients of polynomials in the x_i with coefficients in F. Any permutation of the x_i induces an automorphism of K which fixes F elementwise, and distinct such permutations induce distinct automorphisms of K. Hence S_n, the symmetric group of degree n, is a subgroup of $G(K/F)$. Let S be the fixed field of S_n. We have $F \subset S \subset K$ and $G(K/F) \supset S_n$. The elements of S are called ***symmetric rational functions*** of the x_i. For example, S contains all the symmetric polynomials in the x_i and in particular the elementary symmetric polynomials

$$x_1 + x_2 + \cdots + x_n,$$

$$x_1x_2 + x_1x_3 + \cdots + x_1x_n + x_2x_3 + \cdots + x_{n-1}x_n,$$

$$\vdots$$

$$x_1x_2x_3 \cdots x_n.$$

Let $s(x) = \prod_{i=1}^{n} (x - x_i)$. The coefficients of $s(x)$ are (neglecting sign) these elementary symmetric polynomials. Thus, $s(x) \in S[x]$. Clearly, K is a root field of $s(x)$ over S. This implies that $[K:S] \leq n!$. Since $|G(K/S)| \geq n!$, it follows from 6.3.8 that $[K:S] = n!$ and that $G(K/S) \approx S_n$. At this point, we have of course proved 6.4.7. However, a couple more observations about symmetric polynomials are in order. Let E be the subfield of S generated by F and the ***elementary*** symmetric polynomials in $x_1, x_2, \ldots x_n$. Then $F \subset E \subset S \subset K$. Since the coefficients of $s(x)$ (neglecting sign) are precisely the elementary symmetric polynomials in $x_1, x_2, \ldots, x_n$, then $s(x) \in E[x]$ and K is the root field of $s(x)$ over E. Hence $[K:E] \leq n!$. It follows that $S = E$. In other words, the symmetric rational functions in $x_1, x_2, \ldots, x_n$ are rational functions of the elementary symmetric polynomials in $x_1, x_2, \ldots, x_n$. Actually something stronger is true, but it has nothing to do with field theory. A symmetric polynomial in $x_1, x_2, \ldots, x_n$ is a polynomial in the elementary symmetric polynomials. For example,

$$x_1^2x_2 + x_1x_2^2 = x_1x_2(x_1 + x_2),$$

and

$$\begin{aligned} x_1^2x_2 + x_1x_2^2 + x_1x_3^2 + x_2^2x_3 + x_2x_3^2 + x_1^2x_3 \\ = (x_1x_2 + x_2x_3 + x_2x_3)(x_1 + x_2 + x_3) - 3x_1x_2x_3. \end{aligned}$$

We omit the proof in general, which can be found in Lang, p. 204.

We close this section by exhibiting a specific polynomial in $\mathbb{Q}[x]$ of degree 5 whose Galois group is S_5 and hence is not solvable. Our Theorem 6.4.7 is not applicable. However, its proof does enable one to get such a polynomial over a subfield F of the field of real numbers. One only needs real numbers r_1, r_2, r_3, r_4, r_5 such that if x_1, x_2, x_3, x_4, x_5 are indeterminates, then

$$\mathbb{Q}(x_1, x_2, x_3, x_4, x_5) \approx \mathbb{Q}(r_1, r_2, r_3, r_4, r_5)$$

via an isomorphism α such that $\alpha(x_i) = r_i$ for all i. Then the proof of 6.4.7 shows that the Galois group of $f(x) = \prod_{i=1}^{5} (x - r_i)$ over the field generated by $\mathbb{Q}$ and the coefficients of $f(x)$ is S_5. The existence of such a set of r_i is most easily shown by appealing to cardinality arguments. Pick a real number r_1 which is not algebraic over $\mathbb{Q}$. The field $\mathbb{Q}(r_1)$ is countable, and only countably many real numbers are algebraic over $\mathbb{Q}(r_1)$. Pick a real number r_2 which is not algebraic over $\mathbb{Q}(r_1)$. Proceeding in this way, we get real numbers r_1, r_2, r_3, r_4, r_5 which fill the bill.

A frequently used example of a polynomial in $\mathbb{Q}[x]$ which is not solvable by radicals is $x^5 - 4x + 2$. We will indicate that its Galois group over $\mathbb{Q}$ is S_5. First, we need it to be irreducible over $\mathbb{Q}$. This can be shown using the Eisenstein criterion (8.2.13). The polynomial has exactly three real roots. Just graph it and observe that it crosses the x axis exactly three times. It has no double roots since it is irreducible over $\mathbb{Q}$ and hence separable. Hence, it has exactly two complex roots. Let K be the root field of $x^5 - 4x + 2$ over $\mathbb{Q}$ in the field $\mathbb{C}$ of complex numbers. If a and b are roots of $x^5 - 4x + 2$, then there is an element $\sigma \in G(K/F)$ such that $\sigma(a) = b$. This follows from the irreducibility of the polynomial. The automorphism of $\mathbb{C}$ which takes a complex number $a + bi$ to $a - bi$ induces an automorphism of K which fixes the real roots of $x^5 - 4x + 2$ and interchanges the two complex ones. View $G(K/\mathbb{Q})$ as a subgroup of S_5. We have just seen that this subgroup has the following properties. Given any i and j in $\{1, 2, 3, 4, 5\}$, there is an element α in that subgroup such that $\sigma(i) = j$, and there is a transposition in that subgroup. The only such subgroup of S_5 is S_5 itself, as one can check. Therefore, $G(K/\mathbb{Q})$ is isomorphic to S_5 and $x^5 - 4x + 2$ is not solvable over $\mathbb{Q}$ by radicals.

PROBLEMS

1 Prove that if $F \subset K$ and $f(x) \in F[x]$ is solvable over F, then $f(x)$ is solvable over K.

2 Prove that if $G(K/F)$ is finite and solvable, and if $f(x) \in F[x]$ is solvable over K, then $f(x)$ is solvable over F.

3 Prove that if an irreducible polynomial $f(x) \in F[x]$ has a root in a radical extension of F, then $f(x)$ is solvable by radicals.

4 Prove that if K is a radical extension of $\mathbb{Q}$ of prime degree $p > 2$, then K is not a normal extension of $\mathbb{Q}$.

5 A group G of permutations of a set S is called ***transitive*** if given s and t in S, there is a σ in G such that $\sigma(s) = t$. Let K be the root field over F of an irreducible polynomial $f(x) \in F[x]$. Prove that $G(K/F)$ is transitive when viewed as a permutation of the roots of $f(x)$.

6 Let $f(x)$ be a separable polynomial in $F[x]$, and let K be its root field over F. Prove that $f(x)$ is irreducible in $F[x]$ if and only if $G(K/F)$ is transitive as a group of permutations of the roots of $f(x)$.

7 Determine whether or not the following polynomials are solvable by radicals over $\mathbb{Q}$.

(a) $x^5 - 8x + 3$
(b) $x^6 + x^5 + x^4 + x^3 + x^2 + x + 1$
(c) $x^{18} + x^9 + 1$
(d) $x^6 + x^3 + 1$
(e) $x^5 - 5x^3 - 20x + 5$

8 Let p be a prime ≥ 5. Prove that the polynomial $x^p - 4x + 2$ is not solvable by radicals over $\mathbb{Q}$.

9 Let $f(x)$ be an irreducible polynomial of prime degree over a field F, and let K be its root field. Suppose that $f(x)$ is solvable by radicals over F. Prove that if a and b are two roots of $f(x)$ in K, then $K = F(a, b)$.

10 Let F be a subfield of K, and let $a_1, a_2, \ldots, a_n$ be elements of K. Let $x_1, x_2, \ldots, x_n$ be indeterminates. Prove that if no a_i is algebraic over

$$F(a_1, \ldots, a_{i-1}, a_{i+1}, \ldots, a_n),$$

then $F(a_1, a_2, \ldots, a_n) \approx F(x_1, x_2, \ldots, x_n)$ via an F-isomorphism α such that $\alpha(a_i) = x_i$ for all i.

11 Determine the Galois group over $\mathbb{Q}$ of a polynomial $f(x) \in \mathbb{Q}[x]$ of degree 4 that has a quadratic factor in $\mathbb{Q}[x]$.

7 Topics from group theory

7.1 INTRODUCTION

Group theory is a vast subject which pervades almost every mathematical discipline. This chapter gives a brief introduction to some of the classical topics in non-Abelian group theory. In Chapter 2, we presented the basic notions of homomorphisms, normal subgroup, quotient group, direct sum, and so on. The Fundamental Theorem of Finite Abelian Groups (2.7.7), and the fact that for $n \geq 5$, the alternating group A_n is simple (2.5.7) are the deepest results in that chapter. In Section 6.4, we introduced solvable groups (6.4.2) and derived some of their properties. We will return to that topic in this chapter.

Since our discussion here is with groups that are not necessarily Abelian, we will use multiplicative notation.

7.2 THE JORDAN–HÖLDER THEOREM

Let G be a finite group. If G is not simple, then G has a normal subgroup $G_1 \neq G$ such that G/G_1 is simple. Just let G_1 be a maximal normal subgroup in the finite group G. Similarly, G_1 has a normal subgroup G_2 such that G_1/G_2 is simple. Thus we get a descending chain

$$G = G_0 \supset G_1 \supset G_2 \supset \cdots \supset G_n = \{e\}$$

such that G_{i+1} is normal in G_i, and G_i/G_{i+1} is simple. Now suppose that

$$G = H_0 \supset H_1 \supset H_2 \supset \cdots \supset H_m = \{e\}$$

is another such chain of subgroups. The Jordan-Hölder Theorem asserts that $m = n$, and that there is a one-to-one correspondence between the factor groups G_i/G_{i+1} and H_j/H_{j+1} such that corresponding factor groups are isomorphic. We will prove this remarkable theorem, but first some notation and terminology are needed. *The groups considered are not necessarily finite.*

7.2.1 Definition

Let G be a group. A **normal series** *of G is a chain of subgroups*

$$G = G_0 \supset G_1 \supset G_2 \supset \cdots \supset G_n = \{e\}$$

such that G_{i+1} is a normal subgroup of G_i for each i. The groups G_i/G_{i+1} are the **factor groups** *of the chain. The* **length** *of the chain is the number of strict inclusions in the chain. A normal series is a* **composition series** *if each factor group G_i/G_{i+1} is a simple group $\neq \{e\}$. Two normal series are* **equivalent** *if there is a one-to-one correspondence between their factor groups such that corresponding factor groups are isomorphic.*

7.2.2 Jordan–Hölder Theorem

Any two composition series of a group are equivalent.

Proof If a group has a composition series of length 1, then the group is simple and any two composition series are certainly equivalent. Now suppose that a group G has a composition series

$$G = G_0 \supset G_1 \supset G_2 \supset \cdots \supset G_n = \{e\} \tag{1}$$

of length $n > 1$, and that if a group has a composition series of length less than n, then any two composition series of that group are equivalent. Let

$$G = H_0 \supset H_1 \supset H_2 \supset \cdots \supset H_m = \{e\} \tag{2}$$

be any composition series of G. Consider the series

$$G = G_0 \supset G_1 \supset G_1 \cap H_1 \supset G_2 \cap H_1 \supset \cdots \supset G_n \cap H_1 = \{e\} \tag{3}$$

and

$$G = H_0 \supset H_1 \supset H_1 \cap G_1 \supset H_2 \cap G_1 \supset \cdots \supset H_m \cap G_1 = \{e\}. \tag{4}$$

Since $G_{i+1} \cap H_1$ is a normal subgroup of $G_i \cap H_1$ and $G_i \supset G_{i+1}$, the Third Isomorphism Theorem (2.3.12) yields

$$\begin{aligned}(G_i \cap H_1)/(G_{i+1} \cap H_1) &= (G_i \cap H_1)/(G_{i+1} \cap (G_i \cap H_1)) \\ &\approx G_{i+1}(G_i \cap H_1)/G_{i+1},\end{aligned}$$

and $G_{i+1}(G_i \cap H_1)$ is a normal subgroup of G_i since it is a product of two normal subgroups. Since G_i/G_{i+1} is a simple group, $(G_{i+1}(G_i \cap H_1))/G_{i+1}$ is either G_i/G_{i+1} or G_{i+1}/G_{i+1}. That is, $G_{i+1}(G_i \cap H_1)$ is either G_{i+1} or G_i. Therefore, if we remove repetitions from

$$G_1 \supset (G_1 \cap H_1) \supset (G_2 \cap H_1) \supset \cdots \supset (G_n \cap H_1) = \{e\},$$

we get a composition series for G_1. By our induction hypothesis, the resulting composition series is equivalent to the composition series

$$G_1 \supset G_2 \supset \cdots \supset G_n = \{e\},$$

and hence (*1*) and (*3*) (with repetitions removed) are equivalent. If $G_1 = H_1$, then (*1*) and (*2*) are certainly equivalent. If $G_1 \neq H_1$, then G_1H_1 is a normal subgroup of G properly containing G_1, so $G_1H_1 = G$. Thus $G_1/(G_1 \cap H_1) \approx (G_1H_1)/H_1$, and $H_1/(H_1 \cap G_1) \approx (G_1H_1)/G_1 = G/G_1$. Therefore (*3*) (with repetitions removed) and (*4*) (with repetitions removed) are equivalent, and the theorem is proved.

PROBLEMS

1 Prove that if G has a composition series, then any normal subgroup of G has a composition series.

2 Prove that if N is normal in G, and if G has a composition series, then so does G/N.

3 Suppose that G has a composition series and that N is normal in G. Prove that G has a composition series of which N is a member.

4 Suppose that G has a composition series. Prove that any normal series of G has a refinement that is a composition series.

5 (***Zassenhaus's Lemma***) Let A and B be subgroups of a group G, and let M and N be normal in A and B, respectively. Prove that
(a) $M(A \cap N)$ is a normal subgroup of $M(A \cap B)$.
(b) $N(M \cap B)$ is a normal subgroup of $N(A \cap B)$.
(c) $(M(A \cap B))/(M(A \cap N)) \approx (N(A \cap B))/(N(M \cap B))$.

6 Derive the ***Third Isomorphism Theorem*** (2.3.12) from Problem 5.

7 Prove from Problem 5 that any two normal series of a group G have refinements that are equivalent.

8 Prove the ***Jordan–Hölder Theorem*** from Problem 7.

9 Find all composition series of the cyclic group of order 6; of order 15; of order 30.

10 Find all composition series of the cyclic group of order 8; of order 16; of order 27.

11 Find all composition series of S_3; of S_4; of S_n.

12 Find all composition series of the quaternions $\mathbb{Q}_8$.

13 Find all composition series of the dihedral group D_4.

14 Find all composition series of the holomorph of $\mathbb{Z}(5)$.

15 Find all composition series of the holomorph of $\mathbb{Z}(8)$.

16 Find a group that does not have a composition series.

7.3 THE SYLOW THEOREMS

Let G be a finite group, and let p be a prime. If p^n is the highest power of p that divides $o(G)$, then G has a subgroup S such that $o(S) = p^n$. Any two such subgroups of G are conjugate, and the number of such subgroups is of the form $pq + 1$, that is, is congruent to modulo p, written $\equiv 1 \pmod{p}$. Further, that number divides $o(G)$. This is the essence of the Sylow Theorems, and we begin now to prove these theorems. We assume throughout that all our groups are *finite*.

Let G be a group. Two subsets S and T of G are ***conjugate*** if there is an element $g \in G$ such that $g^{-1}Sg = T$. Conjugacy is an equivalence relation on the set of subsets of G. The equivalence class determined by a set S is denoted $\mathrm{Cl}(S)$. The ***normalizer*** of a subset S of G is the set

$$N(S) = \{g \in G : g^{-1}Sg = S\}.$$

The set $N(S)$ is clearly a subgroup of G.

7.3.1 Lemma

Let S be a subset of G. Then $G{:}N(S) = o(\mathrm{Cl}(S))$.

Proof If $xN(S) = yN(S)$, then $y^{-1}x \in N(S)$, and hence $(y^{-1}x)^{-1}Sy^{-1}x = S$. Thus $x^{-1}ySy^{-1}x = S$, and so $ySy^{-1} = xSx^{-1}$. Similarly, if $ySy^{-1} = xSx^{-1}$, then $xN(S) = yN(S)$. The lemma follows.

We now apply 7.3.1 to the case where S consists of 1 element. Conjugacy is clearly an equivalence relation on the set of 1-element subsets of G, that is, on the set G. Hence, G is the union of the set of equivalence classes $\mathrm{Cl}(a)$ of elements a of G. The number of elements in each $\mathrm{Cl}(a)$ is the index of $N(a)$ in G. Thus

$$o(G) = \sum o(\mathrm{Cl}(a)) = \sum G{:}N(a) = \sum o(G)/o(N(a)),$$

where the sum ranges over a set of representatives of the equivalence classes. If an element a is in the center

$$\mathbb{Z}(G) = \{g \in G: gx = xg \text{ for all } x \in G\}$$

of G, then $\mathrm{Cl}(a) = \{a\}$. The equation in the next theorem is called the ***class equation***, and it follows from the equations above.

7.3.2 Theorem

$o(G) = o(\mathbb{Z}(G)) + \Sigma\, o(G)/o(N(a))$, where the sum is over a set of representatives a such that $a \notin \mathbb{Z}(G)$.

7.3.3 Corollary

Let p be a prime. If $o(G) = p^n$ and $n \geq 1$, then $\mathbb{Z}(G) \neq \{e\}$.

Proof $o(G) = o(\mathbb{Z}(G)) + \Sigma\, o(G)/o(N(a))$ as in 7.3.2. Now $o(G)/o(N(a))$ is divisible by p since $N(a) \neq G$. Since $o(G)$ is divisible by p, $o(\mathbb{Z}(G))$ must be divisible by p.

7.3.4 Corollary

If p is a prime, and if $o(G) = p^2$, then G is Abelian.

Proof If $\mathbb{Z}(G) \neq G$, then $o(G/\mathbb{Z}(G)) = p$, so $G/\mathbb{Z}(G)$ is cyclic. Let $g\mathbb{Z}(G)$ generate $G/\mathbb{Z}(G)$. Then every element of G is of the form $g^n a$, with $a \in \mathbb{Z}(G)$. Any two such elements commute. Hence, G is Abelian.

The next corollary is crucial in showing the existence of Sylow subgroups of groups. If p^n is the highest power of the prime p that divides $o(G)$, and if $n \geq 1$, then for G to have a subgroup of order p^n, it certainly must have an element of order p. This is how we get started showing the existence of such a subgroup.

7.3.5 Corollary (Cauchy's Theorem)

If p is a prime and p divides $o(G)$, then G has an element of order p.

Proof The proof is by induction on $o(G)$. If $o(G) = 1$, the assertion is trivial. Suppose that $o(G) > 1$ and that every group of smaller order than $o(G)$ has an element of order p whenever p divides its order. Consider the class equation

$$o(G) = o(\mathbb{Z}(G)) + \sum G{:}N(a)$$

of G. If p divides some $o(N(a))$, then since $a \notin \mathbb{Z}(G)$, $o(N(a)) < o(G)$, and $N(a)$ has an element of order p. So we may suppose that p divides no $o(N(a))$. Thus p divides each $G{:}N(a)$, and since p divides $o(G)$, it divides $o(\mathbb{Z}(G))$. Let $g \in \mathbb{Z}(G)$, $g \neq e$. If p does not divide $o(\langle g \rangle)$, then p divides $o(\mathbb{Z}(G)/\langle g \rangle)$, so that $\mathbb{Z}(G)/\langle g \rangle$ has an element $x\langle g \rangle$ of order p since $o(\mathbb{Z}(G)/\langle g \rangle) < o(G)$. Thus, $x^p = g^n$ for some n. If $o(g^n) = m$, then $o(x) = p \cdot m$, and it follows that $o(x^m) = p$. If p divides $o(\langle g \rangle)$, then $o(g) = p \cdot n$ for some n, whence $o(g^n) = p$.

Now we can prove that Sylow subgroups exist.

7.3.6 Theorem (Sylow)

Let p^n divide $o(G)$. Then G has a subgroup of order p^n.

Proof We induct on $o(G)$. If $o(G) = 1$, then the proof is trivial. Now suppose that $o(G) > 1$ and that the theorem is true for all groups whose orders are less than $o(G)$. If G has a proper subgroup H such that $G{:}H$ is prime to p, then p^n divides $o(H)$, so that H has a subgroup of order p^n. So we may suppose that p divides $G{:}H$ for all proper subgroups H of G. Now $o(G) = o(\mathbb{Z}(G)) + \sum G{:}N(a)$, so p divides $o(\mathbb{Z}(G))$. Thus, $\mathbb{Z}(G)$ has a subgroup S such that $o(S) = p$. The subgroup S is normal, $o(G/S) < o(G)$, $o(G/S)$ is divisible by p^{n-1}, and so G/S has a subgroup H/S whose order is p^{n-1}. It follows that $o(H)$ is p^n.

7.3.7 Definition

Let p^n be the highest power of the prime p that divides $o(G)$. A subgroup S of G such that $o(S) = p^n$ is called a **Sylow p-subgroup** *of G.*

Now we want to find the number of Sylow p-subgroups of a group and to show that any two are conjugate. We need a preliminary definition and lemma.

7.3.8 Definition

*Let A and B be subgroups of a group G. An A-***conjugate** *of B is a subgroup of the form $a^{-1}Ba$ with $a \in A$. (So a G-conjugate of B is just a conjugate of B.)*

7.3.9 Lemma

Let A and B be subgroups of a group G. Then the number of A-conjugates of B is $A{:}(N(B)\cap A)$.

Proof $a^{-1}Ba = a_1^{-1}Ba_1$ if and only if $a_1a^{-1}Baa_1^{-1} = B$ if and only if $a_1a^{-1} \in N(B)$ if and only if $(N(B)\cap A)a_1 = (N(B)\cap A)a$. Therefore,

$$a^{-1}Ba \rightarrow (N(B)\cap A)a$$

is a one-to-one mapping from the set of A-conjugates of B onto the set of right cosets of $N(B)\cap A$ in A.

7.3.10 Theorem

Let G be a finite group, and let p be a prime.

(a) *If H is a subgroup of G and if $o(H)$ is a power of p, then H is contained in a Sylow p-subgroup of G.*
(b) *Any two Sylow p-subgroups of G are conjugate.*
(c) *The number of Sylow p-subgroups of G is $\equiv 1(\mathrm{mod}\, p)$, and divides $o(G)$.*

Proof Let H be a subgroup of G, and suppose that $o(H) = p^n$. Let S be any Sylow p-subgroup of G. The number of conjugates of S is $G{:}N(S)$, which is prime to p. Now H induces a partition of the set of conjugates of S; two conjugates of S are in the same member of the partition if they are H-conjugate. Let S_1 be any conjugate of S. By 7.3.9, the number of H-conjugates of S_1 is $H{:}(N(S_1)\cap H)$, and this is a power of p since $o(H) = p^n$. Thus, the number of conjugates of S in every member of this partition of the conjugates of S is a power of p and the total number of conjugates is prime to p. Therefore, some member of this partition has exactly 1 element. Let that element be T. Then $H{:}(N(T)\cap H) = 1$, so $N(T) \supset H$. Therefore TH is a subgroup, T is normal in TH, and $TH/T \cong H/(H\cap T)$ is a p-group. It follows that TH is a p-group, so that $TH = T$. Hence $T \supset H$. Hence, H is contained in a Sylow p-subgroup. This proves (a). If H is already a Sylow p-subgroup, then $H = T$, so H is conjugate to S, and (b) follows. If $H = S$, then S partitions the set of conjugates of S (that is, the set of all Sylow p-subgroups of G) into subsets with a power of p elements, and one member of the partition has 1 element. But whenever one member of the partition has only 1 element, that element is $H = S$. Therefore, exactly one member of the partition of the conjugates of S has exactly 1 element. Since they all have a power of p elements, the total number of conjugates of S is

$\equiv 1 \pmod{p}$. Since $o(\mathrm{Cl}(S)) = G{:}N(S)$, the number of Sylow p-subgroups divides $o(G)$, and (c) follows.

7.3.11 Corollary

Let S be a Sylow p-subgroup of the finite group G. Then S is the only Sylow p-subgroup of $N(S)$.

7.3.12 Corollary

Let S be a Sylow p-subgroup of the finite group G. Then $N(N(S)) = N(S)$.

Proof If $x^{-1}N(S)x = N(S)$, then $x^{-1}Sx = S$ since S is the only Sylow p-subgroup of $N(S)$. Thus, $x \in N(S)$. Hence $N(N(S)) \subset N(S)$, so $N(N(S)) = N(S)$.

PROBLEMS

1. Find all the Sylow 2-subgroups and all the Sylow 3-subgroups of S_3, S_4, and S_5.
2. Let H be a proper subgroup of the finite group G. Prove that G is not the set union of the conjugates of H.
3. Let G be a finite group, let N be a normal subgroup of G, and let S be a Sylow p-subgroup of G. Prove that $S \cap N$ is a Sylow p-subgroup of N, and that SN/N is a Sylow p-subgroup of G/N.
4. Prove that a group of order 28 has a normal subgroup of order 7. Prove that a group of order 28 that does not have exactly 7 Sylow 2-subgroups has a normal subgroup of order 4, and is Abelian.
5. Let $o(G) = pq$ with p and q primes, and $p < q$. Let S be a subgroup of G of order p, and let T be a subgroup of G of order q.
   (a) Prove that T is normal in G.
   (b) Prove that G is a split extension of T by S.
   (c) Prove that G is the direct product of S and T, and hence is cyclic, unless p divides $q - 1$.
   (d) Suppose that p divides $q - 1$. Prove that there is a non-Abelian group of order pq.
6. Prove that any group of order less than 60 is either of prime order or has a proper normal subgroup.
7. Prove that a group of order 108 has a normal subgroup of order 9 or 27.

8 Let G be finite, and let S be a Sylow p-subgroup of G. If S is normal in N, and if N is normal in G, prove that S is normal in G.

9 Prove that any simple group of order 60 is isomorphic to A_5.

10 Prove that if $o(G) = p^n$, p a prime, and if H is a proper subgroup of G, then $N(H) \neq H$.

7.4 SOLVABLE AND NILPOTENT GROUPS

Solvable groups have been defined in Chapter 6 (6.4.2) in connection with Galois theory, and a few of their properties were noted there. Recall that a group G is ***solvable*** if it has a normal series

$$G = G_0 \supset G_1 \supset G_2 \supset \cdots \supset G_n = \{e\}$$

such that G_i/G_{i+1} is Abelian. Such a series will be called a ***solvable series***. We showed that subgroups and quotient groups of solvable groups are solvable, and that if N is normal in G, then N and G/N solvable imply that G is solvable. We noted further that a solvable group has a normal chain whose factors are cyclic of prime order. Since a nontrivial p-group G has nontrivial center $\mathbb{Z}(G)$, it follows by induction on $o(G)$ that G is solvable. Indeed, $\mathbb{Z}(G)$ is a solvable normal subgroup of G and $G/\mathbb{Z}(G)$ is solvable since it is a p-group smaller than G. It follows that G is solvable. This fact is important enough to exhibit as a theorem.

7.4.1 Theorem

Finite p-groups are solvable.

Actually, finite p-groups satisfy a much stronger condition than that of solvability. They are nilpotent, a concept we will come to shortly.

Recall that the ***commutator subgroup*** G' of a group G is the subgroup generated by all its commutators $a^{-1}b^{-1}ab$, where a and b are in G. The crucial facts about G' are that G' is a fully invariant subgroup of G, G/G' is Abelian, and G' is contained in every normal subgroup N such that G/N is Abelian. We now define the ***higher commutator subgroups*** of a group G inductively by

$$G^{(0)} = G, \quad \text{and} \quad G^{(i+1)} = (G^{(i)})'.$$

Then we have a normal chain

$$G = G^{(0)} \supset G^{(1)} \supset G^{(2)} \supset \cdots.$$

This chain is called the ***derived series*** of G. In the derived series, either each inclusion is proper, or for some i, $G^{(i)} = G^{(i+n)}$ for all n. Either situation may occur. If G is finite, the latter must occur, of course.

7.4.2 Theorem

A group G is solvable if and only if $G^{(n)} = \{e\}$ for some n.

Proof If $G^{(n)} = \{e\}$, then the normal series

$$G = G^{(0)} \supset G^{(1)} \supset \cdots \supset G^{(n)} = \{e\}$$

has Abelian quotients, whence G is solvable. Now suppose that G is solvable. Let

$$G = G_0 \supset G_1 \supset \cdots \supset G_n = \{e\}$$

be a solvable series for G. Then since G_0/G_1 is Abelian, $G_1 \supset G^{(1)}$. Suppose that $G_i \supset G^{(i)}$. Then $(G_i)' \supset (G^{(i)})' = G^{(i+1)}$, and since G_i/G_{i+1} is Abelian, $G_{i+1} \supset (G_i)'$. Thus $G_{i+1} \supset G^{(i+1)}$, and we have by induction that $G_i \supset G^{(i)}$ for all i. Since $G_n = \{e\}$, $G^{(n)} = \{e\}$, and the theorem is proved.

There are two other series of a group we will now define. If H and K are subgroups of G, $[H, K]$ is the subgroup generated by all elements of the form $h^{-1}k^{-1}hk$ with $h \in H$ and $k \in K$. The ***lower central series*** (or ***descending central series***) of a group G is the chain

$$G = G^0 \supset G^1 \supset G^2 \supset \ldots$$

where $G^{i+1} = [G^i, G]$. Recall that a subgroup H of G is ***fully invariant*** if it is taken into itself by every endomorphism of G. It is an easy exercise to show that *every member of the lower central series of G is a fully invariant subgroup of G.* In particular, each member of the lower central series is a normal subgroup of G. The factors of the series are certainly Abelian, since $G^{i+1} = [G^i, G] \supset [G^i, G^i] = (G^i)'$.

Let $\mathbb{Z}^0(G) = \{e\}$, $\mathbb{Z}(G)$ be the center of G, and in general $\mathbb{Z}^{i+1}(G)$ be given by $\mathbb{Z}(G/\mathbb{Z}^i(G)) = \mathbb{Z}^{i+1}(G)/\mathbb{Z}^i(G)$. The chain

$$\{e\} = \mathbb{Z}^0(G) \subset \mathbb{Z}^1(G) \subset \mathbb{Z}^2(G) \subset \cdots$$

is called the ***upper central series*** (or ***ascending central series***) of G. A subgroup H of a group G is called ***characteristic*** if $f(H) = H$ for every automorphism f of G. It is straightforward to verify that *each member of the upper central series of G is a characteristic subgroup of G.* In particular, each member of the upper central series is a normal subgroup of G. Again, the factors $\mathbb{Z}^{i+1}(G)/\mathbb{Z}^i(G)$ of the series are certainly Abelian.

A fundamental fact about the upper and lower central series of a group G is that the upper central series reaches G in a finite number of steps if and only if the lower central series reaches $\{e\}$ in that same finite number of steps. To prove this, it is convenient to have the following two lemmas, whose proofs are left as exercises (Problems 13 and 14).

7.4.3 Lemma

Let N and K be subgroups of G with N normal in G and with $N \subset K$. Then $K/N \subset \mathbb{Z}(G/N)$ if and only if $[K, G] \subset N$.

7.4.4 Lemma

If $f\colon G \to H$ is an epimorphism, and if $S \subset \mathbb{Z}(G)$, then $f(S) \subset \mathbb{Z}(H)$.

7.4.5 Theorem

$\mathbb{Z}^m(G) = G$ if and only if $G^m = \{e\}$. In this case, $G^i \subset \mathbb{Z}^{m-i}(G)$ for all i.

Proof Note that the theorem is trivial if $m = 1$. So we assume $m \geq 2$. Suppose that $\mathbb{Z}^m(G) = G$. We will show that $G^i \subset \mathbb{Z}^{m-i}(G)$ by induction on i. If $i = 0$, then $G^i = G = \mathbb{Z}^m(G)$, so the inclusion holds. If $G^i \subset \mathbb{Z}^{m-i}(G)$, then

$$G^{i+1} = [G^i, G] \subset [\mathbb{Z}^{m-i}(G), G].$$

Letting $N = \mathbb{Z}^{m-(i+1)}(G)$ and $K = \mathbb{Z}^{m-i}(G)$ in 7.4.3, we get $[\mathbb{Z}^{m-i}(G), G] \subset \mathbb{Z}^{m-(i+1)}(G)$, whence $G^{i+1} \subset \mathbb{Z}^{m-(i+1)}$. Hence, if $\mathbb{Z}^m(G) = G$, then setting $i = m$, we get $G^m \subset \mathbb{Z}^0 = \{e\}$.

Conversely, suppose that $G^m = \{e\}$. We will show that $G^i \subset \mathbb{Z}^{m-i}(G)$ by induction on $m - i$. If $m - i = 0$, then $G^i = G^m = \{e\} \subset \mathbb{Z}^0(G)$, and our inequality holds. If $G^i \subset \mathbb{Z}^{m-i}(G)$, then there is a natural epimorphism

$$G/G^i \to G/\mathbb{Z}^{m-i}(G)$$

given by $gG^i \to g\mathbb{Z}^{m-i}(G)$. Since $G^{i-1}/G^i \subset \mathbb{Z}(G/G^i)$ by 7.4.3, then by 7.4.4, the image $G^{i-1}\mathbb{Z}^{m-i}(G)/\mathbb{Z}^{m-i}(G)$ of G^{i-1}/G^i under the epimorphism is contained in the center of

$$\mathbb{Z}(G/\mathbb{Z}^{m-i}(G)) = \mathbb{Z}^{m-(i-1)}(G)/\mathbb{Z}^{m-i}(G).$$

Therefore, $G^{i-1} \subset \mathbb{Z}^{m-(i-1)}(G)$. This completes the induction. If $G^m = \{e\}$, then setting $i = 0$, we get $G = G^0 \subset \mathbb{Z}^m(G)$, whence $\mathbb{Z}^m(G) = G$. This completes the proof of the theorem.

7.4.6 Definition

A group G is **nilpotent** *if $G^n = \{e\}$ for some n.*

We could equally as well have defined G to be nilpotent if $\mathbb{Z}^n(G) = G$ for some integer n. It is very easy to show that *subgroups and quotient groups of nilpotent groups are nilpotent.* It follows from the definition that if G is nilpotent, then G is solvable. However, not every solvable group is nilpotent. The symmetric group S_3 is solvable but is not nilpotent. From 7.3.3 and 7.4.5, we get the following important fact.

7.4.7 Theorem

Every finite p-group is nilpotent.

7.4.8 Lemma

If H is a proper subgroup of a nilpotent group G, then $N(H) \neq H$.

Proof Let n be the smallest integer such that $G^n \subset H$. Then $[G^{n-1}, H] \subset [G^{n-1}, G] = G^n \subset H$, whence $G^{n-1} \subset N(H)$ by Problem 15. Therefore, $N(H)$ properly contains H.

The next theorem is a basic characterization of finite nilpotent groups. It reduces the study of such groups to that of finite p-groups.

7.4.9 Theorem

A finite group is nilpotent if and only if it is the direct product of its Sylow subgroups.

Proof If a finite group is the direct product of its Sylow subgroups, it is nilpotent by Problem 17 since each Sylow subgroup is nilpotent by 7.4.7. Conversely, suppose that a finite group G is nilpotent. Let S_p be a Sylow p-subgroup of G. Since $N(N(S_p)) = N(S_p)$, then $N(S_p) = G$ by 7.4.8. That is, every Sylow subgroup of G is normal. Since any two Sylow p-subgroups are conjugate, for each p dividing $o(G)$ there is exactly one Sylow p-subgroup. The rest of the proof is routine.

PROBLEMS

1 Prove that $(G \times H)' = G' \times H'$.

2 Prove that if $G_1, G_2, \ldots, G_n$ are solvable, then so is the direct product $\prod G_i$.

3 Prove that if m and n are integers such that $m \leq n$, then $(G/G^{(n)})^{(m)} = G^{(m)}/G^{(n)}$.

4 Prove that any group of order pq is solvable, where p and q are primes.

5 Prove that any group of order p^2q is solvable, where p and q are primes.

6 Prove that a solvable group $G \neq \{e\}$ has a fully invariant Abelian subgroup $\neq \{e\}$.

7 Prove that a finite solvable group $G \neq \{e\}$ has a fully invariant Abelian p-subgroup $\neq \{e\}$.

8 Prove that if H and K are solvable normal subgroups of G, then HK is a solvable normal subgroup of G.

9 Let A and B be two finite solvable groups of the same order. Prove that there is a one-to-one correspondence between the factors of a composition series of A and the factors of a composition series of B such that the corresponding factors are isomorphic.

10 Prove that S_3 is solvable but not nilpotent.

11 Prove that every member of the lower central series of a group G is a fully invariant subgroup of G.

12 Prove that every member of the upper central series of a group G is a characteristic subgroup of G.

13 Prove that if N is a normal subgroup of G, and if K is a subgroup of G containing N, then $K/N \subset \mathbb{Z}(G/N)$ if and only if $[K, G] \subset N$.

14 Prove that if $f: G \to H$ is an epimorphism and if $S \subset \mathbb{Z}(G)$, then $f(S) \subset \mathbb{Z}(H)$.

15 Prove that if A and B are subgroups of G, then $[A, B] \subset B$ if and only if $A \subset N(B)$.

16 Prove directly from the definition of nilpotent groups that subgroups and quotient groups of nilpotent groups are nilpotent.

17 Prove that if $G_1, G_2, \ldots, G_n$ are nilpotent, then so is the direct product $\prod G_i$.

18 Prove that if $H \subset \mathbb{Z}(G)$ and if G/H is nilpotent, then G is nilpotent.

19 Let H be a subgroup of a finite nilpotent group G. Prove that $H = \prod (H \cap S_p)$, where S_p ranges over the Sylow subgroups of G.

7.5 THE WEDDERBURN THEOREM FOR FINITE DIVISION RINGS

A theorem proved by Wedderburn in 1905 asserts that a finite division ring is a field. That is, if D is a finite division ring, then the multiplicative group $D^* = \{d \in D: d \neq 0\}$ is commutative. We prove this theorem here instead of in Chapter 8 because its proof involves only the class equation (7.3.2) and some number theory. Thus its proof is largely group theoretic, even though one thinks of the theorem as a theorem about rings. The number theoretic fact needed is that the nth cyclotomic polynomial

$$\varphi_n(x) = \prod_{i=1}^{\varphi(n)} (x - \omega_i),$$

where the product ranges over the primitive nth roots of 1, is in $\mathbb{Z}[x]$ (6.3.13).

7.5.1 Wedderburn's Theorem

A finite division ring is a field.

Proof Let D be a finite division ring. Its center $\mathbb{Z}(D)$ is a field F, and since D is a finite-dimensional vector space over F, D has q^n elements, where $o(F) = q$ and n is the dimension of D over F. Let $a \in D$. Then it is easy to check that

$$N(a) = \{d \in D: da = ad\}$$

is a subdivision ring of D. Since $N(a) \supset F$, it is a vector space over F, so that $o(N(a)) = q^{n(a)}$ for some integer $n(a)$. Now consider the class equation for the group D^* of nonzero elements of D. It is

$$q^n - 1 = q^{n(a)} - 1 + \sum_a (q^n - 1)/(q^{n(a)} - 1)$$

where the sum is over representatives a of each conjugate class for a not in the center $\mathbb{Z}(D^*)$ of the group D^*. Since $q^{n(a)} - 1$ divides $q^n - 1$, $n(a)$ must divide n. In fact, if $n = n(a)m + r$ with $0 \leq r < n(a)$, then

$$q^n - 1 = (q^{n(a)} - 1)(q^{n-n(a)} + q^{n-2n(a)} + \cdots + q^{n-mn(a)}) + q^{n-mn(a)} - 1,$$

and since $q^{n-mn(a)} - 1 < q^{n(a)} - 1$ and $q^{n(a)} - 1$ divides $q^n - 1$, it follows that $n - mn(a) = 0$. Therefore, in each term $(q^n - 1)/(q^{n(a)} - 1)$ in the class equation, $n(a)$ divides n. Suppose $n(a) < n$. Consider the polynomials

$$x^n - 1 = \prod (x - \alpha_i)$$

and

$$x^{n(a)} - 1 = \prod (x - \beta_j),$$

where the α_i range over the nth roots of 1 and the β_j range over the $n(a)$th roots of 1. Since no β_j is a primitive nth root of 1, $\varphi_n(x) = \prod (x - \alpha_i)$, where α_i ranges over the primitive nth roots of 1, is relatively prime to $x^{n(a)} - 1$. Since all these polynomials are in $\mathbb{Z}[x]$, $\varphi_n(x)$ divides $(x^n - 1)/(x^{n(a)} - 1)$ in $\mathbb{Z}[x]$. Hence the integer $\varphi_n(q)$ divides $(q^n - 1)/(q^{n(a)} - 1)$ in $\mathbb{Z}$. From the class equation, this forces the integer $\varphi_n(q)$ to divide $q - 1$. But $\varphi_n(q) = \prod (q - \alpha_i)$, where α_i ranges over the primitive nth roots of 1. Each $q - \alpha_i$ has absolute value $|q - \alpha_i|$ strictly greater than $q - 1$ if $n > 1$. Indeed, if $\alpha_i = a + bi$, then since $a_i^n = 1$, α_i has absolute value 1, so $a^2 + b^2 = 1$. Thus

$$|q - (a + bi)|^2 = (q - a)^2 + b^2 = q^2 - 2qa + a^2 + b^2 = q^2 - 2qa + 1,$$

and

$$|q - 1|^2 = q^2 - 2q + 1.$$

If $n > 1$, then $a < 1$. Since q is positive, it follows that

$$q^2 - 2qa + 1 > q^2 - 2q + 1,$$

whence

$$|q - \alpha_i| > q - 1.$$

Therefore $\varphi_n(q)$ is strictly larger than $q - 1$. It follows that $n = 1$, that is, that $n = n(a)$, and that $F = D$.

7.6 THE FUNDAMENTAL THEOREM OF ALGEBRA

The existence of Sylow subgroups (7.3.6) together with Galois theory (Section 6.3) enables us to prove the Fundamental Theorem of Algebra—the field of complex numbers $\mathbb{C}$ is algebraically closed. That is, every polynomial in $\mathbb{C}[x]$ of degree ≥ 1 factors into the product of linear factors. Equivalently, every polynomial in $\mathbb{C}[x]$ of degree ≥ 1 has a root in $\mathbb{C}$. We take as the definition of $\mathbb{C}$ the field of all numbers $a + bi$ with a and b in the field $\mathbb{R}$ of real numbers, and with addition and multiplication given by that of $\mathbb{R}$ and by $i^2 = -1$. We will not go into the definition of $\mathbb{R}$, but the properties of $\mathbb{R}$ that we must use are these. First, every polynomial $f(x)$ in $\mathbb{R}[x]$ of odd degree has a root in $\mathbb{R}$. To see this, visualize the graph of a monic polynomial $f(x)$ of odd degree. For sufficiently large positive a, $f(a)$ is positive, and for a negative and of sufficiently large absolute value, $f(a)$ is negative. From the completeness property of the real numbers, the polynomial $f(x)$ must assume every value in between, so $f(x)$ must have a root. Second, we need the fact that positive real numbers have a square root. This can be seen in the same way by considering the polynomial $x^2 - a$ for $a > 0$. Finally, we need to know that every complex number has a square root. Let $a + bi$ be any complex number. We need to solve the equation $(x + yi)^2 = a + bi$. Using the fact that positive real numbers have square roots, this is an easy exercise.

7.6.1 The Fundamental Theorem of Algebra

The field of complex numbers is algebraically closed.

Proof Let $\mathbb{C}$ be the field of complex numbers, and suppose that $\mathbb{C}_1$ is a proper finite extension of $\mathbb{C}$. By 6.2.15, $\mathbb{C}_1 = \mathbb{C}(a)$ with $a \in \mathbb{C}_1$ and with a a root of an irreducible $p(x) \in \mathbb{C}[x]$. The root field K of $p(x)$ is a finite normal separable extension of $\mathbb{C}$, and $K \supset \mathbb{C}_1$ (6.3.2). We need to show that such a proper extension K of $\mathbb{C}$ does not exist. Note that K is the root field of $p(x)(x^2+1)$ over $\mathbb{R}$, so that K is a finite normal separable extension of $\mathbb{R}$. Let F be the fixed field of a Sylow 2-subgroup S of $G(K/\mathbb{R})$. Then $K \supset F \supset \mathbb{R}$, $S = G(K/F)$, $|S| = [K:F]$ (6.3.8), and since $[K:\mathbb{R}] = [K:F][F:\mathbb{R}]$ we get that $[F:\mathbb{R}]$ is odd. Therefore, $F = \mathbb{R}(a)$ with a satisfying an irreducible polynomial $f(x) \in \mathbb{R}[x]$ of odd degree. But every such polynomial of odd degree has a root in $\mathbb{R}$. Thus, $F = \mathbb{R}$. Therefore, $G(K/\mathbb{R})$ is a 2-group, and hence $G(K/\mathbb{C})$ is a 2-group. Since $K \neq \mathbb{C}$, $G(K/\mathbb{C})$ has at least 2 elements. Let L be the fixed field of a subgroup of index 2 in $G(K/\mathbb{C})$ (7.3.6). Then $[L:\mathbb{C}] = 2$, and $L = \mathbb{C}(d)$ with d satisfying an irreducible polynomial $x^2 + bx + c$ in $\mathbb{C}[x]$. But this polynomial has its roots $x = (-b \pm (b^2 - 4c)^{1/2})/2$ in $\mathbb{C}$, since $b^2 - 4c$ is in $\mathbb{C}$. We have reached a contradiction, and must conclude that no such $\mathbb{C}_1$ exists. This concludes the proof.

Complex variable theory offers an elegant proof of the Fundamental Theorem of Algebra. It involves complex integration and matters too far removed from our topic to discuss here. That such disjoint topics, complex variable theory on the one hand, and Galois theory and Sylow theory on the other, should yield such a theorem as the Fundamental Theorem of Algebra is one of the beautiful and intriguing aspects of mathematics.

PROBLEMS

1. Assume that every positive real number has a square root. Prove that every complex number has a square root.
2. Prove that every polynomial in $\mathbb{R}[x]$ factors in $\mathbb{R}[x]$ into a product of linear and quadratic factors.
3. Let A be the set of all elements in $\mathbb{C}$ that are algebraic over $\mathbb{Q}$. Use 7.6.1 to prove that A is algebraically closed. Generalize.

8 Topics in ring theory

8.1 INTRODUCTION

Ring theory is broadly divided into two areas—commutative ring theory and noncommutative ring theory. Commutative ring theory has developed from algebraic geometry and algebraic number theory. The prototype of the rings studied in algebraic geometry is the ring $F[x_1, x_2, \ldots, x_n]$ of polynomials in n indeterminates $x_1, x_2, \ldots, x_n$ with coefficients from a field F. In algebraic number theory, the prototype is the ring $\mathbb{Z}$ of integers. Both of these areas are vast and important mathematical topics, and they are part of what is generally called commutative algebra. Commutative algebra is the study of commutative rings.

The first topics we take up are from commutative ring theory. Section 8.2 discusses unique factorization domains, and the goal there is to show that if R is a unique factorization domain, then so is $R[x]$ (8.2.10). This is one of several important theorems in ring theory which asserts that if R satisfies a certain property, then so does $R[x]$. Section 8.3 is in that same spirit. There we prove one of the most famous of these theorems—the Hilbert Basis Theorem (8.3.2). That theorem says that if R is a commutative Noetherian ring, then so is $R[x]$. Section 8.4 presents a classical topic from commutative Noetherian rings, the Noether–Lasker decomposition theory. These three topics from commutative ring theory were chosen because they require minimum background, and yet they impart some of the flavor of commutative ring theory.

The prototype of the rings studied in noncommutative ring theory is the ring F_n of all $n \times n$ matrices over a field F, or more generally the ring D_n of all such matrices over a division ring D. In fact, our only topic (8.5) from noncommutative ring theory deals essentially with just these rings. We begin that topic with an examination of the rings D_n. One of the best theorems in noncommutative ring theory is the Wedderburn–Artin Theorem (8.5.9), which gives elegant necessary and sufficient conditions for a ring to be isomorphic to some D_n. We will prove this theorem via the more general Jacobson Density Theorem (8.5.8). It is hoped that this topic will impart some of the flavor of noncommutative ring theory.

8.2 UNIQUE FACTORIZATION DOMAINS

We will be using the notation and terminology of Section 4.4.

8.2.1 Definition

A **unique factorization domain** (UFD) *is an integral domain in which every element not* 0 *or a unit is a product of irreducible elements, and these irreducible elements are unique up to order and associates.*

Thus if R is a UFD, a is in R, and a is not 0 or a unit, then $a = a_1 a_2 \cdots a_n$ with each a_i irreducible. If $a = b_1 b_2 \cdots b_n$ with each b_i irreducible, then $m = n$, and there is a one-to-one correspondence between the a_i's and the b_i's such that corresponding elements are associates.

From Section 4.4, we know that any PID is a UFD. However, there are UFD's that are not PID's. For example, if F is a field, then $F[x, y]$ is a UFD but is not a PID, and $\mathbb{Z}[x]$ is a UFD but not a PID. Our aim is to prove that if R is a UFD, then so is $R[x]$.

First we dispose of some easy preliminaries. Suppose that R is a UFD and that a is in R with a not 0 or a unit. Then $a = a_1 a_2 \cdots a_n$ with each a_i irreducible. If a_i is an associate of a_1, write $a_i = u_i a_1$. In this manner, we can express a as a product

$$a = u a_1^{m_1} a_2^{m_2} \cdots a_k^{m_k},$$

where the a_i's are nonassociate irreducible elements, $m_i > 0$, and u is a unit. If a is also another such product

$$a = v b_1^{n_1} b_2^{n_2} \cdots b_s^{n_s},$$

then $s = k$, and after renumbering, a_i and b_i are associates and $m_i = n_i$ for all i. If a and b are two nonzero elements of R, then we can write

$$a = u a_1^{m_1} a_2^{m_2} \cdots a_r^{m_r},$$

and

$$b = v a_1^{n_1} a_2^{n_2} \cdots a_r^{n_r},$$

with u and v units, a_i irreducible, and m_i and $n_i \geq 0$. If this is done, then a divides b if and only if $m_i \leq n_i$ for all i. Finally, note that an element a in R is irreducible if and only if it is prime (Problem 1).

8.2.2 Lemma

In a UFD, *any* 2 *elements have a greatest common divisor* (gcd).

Proof Let a and b be in a UFD R. The only nontrivial case is the one where neither a nor b is 0 or a unit. Write a and b as above, and let k_i be the minimum of m_i and n_i. We claim that

$$d = a_1^{k_1} a_2^{k_2} \cdots a_r^{k_r}$$

is a gcd of a and b. It certainly divides both. Suppose that e divides a and b. Then by earlier remarks,

$$e = a_1^{j_1} a_2^{j_2} \cdots a_r^{j_r},$$

with $j_i \leq m_i$ and $j_i \leq n_i$, and hence $j_i \leq k_i$. Thus e divides d, so that d is a gcd of a and b.

Similarly, we can show that a finite subset (or, in fact, any subset) of elements of a UFD has a gcd.

8.2.3 Definition

Let R be a UFD, *and let $f(x)$ be in $R[x]$. Then $f(x)$ is* **primitive** *if a* gcd *of the coefficients of $f(x)$ is* 1. *A* gcd *of the coefficients of $f(x)$ is called the* **content** *of $f(x)$, and it is denoted by $c(f(x))$.*

The content $c(f(x))$ is unique only up to multiplication by a unit. Note that if $f(x)$ is a nonzero element of $R[x]$ where R is a UFD, then $f(x) = c(f(x))f_1(x)$ with $f_1(x)$ primitive. Also note that if $f(x) = c \cdot g(x)$ with $g(x)$ primitive, then c is a content of $f(x)$. The following classical lemma is crucial.

8.2.4 Gauss' Lemma

If $f(x)$ and $g(x)$ are primitive polynomials in $R[x]$ and R is a UFD, *then $f(x)g(x)$ is primitive.*

Proof Let $f(x) = \sum_{i=0}^{m} a_i x^i$ and $g(x) = \sum_{i=0}^{n} b_i x^i$. Suppose that p is a prime dividing all the coefficients of $f(x)g(x)$. Let i be the smallest integer such that p does not divide a_i, and let j be the smallest integer such that p does not divide b_j. The coefficient of x^{i+j} in $f(x)g(x)$ is

$$c_{i+j} = a_0 b_{i+j} + a_1 b_{i+j-1} + \cdots + a_i b_j + \cdots + a_{i+j} b_0.$$

(If $i+j>m$, take $a_{i+j}=0$, and similarly for b_{i+j}.) Since p divides $a_0, a_1, \ldots, a_{i-1}$ and $b_0, b_1, \ldots, b_{j-1}$, then p divides every term except possibly $a_i b_j$. But p divides c_{i+j}, whence p divides $a_i b_j$. Thus p divides either a_i or b_j, which is impossible. Thus, $f(x)g(x)$ is primitive.

There are a number of corollaries of interest.

8.2.5 Corollary

Let R be a UFD, *and let $f(x)$ and $g(x)$ be in $R[x]$. Then $c(f(x)g(x)) = c(f(x))c(g(x))$, up to units.*

Proof Write $f(x) = c(f(x))f_1(x)$, and $g(x) = c(g(x))g_1(x)$ with $f_1(x)$ and $g_1(x)$ primitive. Then

$$f(x)g(x) = c(f(x))c(g(x))f_1(x)g_1(x).$$

Since $f_1(x)g_1(x)$ is primitive, $c(f(x))c(g(x))$ is the content of $f(x)g(x)$.

8.2.6 Corollary

Let R be a UFD, *and let F be its quotient field. If $f(x)$ is in $R[x]$ and if $f(x) = g(x)h(x)$ in $F[x]$, then $f(x) = g_1(x)h_1(x)$ with $g_1(x)$ and $h_1(x)$ in $R[x]$ and with* $\deg(g_1(x)) = \deg(g(x))$.

Proof Let r and s be nonzero elements of R such that $rg(x)$ and $sh(x)$ are in $R[x]$. Then $rsf(x) = (rg(x))(sh(x))$, $rg(x) = c_1g_1(x)$, and $sh(x) = c_2h_1(x)$ with c_i in R and $g_1(x)$ and $h_1(x)$ primitive elements of $R[x]$. Thus, $rsf(x) = c_1c_2g_1(x)h_1(x)$. Since $g_1(x)$ and $h_1(x)$ are primitive, rs divides c_1c_2. Thus, $f(x) = (cg_1(x))h_1(x)$. Since the degree of $g_1(x)$ is the degree of $g(x)$, the corollary follows.

The following two corollaries are immediate consequences of 8.2.6.

8.2.7 Corollary

Let R be a UFD, *and let F be its quotient field. A primitive polynomial $f(x)$ in $R[x]$ is irreducible in $F[x]$ if and only if it is irreducible in $R[x]$.*

8.2.8 Corollary

Let R be a UFD, *and let F be its quotient field. If $f(x)$ is monic, $f(x)$ is in $R[x]$, and $f(x) = g(x)h(x)$ in $F[x]$, then $f(x) = g_1(x)h_1(x)$ in $R(x)$ with $g_1(x)$ and $h_1(x)$ monic and having the same degrees as $g(x)$ and $h(x)$, respectively.*

Corollaries 8.2.6, 8.2.7, and 8.2.8 are of interest even for the case $R = \mathbb{Z}$. Here is the main theorem.

8.2.9 Theorem

If R is a UFD, *then $R[x]$ is a* UFD.

Proof Suppose that $f(x)$ is a polynomial in $R[x]$ of positive degree. Let F be the quotient field of R. In $F[x]$, $f(x) = f_1(x)f_2(x) \cdots f_n(x)$ with the $f_i(x)$ prime in $F[x]$. Let r_i be a nonzero element of R such that $r_i f_i(x)$ is in $R[x]$. Then $r_1 r_2 \cdots r_n f(x) = (r_1 f_1(x))(r_2 f_2(x)) \cdots (r_n f_n(x))$. Write each $r_i f_i(x) = c_i g_i(x)$ with c_i in R and $g_i(x)$ a primitive polynomial in $R[x]$. Then $g_i(x)$ is irreducible in $R[x]$ by 8.2.7, and we have

$$r_1 r_2 \cdots r_n f(x) = c_1 c_2 \cdots c_n g_1(x) g_2(x) \cdots g_n(x).$$

Now $c = c_1 c_2 \cdots c_n$ is the content of $r_1 r_2 \cdots r_n f(x)$ since $g_1(x)g_2(x) \cdots g_n(x)$ is primitive. Thus, $r_1 r_2 \cdots r_n$ divides c. Hence $f(x) = r g_1(x) g_2(x) \cdots g_n(x)$ with r in R, and with r a product $u a_1 a_2 \cdots a_k$ of irreducible elements a_i in R and a unit u in R. Since any element in R not 0 or a unit is a product of irreducible elements in R, and since every element irreducible in R is irreducible in $R[x]$, we have shown that every element in $R[x]$ is a product of irreducible elements. Suppose that

$$a_1 a_2 \cdots a_j f_1(x) f_2(x) \cdots f_m(x) = b_1 b_2 \cdots b_k g_1(x) g_2(x) \cdots g_n(x)$$

with each factor irreducible in $R[x]$, with the a_i and b_i in R, and the $f_i(x)$ and $g_i(x)$ polynomials of positive degree. Each $f_i(x)$ and $g_i(x)$ is primitive and hence irreducible in $F[x]$ by 8.2.7. In $F[x]$, the a_i and b_i are units. Therefore, viewing the equality in the PID $F[x]$, $m = n$, and after rearrangement, $f_i(x)$ and $g_i(x)$ are associates in $F[x]$. Hence $(r_i/s_i)f_i(x) = g_i(x)$ with r_i and s_i in R, and so $r_i f_i(x) = s_i g_i(x)$. Since $f_i(x)$ and $g_i(x)$ are primitive, r_i and s_i are associates in R. Thus, we have $a_1 a_2 \cdots a_j = b_1 b_2 \cdots b_k v$ with v a unit. Since R is a UFD, $j = k$, and after renumbering, a_i and b_i are associates in R. This completes the proof.

8.2.10 Corollary

If R is a UFD, *then $R[x_1, x_2, \ldots, x_n]$ is a UFD.*

8.2.11 Corollary

$\mathbb{Z}[x_1, x_2, \ldots, x_n]$ *is a* UFD.

8.2.12 Corollary

If F is a field, then $F[x_1, x_2, \ldots, x_n]$ is a UFD.

Let R be a UFD. It is usually quite difficult to tell whether a polynomial in $R[x]$ is irreducible or not. One test that is easy to apply is the following.

8.2.13 Theorem (The Eisenstein Criterion)

Let R be a UFD, *and let $f(x) = a_0 + a_1x + \cdots + a_nx^n$ be in $R[x]$. If p is a prime in R such that p does not divide a_n, p divides each of $a_0, a_1, \ldots, a_{n-1}$, and p^2 does not divide a_0, then $f(x)$ is irreducible over $F[x]$, where F is the quotient field of R.*

Proof Write $f(x) = c(f(x))f_1(x)$. Then $f(x)$ satisfies the hypothesis on its coefficients if and only if $f_1(x)$ does, and $f(x)$ is irreducible in $F[x]$ if and only if $f_1(x)$ is irreducible in $F[x]$. Thus, we may assume that $f(x)$ is primitive. Suppose that $f(x)$ factors properly in $F[x]$. Then by 8.2.7, $f(x) = (b_0 + b_1x + \cdots + b_rx^r)(c_0 + c_1x + \cdots + c_sx^s)$ with the b_i and c_i in R, r, and s positive, and b_r and c_s not 0. Since p divides a_0, p divides $b_0c_0 = a_0$. Since p^2 does not divide a_0, p divides exactly one of b_0 and c_0, say b_0. If p divides all the b_i, then p divides a_n. Thus p does not divide some b_k, and let k be the smallest such integer. Now $a_k = b_kc_0 + b_{k-1}c_1 + \cdots + b_0c_k$, p divides $b_0, b_1, \ldots, b_{k-1}$, and since $k < n$, p divides a_k. Thus, p divides b_kc_0. But p divides neither b_k nor c_0. This contradiction establishes the theorem.

Eisenstein's Criterion is of special interest for the case $R = \mathbb{Z}$.

An easy example of an integral domain that is not a UFD is this. Let F be any field, and let $R = F[x^2, x^3]$ be the subring of $F[x]$ generated by F, x^2, and x^3. Then in R, $x^6 = x^2x^2x^2 = x^3x^3$, and both x^2 and x^3 are irreducible elements. Neither element is prime. For example, x^2 divides x^3x^3, but does not divide x^3.

A less apparent example is the integral domain $R = \{m + n(-5)^{1/2}: m, n \in \mathbb{Z}\}$. To show that R is not a UFD, it is convenient to introduce the map $N: R \to \mathbb{Z}$ given by

$$N(m + n(-5)^{1/2}) = (m + n(-5)^{1/2})(m - n(-5)^{1/2}) = m^2 + 5n^2.$$

One can readily check that N is multiplicative, that is, that if r and s are in R, then $N(rs) = N(r)N(s)$. Therefore, if u is a unit in R, then $N(uu^{-1}) = N(u)N(u^{-1}) = N(1)$, whence $N(u) = 1$. Thus, the only units in R are ± 1. Now 21 can be factored in three ways into the product of irreducible factors:

$$21 = 3 \cdot 7 = (1 + 2(-5)^{1/2})(1 - 2(-5)^{1/2}) = (4 + (-5)^{1/2})(4 - (-5)^{1/2}).$$

The factors are all irreducible. For example, $N(3) = 9$, so for any factor r of 3, $N(r) = 1$ or $N(r) = 3$. But $N(r) = 1$ implies that $r = \pm 1$, and $N(m + n(-5)^{1/2}) = m^2 + 5n^2 = 3$ is impossible. Similarly, one can show that the other factors are irreducible. Therefore, R is not a UFD.

PROBLEMS

1 Prove that an element in a UFD is irreducible if and only if it is prime.

2 Prove that any finite subset of a UFD has a gcd.

3 Let R be a UFD, let

$$a = ua_1^{m_1}a_2^{m_2} \cdots a_r^{m_r},$$

and let

$$b = va_1^{n_1}a_2^{n_2} \cdots a_r^{n_r},$$

with the a_i irreducible, m_i and n_i positive, and u and v units. Prove that a divides b if and only if $m_i \leq n_i$ for all i.

4 In Problem 3, let k_i be the minimum of m_i and n_i. Prove that

$$\gcd(a, b) = a_1^{k_1}a_2^{k_2} \cdots a_r^{k_r}.$$

5 Let R be a UFD, and let $f(x) = ag(x) = bh(x)$ with a and b in R, and $g(x)$ and $h(x)$ primitive elements of $R[x]$. Prove that a and b are associates in R, and prove that $g(x)$ and $h(x)$ are associates in $R[x]$.

6 Let R be a UFD. Prove that

$$x^4 + 2y^2x^3 + 3y^3x^2 + 4yx + y + 8y^2$$

is irreducible in $R[x, y]$.

7 Let R be a UFD. Prove that if $f(x)$ is in $R[x]$ and $f(x)$ is reducible, then so is $f(x + a)$ for each a in R.

8 Prove that if p is a prime, then $x^{p-1} + x^{p-2} + \cdots + x + 1$ is irreducible in $\mathbb{Z}[x]$.

9 Prove that $x^4 + 1$ is irreducible over $\mathbb{Q}[x]$.

10 Prove that if R is a UFD and if $f(x)$ is a monic polynomial with a root in the quotient field of R, then that root is in R.

8.3 THE HILBERT BASIS THEOREM

In this section we will prove the Hilbert Basis Theorem: If R is a commutative Noetherian ring, then so is $R[x]$. This theorem is fundamental in the ideal theory of commutative rings.

A ring R is Noetherian (4.4.5) if every infinite chain

$$I_1 \subset I_2 \subset I_3 \cdots$$

of ideals becomes constant. That is, it is impossible to have an infinite strictly increasing chain of ideals of R. This is also expressed by saying that R satisfies the ***ascending chain condition*** (acc) on ideals. A ring R satisfies the ***maximum condition*** on ideals if every nonempty set of ideals of R has a member that is not properly contained in any other member of that set. Such a member is called a ***maximal element*** of that set. An ideal I of R is ***finitely generated*** if there are finitely many elements $r_1, r_2, \ldots, r_n$ of R such that I is the smallest ideal of R containing all the r_i. If R is commutative, this simply means that $I = Rr_1 + Rr_2 + \cdots + Rr_n$.

The Hilbert Basis Theorem is concerned with commutative rings. However, we will need in later sections the notion of ***left Noetherian ring*** (and right Noetherian ring). A ring R is ***left Noetherian*** if every ascending chain of ***left*** ideals becomes constant. Similarly, R satisfies the ***maximum condition on left ideals*** if every nonempty set of left ideals has a maximal element.

8.3.1 Lemma

Let R be a ring. The following are equivalent.
(a) *R is left Noetherian.*
(b) *R satisfies the maximum condition on left ideals.*
(c) *Every left ideal of R is finitely generated.*

Proof Assume (a), and let S be a nonempty set of left ideals of R. Let I_1 be in S. If I_1 is not a maximal element of S, then there is an element I_2 of S such that I_1 is properly contained in I_2. If I_2 is not maximal, then there is an I_3 in S such that I_2 is properly contained in I_3, and so on. Since R is left Noetherian, we eventually reach an I_n that is maximal in S. Thus, (a) implies (b).

Assume (b), and let I be a left ideal of R. Let r_1 be in I, and let I_1 be the left ideal generated by r_1, that is, $I_1 = Rr_1$. If $I_1 \neq I$, then there is an element r_2 in I and not in I_1. Let I_2 be the ideal $Rr_1 + Rr_2$ generated by r_1 and r_2. Then I_2 properly contains I_1. If $I_2 \neq I$, then there is an element r_3 in I and not in I_2. This process must stop. Thus we must come to an I_n such that $I_n = I$. Otherwise, we get a nonempty set $\{I_1, I_2, I_3, \ldots\}$ of left ideals which has no maximal element. Thus $I = I_n$ for some n, whence I is generated by the finite set $\{r_1, r_2, \ldots, r_n\}$. Thus, (b) implies (c).

Assume (c), and let $I_1 \subset I_2 \subset I_3 \subset \cdots$ be a chain of left ideals of R. By (c), the left ideal $I = \bigcup_j I_j$ is generated by a finite subset $\{r_1, r_2, \ldots, r_n\}$ of I. Each r_i is in some member of the chain of left ideals, and hence there exists an n such that all r_i are in I_n. Thus $I_n = I$, and so

$$I_n = I_{n+1} = I_{n+2} \cdots.$$

Hence (c) implies (a), and the proof is complete.

If R is a commutative ring, then left ideals are the same as ideals. Hence, if R is a commutative ring, then R is Noetherian if and only if every ideal of R is finitely generated. This is the fact we need in the proof below of the Hilbert Basis Theorem.

8.3.2 The Hilbert Basis Theorem

If R is a commutative Noetherian ring, then so is $R[x]$.

Proof Let A be an ideal of $R[x]$. Let A_i be the set of elements a in R which appear as the leading coefficient in some polynomial $a_0 + a_1x + a_2x^2 + \cdots + ax^j$ in A of degree $j \leq i$. If a and b are in A_i, then a is the leading coefficient of a polynomial $f(x)$ in A of degree $j \leq i$, and b is the leading coefficient of a polynomial $g(x)$ in A of degree $k \leq i$. Therefore, $a + b$ is the leading coefficient of the polynomial $x^{i-j}f(x) + x^{i-k}g(x)$ in A of degree i. Since ra is the leading coefficient of $rf(x)$, it follows that A_i is an ideal of R. The ideals A_i form an increasing chain

$$A_0 \subset A_1 \subset A_2 \subset \cdots$$

and since R is Noetherian, this chain stops, say at A_n. So $A_n = A_{n+1} = A_{n+2} = \cdots$. Each A_i is finitely generated, and let $a_{i1}, a_{i2}, \ldots, a_{ir_i}$ generate A_i for $i = 0, 1, \ldots, n$. For each i, let f_{ij} be a polynomial of A of degree i with leading coefficient a_{ij}, $j = 1, 2, \ldots, r_i$. We will now show that the f_{ij} generate A. Let f be in A, $f \neq 0$. We induct on the degree d of f. If $d > n$, then the leading coefficients of $x^{d-n}f_{n1}, \ldots, x^{d-n}f_{nr_n}$ generate A_d. Hence there exist elements $s_1, s_2, \ldots, s_{r_n}$ in R such that

$$g = f - s_1x^{d-n}f_{n1} - \cdots - s_{r_n}x^{d-n}f_{nr_n}$$

has degree less than d, and g is in A. Similarly, if $d \leq n$ we get an element $g = f - s_1f_{d1} - \cdots - s_{r_n}f_{dr_n}$ of degree less than d with g in A. ($g = 0$ if $d = 0$.) By induction, we can subtract from f a polynomial h in the ideal generated by the f_{ij} such that $f - h$ is again in the ideal generated by the f_{ij}. This completes the proof.

8.3.3 Corollary

If R is a commutative Noetherian ring, then so is $R[x_1, x_2, \ldots, x_n]$.

8.3.4 Corollary

If F is a field, then $F[x_1, x_2, \ldots, x_n]$ is Noetherian.

8.4 THE LASKER–NOETHER DECOMPOSITION THEOREM

Throughout this section all our rings will be ***commutative***. An ideal I in a ring R is ***prime*** if whenever a product $a \cdot b$ is in I, then either a or b is in I. The concept corresponds to the concept of a prime number in ordinary arithmetic. In fact, in a PID, an ideal I such that $I \neq 0$ and $I \neq R$ is prime if and only if $I = Rp$ where p is prime. An ideal I is ***primary*** if whenever a product $a \cdot b$ is in I and a is not in I, then b^n is in I for some positive integer n. This concept corresponds to that in ordinary arithmetic of being a power of a prime number. In a PID, an ideal I not zero or R is primary if and only if $I = Rp^n$, where p is prime. In a PID, every element a, not a unit or 0, is uniquely a product

$$a = p_1^{n_1} p_2^{n_2} \cdots p_r^{n_r}$$

of powers of distinct primes. This can be expressed by the equation

$$Ra = Rp_1^{n_1} \cap Rp_2^{n_2} \cap \cdots \cap Rp_r^{n_r}.$$

In an arbitrary Noetherian ring, it is not necessarily true that an element is a product of powers of primes. However, it is true that every ideal in a Noetherian ring is the intersection of primary ideals, and there is a certain amount of uniqueness to the set of primary ideals involved. That is the gist of the Lasker–Noether decomposition theorem. We proceed now to the precise formulation and proof of this theorem. Some further preliminaries are needed.

An ideal in a ring is ***irreducible*** if it is not the intersection of a finite number of ideals strictly containing it.

8.4.1 Lemma

Let R be a Noetherian ring. Then every ideal of R is the intersection of finitely many irreducible ideals.

Proof Let Ω be the set of all ideals of R which are not intersections of finitely many irreducible ideals. We need to show that Ω is empty. If Ω is not empty, then it has a maximal element I because R is Noetherian. Since I is not irreducible, $I = J \cap K$, where J and K are ideals properly containing I. Since J and K properly contain I and I is a maximal element of Ω, neither J nor K is in Ω. Hence J and K are finite intersections of irreducible ideals, whence I is a finite intersection of irreducible ideals. This is a contradiction. Therefore Ω is empty, and the lemma is proved.

The next step is to show that in a Noetherian ring, irreducible ideals are primary. Let I be an ideal, and let S be a subset of R. Then $I{:}S$ is defined to be the set $\{r \in R : rS \subset I\}$. That is, $I{:}S$ is the set of all r in R such that $r \cdot s$ is in I for all s in S. It is easy to check that $I{:}S$ is an ideal of R.

8.4.2 Lemma

Let R be a Noetherian ring. Then irreducible ideals of R are primary.

Proof Let I be an ideal of R. If I is not primary, then there are elements a and b in R with $a \cdot b$ in I, a not in I, and no power of b in I. Since R is Noetherian, the chain

$$I{:}b \subset I{:}b^2 \subset I{:}b^3 \subset \cdots$$

of ideals must terminate. Thus, there is an integer n such that $I{:}b^n = I{:}b^{n+1}$. The ideals $I + Ra$ and $I + Rb^n$ properly contain I since neither a nor b^n is in I. We will show that

$$I = (I + Ra) \cap (I + Rb^n).$$

Let x be in $(I + Ra) \cap (I + Rb^n)$. Then $x = u + ra = v + sb^n$, with u and v in I, and r and s in R. Then

$$bx = bu + rab = bv + sb^{n+1}$$

is in I since bu and rab are in I. Since bv is also in I, we get that sb^{n+1} is in I. Thus s is in $I{:}b^{n+1}$, whence s is in $I{:}b^n = I{:}b^{n+1}$. Hence sb^n is in I, and so x is in I. It follows that $I = (I + Ra) \cap (I + Rb^n)$, and that I is not irreducible. This concludes the proof.

From 8.4.1 and 8.4.2, we know that every ideal in a Noetherian ring is the intersection of finitely many primary ideals. We will do much better than this, but another concept is needed. Let I be an ideal. Then the ***radical*** of I, denoted $\sqrt{I}$, is defined to be the set $\{r \in R : r^n \in I \text{ for some positive integer } n\}$. It is not difficult to show that $\sqrt{I}$ is an ideal.

8.4.3 Lemma

The radical of a primary ideal is a prime ideal.

Proof Let I be a primary ideal of R, and suppose that $a \cdot b$ is in $J = \sqrt{I}$ with a not in J. Since $a \cdot b$ is in J, then $(a \cdot b)^n = a^n b^n$ is in I for some positive integer n. Since a is not in J, a^n is not in I. Thus $(b^n)^m$ is in I for some m, and so b is in J. Hence, J is prime.

If I is a primary ideal, then the prime ideal $\sqrt{I}$ is called the ***associated prime ideal of*** I, I is said to ***belong to*** the prime ideal $\sqrt{I}$, and I is said to be ***primary for*** $\sqrt{I}$.

8.4.4 Lemma

Let Q_i be primary, $i = 1, 2, \ldots, n$, with the same associated prime ideal P. Then $\bigcap_i Q_i = Q$ is primary with associated prime ideal P.

Proof Suppose that $a \cdot b$ is in Q, and a is not in Q. Then a is not in Q_j for some j. Since $a \cdot b$ is in Q_j, b^m is in Q_j for some m. Hence, b is in P. Since $\sqrt{Q_i} = P$ for all i, b^{m_i} is in Q_i for some m_i, $i = 1, 2, \ldots, n$, whence b^k is in Q for $k \geq \sum_i m_i$. Therefore, Q is primary. We need $\sqrt{Q} = P$. But $\sqrt{Q} \subset \sqrt{Q_i} = P$, and we have just seen that if an element is in P, then a power of it is in Q. Thus, $\sqrt{Q} = P$.

Let $I = \bigcap_{i=1}^{n} Q_i$ with the Q_i primary. This representation of I as the intersection of primary ideals is called ***irredundant*** if no Q_i contains the intersection of the rest of the Q_i's, and the Q_i's have distinct associated primes.

8.4.5 The Lasker–Noether Decomposition Theorem

Let R be a Noetherian ring. Then every ideal I of R admits an irredundant representation as a finite intersection of primary ideals. If

$$I = Q_1 \cap Q_2 \cap \cdots \cap Q_m = Q_1' \cap Q_2' \cap \cdots \cap Q_n'$$

are two such representations, then $m = n$ and the two sets of associated prime ideals $P_i = \sqrt{Q_i}$, $i = 1, 2, \ldots, m$ and $P_i' = \sqrt{Q_i'}$, $i = 1, 2, \ldots, n$ are the same.

Proof That an ideal I of R admits such a representation follows immediately from 8.4.1, 8.4.2, and 8.4.4. Indeed, by 8.4.1, $I = Q_1 \cap Q_2 \cap \cdots \cap Q_m$ with Q_i irreducible. By 8.4.2, each Q_i is primary. If some Q_i contains the intersection of the rest, it can be discarded. If Q_i and Q_j have the same associated prime, then $Q_i \cap Q_j$ is primary with that same associated prime by 8.4.4. Replace Q_i and Q_j by the single primary ideal $Q_i \cap Q_j$. This clearly leads to an irredundant representation of I as the intersection of finitely many primary ideals.

Now to the uniqueness part of the theorem. If the set of associated primes of an ideal I is unique, then it should be describable in terms of I alone, that is, independent of any representation of I as an intersection of other ideals. The following lemma does this, and proves the uniqueness part of 8.4.5.

8.4.6 Lemma

Let I be an ideal of R, $I \neq R$, and suppose that $I = \bigcap_{i=1}^{m} Q_i$ is an irredundant representation of I as the intersection of finitely many primary ideals. Then a prime ideal P of R is the associated prime of one of the Q_i if and only if there is an element r in R not in I such that $I{:}r$ is primary for P.

Proof There is an element r in

$$Q_1 \cap \cdots \cap Q_{i-1} \cap Q_{i+1} \cap \cdots \cap Q_m$$

which is not in Q_i. If a is in $I{:}r$, then $a \cdot r$ is in I, whence $a \cdot r$ is in Q_i. But since r is not in Q_i, a^k is in Q_i for some k, whence a is in P_i. Therefore $I{:}r \subset P_i$, and clearly $Q_i \subset I{:}r = Q_i{:}r$. If $x \cdot y$ is in $I{:}r$, and x is not in $I{:}r$, then $x \cdot y \cdot r$ is in Q_i, $x \cdot r$ is not in Q_i, so y^n is in Q_i and hence in $I{:}r$ for some n. Therefore, the ideal $I{:}r$ is primary. Since

$$Q_i \subset Q_i{:}r = I{:}r \subset P_i, \quad \sqrt{Q_i} = P_i \subset \sqrt{I{:}r} \subset \sqrt{P_i} = P_i,$$

$I{:}r$ is primary for P_i.

Now suppose that there is an element r in R not in I such that $I{:}r$ is primary for the prime ideal P. Since

$$I{:}r = \left(\bigcap_i Q_i\right){:}r = \bigcap_i (Q_i{:}r),$$

we get $\sqrt{I{:}r} = P = \sqrt{\bigcap_i (Q_i{:}r)} = \bigcap_i \sqrt{(Q_i{:}r)}$. Applying the first part of the proof to the case $I = Q_i$, for r not in Q_i we get $Q_i \subset Q_i{:}r \subset P_i$. Thus $\sqrt{Q_i{:}r} = P_i$ if r is not in Q_i, and $\sqrt{Q_i{:}r} = R$ if r is in Q_i. Thus, P is the intersection of some of the prime ideals P_i. This forces P to be one of the P_i (Problem 2).

PROBLEMS

1 Let $I_1, I_2, \ldots, I_n$ and J be ideals of R, and let $J_i = \sqrt{I_i}$. Prove that
(a) $(\bigcap_i I_i):J = \bigcap_i (I_i:J)$,
(b) $J:(\sum_i I_i) = \bigcap_i (J:I_i)$,
(c) $I_1:(I_2 I_3) = (I_1:I_2):I_3$,
(d) $\sqrt{I_1 I_2} = \sqrt{I_1 \cap I_2} = \sqrt{I_1} \cap \sqrt{I_2}$,
(e) $\sqrt{I_1 + I_2} = \sqrt{J_1 + J_2}$, and
(f) $\sqrt{J_i} = J_i$.

2 Prove that if $P = \bigcap_i P_i$, where P and the P_i are prime ideals of R, then $P = P_i$ for some i.

3 Prove that if Q is primary for P, and if I and J are ideals such that $IJ \subset Q$ and I is not contained in Q, then $J \subset P$.

4 Let P and Q be ideals of R. Prove that Q is primary and $\sqrt{Q} = P$ if and only if
(a) $Q \subset P$,
(b) if a is in P, then a^n is in Q for some n, and
(c) if $a \cdot b$ is in Q and a is not, then b is in P.

5 Prove that if Q is primary for P, and if I is an ideal not contained in Q, then $Q:I$ is primary for P.

6 Let F be a field, and let $R = F[x, y]$.
(a) Prove that $P = Rx + Ry$ is prime (in fact, maximal).
(b) Prove that $Q = Rx + Ry^2$ is primary.
(c) Prove that $\sqrt{Q} = P$.
(d) Prove that $Q \neq P^2$.
(e) Prove that Q is not a power of any prime ideal of R.
(f) Prove that the ideal $A = Rx^2 + Rxy$ is not primary.
(g) Prove that $\sqrt{A}$ is prime.

8.5 SEMISIMPLE RINGS

Let D be a division ring. We want to examine the ring D_n of all $n \times n$ matrices over D, but to do so we will need to consider modules over the ring D_n. Modules over a division ring D are called ***vector spaces over*** D, and many of the elementary facts about vector spaces over fields carry over to vector spaces over division rings. Linear independence, and bases are defined in exactly the same way, and the proof of invariance of dimension (3.3.6) goes through word for word. In fact, all of section 3.3 goes through for vector spaces over division rings. Suppose now that V is a right vector space over D, $\{v_1, v_2, \ldots, v_n\}$ is a basis of V, and $w_1, w_2, \ldots, w_n$ are in V. Then there is exactly one linear transformation α on V such that $\alpha(v_i) = w_i$. Therefore, as in the case for vector spaces over fields, α is uniquely determined by its image $\alpha(v_i) = w_i = \sum_j v_j a_{ij}$, that is, by the matrix (a_{ij}). Associating α with its matrix (a_{ij}) relative to the basis $\{v_1, v_2, \ldots, v_n\}$ is an isomorphism from the ***ring*** of all linear transformations on V to the ring D_n of all $n \times n$ matrices over D. (Compare with 5.1.3.)

Now to the ring D_n. First, D_n is a ***simple ring***. That is, it has no two-sided ideals except 0 and D_n. To see this, suppose that I is an ideal in D_n. Let E_{ij} be the element of D_n with 1 in the (i, j) position and 0's elsewhere. Now suppose that (a_{ij}) is in I and that $a_{pq} \neq 0$. Then $E_{kp}(a_{ij})E_{qk}$ is the matrix $a_{pq}E_{kk}$. Multiplying $a_{pq}E_{kk}$ by the diagonal matrix $a_{pq}^{-1}(\delta_{ij})$ gets E_{kk} in I. Hence $\sum_k E_{kk} = (\delta_{ij})$ is in I, and so $I = D_n$, and therefore D_n is a simple ring.

A (left or right) module S over any ring is a ***simple module*** if it has no submodules other than 0 and S. Now consider D_n as a left module over itself. Its submodules are the left ideals of D_n, and D_n is not a simple module if $n > 1$. Indeed, let C_m be the set of matrices with 0's off the mth column. Thus, an element of C_m looks like

$$\begin{pmatrix} 0 & 0 & \cdots & 0 & c_{1m} & 0 & \cdots & 0 \\ 0 & 0 & \cdots & 0 & c_{2m} & 0 & \cdots & 0 \\ \vdots & & & & & & & \vdots \\ 0 & 0 & \cdots & 0 & c_{nm} & 0 & \cdots & 0 \end{pmatrix}.$$

Then C_m is a left submodule of D_n, and in fact the module D_n is the direct sum of the submodules C_m. The C_m are ***simple*** modules. To prove that C_m is simple, let (a_{ij}) be any nonzero element of C_m, and let a_{km} be a nonzero entry in (a_{ij}). If (c_{ij}) is any element of C_m, then it is easy to check that $(c_{ij}) = \sum_r c_{rm}a_{km}^{-1}E_{rk}(a_{ij})$. Thus, the left module D_n is a direct sum of simple modules.

A module over any ring is called ***semisimple*** if it is a direct sum $S_1 \oplus S_2 \oplus \cdots \oplus S_m$ of simple modules. (The usual definition of semisimple allows infinite direct sums, but our direct sums are limited to a finite number of summands.) Thus, D_n is semisimple as a left module over itself. If D_n is considered as a right module over itself, then it is also semisimple. It is the direct sum of the right ideals of matrices with zeroes off the ith row. A ring is called ***left (right) semisimple*** if it is semisimple as a left (right) module over itself. Thus, D_n is both left and right semisimple.

The simple left submodules C_i of D_n are all isomorphic. Indeed, the map $C_i \to C_j$ given by taking an element of C_i to the element of C_j whose jth column is the ith column of C_i is a D_n-isomorphism. Thus D_n is not only left semisimple, but it is the direct sum of mutually isomorphic simple modules. Analogous statements hold for D_n considered as a right module over itself.

Now consider all the endomorphisms $E(C_j)$ of one of the simple left D_n modules C_j. Let d be in D. Note that the map $C_j \to C_j$ given by $(c_{ij}) \to (c_{ij})d$ is an element of $E(C_j)$. We will show that every element of $E(C_j)$ is just such a right multiplication by an element of D. It is notationally convenient to write left module homomorphisms on the right and to write right module homomorphisms on the left. Thus, we will be writing our homomorphisms on the opposite side of the scalars. Also, $\alpha \circ \beta$ for left module homomorphisms means α followed by β. Let $\alpha: C_j \to C_j$ be an endomorphism of C_j. Each E_{ij}, $i = 1, 2, \ldots, n$ is in C_j. If $q \neq i$, then $E_{pq}E_{ij} = 0$. Thus $(E_{pq}E_{ij})\alpha = 0 = E_{pq}(E_{ij}\alpha)$, so that $E_{ij}\alpha$ has no nonzero entry off the ith row. Thus, $E_{ij}\alpha = E_{ij}d_i$ for some d_i in D. Since $E_{ki}E_{ij} = E_{kj}$, we have

$$(E_{ki}E_{ij})\alpha = E_{kj}d_k = E_{ki}E_{ij}d_i = E_{kj}d_i.$$

Thus, $d_i = d_k$. Thus, α multiplies each E_{ij} by a fixed element d of D. Now observe that for any a in D,

$$(a(c_{ij}))\alpha = a((c_{ij})\alpha).$$

For any (c_{ij}) in C_j, $(c_{ij}) = \sum_i c_{ij}E_{ij}$, so

$$(c_{ij})\alpha = \left(\sum_i c_{ij}E_{ij}\right)\alpha = \sum_i (c_{ij}(E_{ij}))\alpha = \sum_i c_{ij}E_{ij}d = (c_{ij})d.$$

Hence for any (c_{ij}) in C_j, $(c_{ij})\alpha = (c_{ij})d$. Now it is easy to see that the endomorphism ring of C_j is isomorphic to D. Just associate each α with its corresponding d.

Now consider C_j as a right module over $E(C_j)$, or equivalently, over D. It is then a right vector space of dimension n over D, and its endomorphism ring, that is, the ring of linear transformations on C_j, is just the ring D_n.

We sum up our discussion so far in the following theorem.

8.5.1 Theorem

Let D be a division ring, and let D_n be the ring of all $n \times n$ matrices over D. Then D_n is a simple ring. As a left module over itself, D_n is semisimple. In fact, D_n is the direct sum of n mutually isomorphic simple modules. The endomorphism ring E of each of these simple D_n-modules S is isomorphic to D, and the endomorphism ring of S considered as a module over E is isomorphic to D_n. Analogous statements hold for D_n considered as a right module over itself.

There are still some things about D_n that we would like to know. For example, suppose that S is a simple left D_n-module. Is S isomorphic to the simple modules C_j? That is, does D_n have, up to isomorphism, just one simple left module? The answer is yes, and to see that, let s be any nonzero element of S. Then the map $D_n \to S: r \to rs$ is an epimorphism of left D_n-modules. Its kernel M is a left ideal, and since $D_n/M \approx S$ and S is simple, M is maximal in D_n. That is, there is no left ideal strictly between M and D_n. Since $D_n = C_1 \oplus C_2 \oplus \cdots \oplus C_n$ and $M \neq D_n$, M does not contain some C_j. But since

C_j is simple, $M \cap C_j = 0$. Since M is maximal in D_n, $D_n = M \oplus C_j$. Hence $D_n/M \approx C_j$. But $D_n/M \approx S$. Therefore $S \approx C_j$. Similarly, D_n has only one simple right module (up to isomorphism).

The left module D_n is the direct sum of n simple left modules. It cannot be the direct sum of m simple left modules with $m \neq n$ since any simple left D_n-module has dimension n as a vector space over D. This also follows from the following more general theorem, which we will need later.

8.5.2 Theorem

Let M be a left module over any ring R. Suppose that M is the direct sum $S_1 \oplus S_2 \oplus \cdots \oplus S_n$ of simple modules S_i. If $M = T_1 \oplus T_2 \oplus \cdots \oplus T_m$ with each T_i simple, then $m = n$, and after renumbering, $S_i \approx T_i$ for all i.

Proof We will induct on n. If $n = 1$, the theorem is obviously true. Suppose that $n > 1$ and that it holds for any module which is the direct sum of fewer than n simple modules. If $M = S_1 \oplus S_2 \oplus \cdots \oplus S_n = T_1 \oplus T_2 \oplus \cdots \oplus T_m$, then

$$S = S_1 \oplus S_2 \oplus \cdots \oplus S_{n-1}$$

does not contain some T_j, and we may as well suppose that $j = m$. Since T_m is simple, $S \cap T_m = 0$. Since $M/S \approx S_n$ is simple, $S \oplus T_m = M$. Thus,

$$S \approx M/T_m \approx T_1 \oplus T_2 \oplus \cdots \oplus T_{m-1} \approx S_1 \oplus S_2 \oplus \cdots \oplus S_{n-1}.$$

By induction, $m - 1 = n - 1$, whence $m = n$. Further, after renumbering, $S_i \approx T_i$ for $i < n$. But $M = S \oplus T_m$, $M/S \approx T_m$, and $M/S \approx S_n$. Thus $T_n = T_m \approx S_n$, and the proof is complete.

Let R be any ring. As a left module over R, its submodules are its left ideals. Such a submodule S is a simple module if and only if it is a ***minimal*** left ideal, that is, if there are no left ideals strictly between S and 0.

8.5.3 Corollary

Suppose that R is a ring, and as a left module over itself, R is the direct sum of n minimal left ideals. If R is the direct sum of m minimal left ideals, then $m = n$.

8.5.4 Corollary

If $D_n = S_1 \oplus S_2 \oplus \cdots \oplus S_m$ with each S_i a simple left D_n-module, then $m = n$, and the S_i are mutually isomorphic. Similar statements hold for right D_n-modules.

8.5.5 Theorem

Suppose that D and E are division rings, and that the rings D_n and E_m are isomorphic. Then $m = n$ and $D \approx E$.

Proof Considered as a left module over itself, D_n is the direct sum of n minimal left ideals, and similarly E_m is the direct sum of m minimal left ideals. Since the rings E_m and D_n are isomorphic, $m = n$. The endomorphism ring of any minimal left ideal (or simple left submodule) of D_n is D and of any minimal left ideal (or simple left submodule) of E_m is E. Since the rings D_n and E_m are isomorphic, $D \approx E$.

Now suppose that R is a simple ring and that R is left semisimple. Thus, $R = S_1 \oplus S_2 \oplus \cdots \oplus S_n$ with the S_i minimal left ideals. It turns out that $R \approx D_n$ for some division ring D. Of course, n and D are unique by the last theorem. Actually, n is unique by 8.5.2. Where does one get D?

8.5.6 Theorem (Schur's Lemma)

Let S be a simple left module over any ring R. Then the endomorphism ring $E_R(S)$ of S is a division ring.

Proof We must show that any nonzero endomorphism of S is a unit. Let α be in $E_R(S)$, $\alpha \neq 0$. Then α is onto since $\operatorname{Im}(\alpha) \neq 0$ and S has no submodules except 0 and S. Since $\operatorname{Ker}(\alpha) \neq S$, $\operatorname{Ker}(\alpha)$ must be 0. Thus α is one-to-one and onto, so has an inverse α^{-1} as a mapping of the set S onto itself. But α^{-1} is an endomorphism, as can be easily checked.

8.5.7 Lemma

Let R be a left semisimple ring. Then R is a simple ring if and only if there is, up to isomorphism, only one simple left R-module.

Proof Since R is left semisimple, $R = S_1 \oplus S_2 \oplus \cdots \oplus S_n$ with the S_i simple left R-modules. Suppose that S is a simple left R-module. Then $R/M \approx S$ for some maximal left ideal M. Since M does not contain some S_i, $M \cap S_i = 0$ and $M \oplus S_i = R$. Thus, $R/M \approx S_i \approx S$. Hence, we need to show that the S_i are mutually isomorphic. We have not used yet the fact that R is a simple ring. The set $0{:}S_j = \{r \in R : rS_j = 0\}$ is a ***two-sided*** ideal of R, and thus $0{:}S_j = 0$. For any s_i in S_i, $s_i \neq 0$, there must be an s_j in S_j with $s_i s_j \neq 0$. The map $S_i \to S_j : x \to xs_j$ is a left R-module homomorphism, and is not 0 since $s_i s_j \neq 0$. Since S_i and S_j are simple, this map must be an isomorphism.

To show that R is a simple ring if it has only one simple left module, let I be a two-sided ideal of R with $I \neq R$. Then I does not contain some S_i, say S_1, and since S_1 is a simple left module, $I \cap S_1 = 0$. The S_i are mutually isomorphic. Let φ_i be an isomorphism from S_1 to S_i. Let s_1 be a generator of S_1. Then $s_i = \varphi_i(s_1)$ generates S_i. Since I is a two-sided ideal and S_1 is a left ideal, $IS_1 = 0$, being contained in $I \cap S_1 = 0$. For i in I, $\varphi_i(i \cdot s_1) = \varphi_i(0) = i\varphi_i(s_1) = i \cdot s_i$, so $Is_i = 0$. But $I = IR$, so $(IR)s_i = I(Rs_i) = IS_i = 0$. Thus

$$I(S_1 \oplus \cdots \oplus S_n) = IR = I = 0,$$

and R is simple.

Thus with a simple ring R which is left semisimple, there is naturally associated a division ring D and a number n. The division ring D is the endomorphism ring of any simple left R-module, and n is the number of summands in any decomposition $R = S_1 \oplus S_2 \oplus \cdots \oplus S_n$ of R into a direct sum of such minimal left ideals. We would like to show that $R \approx D_n$ as rings. This is a consequence of the following powerful theorem.

8.5.8 The Jacobson Density Theorem

Let R be a ring, and suppose that R has a simple left module S such that

$$0{:}S = \{r \in R: rS = 0\} = 0.$$

Let D be the endomorphism $E_R(S)$ of S. If

$$\{s_1, s_2, \ldots, s_n\}$$

is a linearly independent subset of the right vector space S over the division ring D, and if $x_1, x_2, \ldots, x_n$ are elements of S, then there is an element r in R such that $rs_i = x_i$ for all i.

Proof Before we start the proof of the theorem, let us see what it says in a special case. If R is a simple ring, then certainly $0{:}S = 0$. If it turns out that S is finite dimensional over D, then taking $\{s_1, s_2, \ldots, s_n\}$ to be a basis of S, we see that every linear transformation of the vector space S over D is just multiplication by an element of r. Since no element r of R annihilates all of S, R is, in effect, the ring of all linear transformations on the n-dimensional vector space S. Thus, $R \approx D_n$. We will draw additional corollaries.

Now to the proof of the theorem. First we will prove the preliminary fact that if T is a subspace of S of finite dimension, and if s is in S and not in T, then there is an element r in R such that $rT = 0$ and $rs \neq 0$. We induct on $\dim(T)$. If $\dim(T) = 0$, let $r = 1$. Suppose that $\dim(T) = n > 0$, and that the assertion is true for all subspaces of dimension less than n. Now $T = V \oplus xD$ with $\dim(V) = n - 1$. Let

$$I = \{r \in R : rV = 0\}.$$

Now I is a left ideal of R, and by the induction hypothesis, there is an element r in R such that $rV = 0$ and $rx \neq 0$. Such an r is in I, so $Ix \neq 0$. But Ix is a nonzero submodule of the simple module S, so $Ix = S$. Let s be in S, s not in T, and assume that every element of R which annihilates T also annihilates s. Letting $i \cdot x$ correspond to $i \cdot s$ defines a map from $Ix = S$ to S since $i \cdot x = 0$ yields $iT = 0$, and thus $i \cdot s = 0$. This map is clearly an R-homomorphism, and therefore is an element d in $D = E_R(S)$. We have, for each i in I,

$$i(xd - s) = ixd - i \cdot s = 0,$$

so by the induction hypothesis, $xd - s$ is in V. Thus s is in $V \oplus xD$, and this is a contradiction. Therefore, some element of R annihilates T and does not annihilate x. This concludes the proof of our preliminary result.

Now let $\{s_1, s_2, \ldots, s_n\}$ be a linearly independent subset of S, and let $x_1, x_2, \ldots, x_n$ be in S. We induct on n. If $n = 1$, there is no problem since $Rs_1 = S$. Assume that the theorem is true for independent sets of size $< n$. Then there are elements r and a in R such that $rs_i = x_i$ for $i = 1, 2, \ldots, n-1$, $as_n \neq 0$, and $as_1 = as_2 = \cdots = as_{n-1} = 0$. Since $Ras_n = S$, there is an element r_1 in R such that $r_1 as_n = xn - rs_n$. We have $(r + r_1 a)s_i = rs_i + r_1 as_i = rs_i = x_i$ for $i < n$, and

$$(r + r_1 a)s_n = rs_n + r_1 as_n = x_n.$$

This concludes the proof of the theorem.

Suppose that R is a simple ring and that S is a simple left R-module. Then the hypotheses of 8.5.8 are satisfied, and as we pointed out, if S is of dimension n over the division ring $D = E_R(S)$, then $R \approx D_n$. How can we tell from R itself whether or not S is finite dimensional? Suppose that S is infinite dimensional. Then there is a subset $\{s_1, s_2, \ldots\}$ of S such that for all n, $\{s_1, s_2, \ldots, s_n\}$ is linearly independent. Let I_n be the left ideal $\{r \in R : rs_i = 0$ for all $i \leq n\}$. We have the chain $I_1 \supset I_2 \supset I_3 \supset \cdots$ of left ideals, and by the proof of 8.5.8, all the inclusions are proper ones. That is, R does not satisfy the descending chain condition on left ideals. Thus, if R satisfies the descending chain condition on left ideals, S would have to be finite dimensional. A ring that satisfies the dcc on left ideals is called ***left Artinian***, and right Artinian rings are similarly defined. Also, a left Artinian ring R has a minimal left ideal S, and S is then a simple left R-module. Thus, we have the following theorem.

8.5.9 The Wedderburn–Artin Theorem

Let R be a left Artinian simple ring. Then $R \approx D_n$ for some division ring D and integer n.

A ring R which has a simple left module S such that $0{:}S = 0$ is called ***left primitive***.

8.5.10 Corollary

A left primitive left Artinian ring is isomorphic to D_n for some division ring D and integer n.

8.5.11 Corollary

Let R be a simple ring that is left semisimple. Then $R \approx D_n$ for some division ring D and integer n.

Proof A ring that is left semisimple is left Artinian (Problem 7). Then 8.5.11 follows from 8.5.9.

8.5.12 Corollary

A simple ring R is left semisimple if and only if it is right semisimple.

Proof If R is left semisimple, then $R \approx D_n$, and D_n is right semisimple. Now 8.5.8 holds with left and right interchanged. There is no difference in the proof. Hence if R is right semisimple, $R \approx D_n$, whence R is left semisimple.

Now suppose that R is a left semisimple ring. Then $R = S_1 \oplus S_2 \oplus \cdots \oplus S_n$ with the S_i minimal left ideals. The S_i are not necessarily isomorphic as in the case when R is also a simple ring (8.5.7) We rewrite

$$R = (S_{11} \oplus S_{12} \oplus \cdots \oplus S_{1n_1}) \oplus (S_{21} \oplus S_{22} \oplus \cdots \oplus S_{2n_2}) \\ \oplus \cdots \oplus (S_{m1} \oplus S_{m2} \oplus \cdots \oplus S_{mn_m})$$

with $S_{ij} \approx S_{pq}$ if and only if $i = p$. Let

$$R_i = S_{i1} \oplus \cdots \oplus S_{in_i}.$$

Thus $R = R_1 \oplus R_2 \oplus \cdots \oplus R_m$, with R_i the direct sum of mutually isomorphic minimal left ideals of R. Let s_{ij} be in S_{ij}. If $s_{ij}s_{pq} \neq 0$ for some s_{ij} and s_{pq}, then the map $S_{ij} \to S_{pq}$ given by $x \to xs_{pq}$ is an R-isomorphism since S_{ij} and S_{pq} are simple left R-modules. Thus $i = p$. Therefore, $S_{ij}S_{pq} = 0$ for $i \neq p$, and more generally, $R_iR_j = 0$ if $i \neq j$. In particular, the R_i are two-sided ideals of R. That is, R is the ***ring direct sum*** of the subrings R_i. Hence, R can be considered as m-tuples $(r_1, r_2, \ldots, r_m)$ with addition and multiplication componentwise. Let $1 = e_1 + e_2 + \cdots + e_m$ with e_i in R_i. Then for r_i in R_i,

$$r_i = 1 \cdot r_i = r_i \cdot 1 = (e_1 + e_2 + \cdots + e_m)r_i = r_i(e_1 + e_2 + \cdots + e_m) = e_ir_i = r_ie_i,$$

so that e_i is the identity of the ring R_i. Now it should be clear that each R_i is a left semisimple ring, $R_i = S_{i1} \oplus \cdots \oplus S_{in_i}$, and the S_{ij}, $j = 1, 2, \ldots, n_i$ are mutually isomorphic. By 8.5.7, R_i is simple, and so is a ring of matrices over a division ring.

8.5.13 Theorem

Let R be left semisimple. Then R is a direct sum $R_1 \oplus R_2 \oplus \cdots \oplus R_m$ of minimal two-sided ideals R_i. The R_i are unique, and each R_i is the ring of $n_i \times n_i$ matrices over a division ring D_i.

Proof Only the uniqueness of the R_i remains to be proved. Suppose that $R = A_1 \oplus A_2 \oplus \cdots \oplus A_k$ with each A_i a minimal two-sided ideal of R. If $R_i \cap A_j = 0$ for all i, then

$$(R_1 \oplus \cdots \oplus R_m)A_j = RA_j = A_j = 0.$$

Thus, $R_1 \cap A_j \neq 0$ for some i. Since R_i and A_j are both minimal two-sided ideals, $R_i = A_j$. Thus each A_j equals some R_i, and the uniqueness of the R_i follows.

8.5.14 Corollary

A ring R is left semisimple if and only if it is right semisimple.

PROBLEMS

1 Let α be an isomorphism from the ring R to the ring S. Prove that if I is a left ideal of R, then $\alpha(I)$ is a left ideal of S. Prove that if I is minimal, then so is $\alpha(I)$. Prove that if the left R-module R is semisimple, then so is the left S-module S. Prove that if R is left Artinian, then so is S.

2 Let R be a ring, and let I be a summand of the left R-module R. Prove that $I = Re$ with $e^2 = e$.

3 Prove that if a ring R has no nonzero nilpotent elements, that is, no element $r \neq 0$ such that $r^n = 0$ for some integer n, then any idempotent in R is in the center of R.

4 Prove that if R and S are left Artinian (Noetherian), then so is $R \oplus S$.

5 Let R be a ring such that every left ideal of R is a summand of the left R-module R. Prove that if R is left Noetherian, then R is semisimple.

6 Suppose that the left R-module $M = S_1 \oplus \cdots \oplus S_n$ with each S_i a simple module. Prove that every submodule of M is a summand of M.

7 Prove that a ring that is left semisimple is left Artinian.

8 Prove that if R is semisimple, then every finitely generated left R-module is a direct sum of simple modules. Prove that every simple left R-module is isomorphic to a minimal left ideal of R.

9 Let e_1 and e_2 be idempotents in the ring R. Prove that the left modules Re_1 and Re_2 are isomorphic if and only if the right modules e_1R and e_2R are isomorphic.

10 Prove that the ring of all upper triangular 2×2 matrices

$$\begin{pmatrix} a & b \\ 0 & c \end{pmatrix},$$

where a is in $\mathbb{Z}$, and b and c are in $\mathbb{Q}$, is right Noetherian but not left Noetherian.

11 Let D be a division ring. Let R be the ring of all upper triangular 2×2 matrices with entries from D. Is R simple? Is R semisimple?

12 Which rings $\mathbb{Z}/n\mathbb{Z}$ are semisimple?

13 Let $G = \{g_1, g_2, \ldots, g_n\}$ be a finite group. Prove that the element $e = 1/n(g_1 + g_2 + \cdots + g_n)$ is an idempotent in $\mathbb{Q}(G)$. Prove that e is in the center of $\mathbb{Q}(G)$. Thus, prove that $\mathbb{Q}(G)$ is the ring direct sum $\mathbb{Q}(G)e \oplus \mathbb{Q}(G)(1-e)$. Prove that the ring $\mathbb{Q}(G)e$ is isomorphic to $\mathbb{Q}$.

14 Let G be the group with 2 elements. Prove that the group ring $\mathbb{Q}(G)$ is semisimple. Do the same thing for the group G with 3 elements. [Actually, $\mathbb{Q}(G)$ is semisimple for every finite group G.]

15 Let I be a minimal left ideal of a ring R. Prove that I is a summand of R if and only if $I^2 \neq 0$.

16 Prove that a ring is semisimple if and only if it is left Artinian, and for all minimal left ideals I, $I^2 \neq 0$.

17 Prove that a simple ring with a minimal left ideal is semisimple.

18 Prove that the intersection J of all maximal left ideals of a ring R is a two-sided ideal. Prove that J is the intersection of all maximal right ideals of R.

19 Let I be a left ideal of a ring R. Prove that if $1 + a$ has a left inverse for all a in I, then I is contained in all the maximal left ideals of R. Prove that if $1 + a$ has a left inverse for all a in I, then $1 + a$ has a right inverse for all a in I. Prove that if R is left Noetherian and I is contained in all maximal left ideals of R, then $1 + a$ has a left inverse for all a in I. (This last statement is true without the assumption that R is left Noetherian. See the Appendix, Problem 9.)

Appendix: Zorn's Lemma

Although we have not hesitated to use the Axiom of Choice in its believable intuitive form, and have done so without explicit mention, we have refrained from using the more technical arguments involving Zorn's Lemma. Such arguments are used extensively in mathematics, especially in algebra. We will illustrate the use of this important tool in this appendix.

Let I be a nonempty set, and let $\{S_i\}_{i \in I}$ be a family of nonempty sets S_i. Is there a function $f: I \to \bigcup_i S_i$ such that $f(i)$ is in S_i for all i in I? The assumption that there is such a function is the ***Axiom of Choice***. The function f "chooses" an element from each set S_i. The Axiom of Choice can be neither proved nor disproved from the usual axioms of set theory. However, it is an extremely useful axiom, and it is used by most mathematicians. There are several mathematical statements logically equivalent to the Axiom of Choice, and on the intuitive level, some of them are not nearly as believable as the Axiom of Choice. Zorn's Lemma is one of these. We need some preliminary definitions before stating it.

Definition

Let S be a set. A relation $\leq$ on S is a **partial ordering** *of S, or a* **partial order** *on S if it satisfies*

(1) $s \leq s$ *for all s in S;*
(2) *if $s \leq t$ and if $t \leq u$, then $s \leq u$;*
(3) *if $s \leq t$ and $t \leq s$, then $s = t$.*

If in addition, for s and t in S, either $s \leq t$ or $t \leq s$, then $\leq$ is a ***linear ordering*** of S, or simply an ***ordering*** of S. If S is a partially ordered set, a subset T of S is called a ***chain*** if for any pair of elements s and t of T, either $s \leq t$ or $t \leq s$. That is, T is a chain if $\leq$ induces an ordering of T. If T is a subset of a partially ordered set S, then an element s in S is an ***upper bound*** of T if $t \leq s$ for all t in T. An element s of a partially ordered set S is ***maximal*** if whenever t is in S and $s \leq t$, then $s = t$. That is, there is no element in S which is strictly bigger than s. Zorn's Lemma pertains to the existence of maximal elements.

Zorn's Lemma

Let S be a partially ordered set. If every chain in S has an upper bound in S, then S has a maximal element.

Zorn's Lemma is logically equivalent to the Axiom of Choice, but this is not obvious.

Note that S cannot be empty. If S were empty, then S is a chain in S, and thus must have an upper bound in S. That is, S must have an element s such that $t \leq s$ for all t in S.

Here is how Zorn's Lemma will often be applied. Let A be a set. Let S be a set of subsets of A. Then S is partially ordered by setting $A_1 \leq A_2$ if A_1 is a subset of A_2. To conclude that S has a maximal element, one must show that every chain C in S has an upper bound in S. This is usually done by showing that the union of the elements of C is in S. Of course, S has to be a very special set of subsets of A for this to be so.

Here are some applications of Zorn's Lemma.

Theorem

Every vector space $\neq \{0\}$ has a basis

Proof Let V be a vector space over F, $V \neq \{0\}$. By a basis of V, we mean a linearly independent subset B of V that generates V. The set B may, of course, be infinite, and an infinite set is defined to be linearly independent if every finite subset of it is linearly independent. Let S be the set of all linearly independent subsets of V. The set S is partially ordered by set inclusion and is not empty. Let C be any chain in S. Thus C is a set of linearly independent subsets of V such that for any two of them, one is contained in the other. We need to show that C has an upper bound in S. Let $U = \bigcup_{T \in C} T$. The set U is linearly independent. Indeed, if $\{u_1, u_2, \ldots, u_n\}$ is a finite subset of U, then there is a T_i in C such that u_i is in T_i. Thus, there is a T in C such that $\{u_1, u_2, \ldots, u_n\}$ is

contained in T, namely the biggest of the T_i's. Since T is linearly independent, so is $\{u_1, u_2, \ldots, u_n\}$, and thus U is linearly independent. By Zorn's Lemma, S has a maximal element B. Then B is a linearly independent subset of V that is not properly contained in any other linearly independent subset of V. We need to show that B generates V. Suppose not. Then there is an element v in V not in the subspace generated by B. It follows readily that $B \cup \{v\}$ is a linearly independent subset of V, contradicting the fact that B is maximal among those subsets of V. Thus B is a basis of V.

Theorem

Let W be a subspace of the vector space V over the field F. Then W is a summand of V.

Proof Let S be the set of all subspaces X of V such that $X \cap W = \{0\}$. The set S is partially ordered by inclusion, and $S \notin \emptyset$. Suppose that C is a chain in S. Let $Y = \bigcup_{X \in C} X$. Then Y is a subspace of V, and $Y \cap W = \{0\}$, so Y is in S. Clearly, Y is an upper bound of C. Therefore Zorn's Lemma applies, and S has a maximal element X. Since $X \cap W = \{0\}$, all we need is that $V = X + W$. Suppose that there is an element v in V such that v is not in $X + W$. Then $(X + Fv) \cap W = \{0\}$ since $x + av = w \neq 0$ implies that $a \neq 0$, whence $v = -a^{-1}x + a^{-1}w$ is in $X + W$. Since $X + Fv$ properly contains S, and $(X + Fv) \cap W = \{0\}$, X is not maximal in S. This contradiction establishes that $V = X \oplus W$.

Theorem

Let R be a ring with identity. Then R has a maximal left ideal.

Proof Recall that a maximal left ideal of R is an ideal M of R such that $M \neq R$ and such that there is no left ideal strictly between M and R.

Let S be the set of all left ideals I of R such that $I \neq R$. The set S is partially ordered by inclusion and $\{0\} \in S$. For any chain C in R, $\bigcup_{I \in C} I = J$ is a left ideal of R and is not R since 1 is not in J. Also J is an upper bound of C. Thus, S has a maximal element.

Theorem

Let R be a semisimple ring. Then any left R-module M is a direct sum of simple modules.

Proof Let M be a left R-module. The module M is a direct sum of simple submodules if there is a set $\{S_i\}_{i \in I}$ of simple submodules of M such that every element m in M can be written uniquely in the form $m = \sum_{i \in I} s_i$ with s_i in S_i. Of course, all but finitely many s_i must be 0. Let S be the set of all sets X of simple submodules of M such that the submodule generated by the submodules in X is the direct sum of the submodules in X. The set S is partially ordered by inclusion, is nonempty, and the union of a chain in S is in S, so that chains in S have upper bounds in S. Thus, S has a maximal element Y. We need to show that M is the direct sum of the elements of Y. The only problem is that the submodule N generated by the elements of Y may not be all of M. Suppose that there is an element m in M such that m is not in N. Then since R is semisimple, Rm is the direct sum $A_1 \oplus \cdots \oplus A_k$ of simple modules. Some A_i is not in Y, and it follows readily that $Y \cup \{A_i\}$ is an element of S. That is, Y is not maximal in S. This completes the proof.

Theorem

Any field F has an algebraic closure, and any two algebraic closures of F are F-isomorphic.

Proof Recall that an algebraic closure of F is an algebraic extension K of F such that every polynomial in $K[x]$ of degree at least 1 factors into linear factors in $K[x]$. Let X be a set in one-to-one correspondence with the set of all polynomials in $F[x]$ of degree at least 1. We denote by $x(f)$ the element in X corresponding to $f(x)$. Now form the ring $F[X]$. This is the ring of all polynomials in indeterminates from X with coefficients from F. Thus, any element in $F[X]$ is an element of

$$F[x(f_1), \ldots, x(f_n)]$$

for some finite number $x(f_1), \ldots, x(f_n)$ of indeterminates. Let I be the ideal of $F[X]$ generated by all the elements $f(x(f))$. If I were $F[X]$, then $1 = \sum_{i=1}^{n} g_i f_i(x(f_i))$ for some g_i in $F[X]$ and generators $f_i(x(f_i))$. The g_i involve only finitely many $x(f)$'s, so each g_i is a polynomial $g_i(x(f_1), \ldots, x(f_m))$, with $m \geq n$. Let F_1 be a finite extension of F in which each $f_i x(f_i))$, $i = 1, \ldots, n$ has a root r_i, and for $i > n$, set $r_i = 0$. Substituting r_i for $x(f_i)$ in our relation

$$1 = \sum_{i=1}^{n} g_i(x(f_1), \ldots, x(f_m)) f_i(x(f_i)),$$

we get $1 = 0$ in F. Therefore, $I \neq F[X]$. By Zorn's Lemma, I is contained in a maximal ideal M. Since M is maximal, $F[X]/M = K_1$ is a field. The

field F is contained in K_1 via the map

$$F \to K_1: a \to a + M.$$

and we view F now as a subfield of K_1. For any polynomial $f(x)$ of degree at least 1 in $F[x]$, $f(x(f) + M) = f(x(f)) + M = M$, which is 0 in K_1. That is, every polynomial in $F[x]$ of degree at least 1 has a root in K_1. The field K_1 is an algebraic extension of F since each $x(f) + M$ is algebraic over F and K_1 is generated by F and the set of all $x(f) + M$. Thus, given a field F, we have constructed an algebraic extension K_1 of F such that every polynomial $f(x)$ in $F[x]$ of degree at least 1 has a root in K_1, or equivalently, a linear factor in $K_1[x]$. We do not know that K_1 is algebraically closed because we do not even know that every such polynomial in $F[x]$ factors into linear factors in $K_1[x]$. However, construct K_2 from K_1 in the same way that K_1 was constructed from F; in general, from K_i construct K_{i+1} such that every polynomial in $K_i[x]$ of degree at least 1 has a root in K_{i+1}. Let $K = \bigcup_i K_i$. Then K is an algebraic extension of F, and if $f(x)$ is in $K[x]$ and has degree at least 1, then $f(x)$ is in $K_i[x]$ for some i, whence has a root in K_{i+1} and hence in K. Therefore, K is algebraically closed.

Now let K and L be two algebraic closures of F. We want an F-isomorphism between K and L. Let Ω be the set of all F-isomorphisms $\alpha: S \to T$ between subfields S of K containing F and subfields T of L containing F. One such is the identity map on F. For $\alpha': S' \to T'$, let $\alpha \leq \alpha'$ if $S \subset S'$ and α' agrees with α on S. This partially orders Ω, and it is easy to see that every chain in Ω has an upper bound in Ω. Therefore by Zorn's Lemma, Ω has a maximal element $\Phi: A \to B$. If $A = K$, then $B \approx K$, so is algebraically closed. But L is an algebraic extension of B. Hence $B = L$. Similarly, if $B = L$, then $A = K$. So we may suppose that $A \neq K$ and $B \neq L$. Let a be in K and not in A, and let $f(x)$ be in $A[x]$ with $\deg(f(x))$ at least 1 and $f(a) = 0$. By 6.2.5, we get an isomorphism $\alpha': A' \to B'$, where A' is a root field of $f(x)$, $B \subset B' \subset L$, and α' extends α. This contradicts the maximality of $\alpha: A \to B$, whence $A = K$ and $B = L$.

PROBLEMS

1 Let L be a linearly independent subset of a vector space V. Prove that V has a basis containing L.

2 Let W be a subspace of a vector space V, and let X be a subspace of V such that $X \cap W = 0$. Prove that $V = A \oplus W$ for some subspace A containing X.

3 Let I be a left ideal of a ring R, with $I \neq R$. Prove that R has a maximal left ideal M such that $I \subset M$.

4 Prove that every ring with identity has a maximal two-sided ideal.

5 Let M be a direct sum of simple modules. Prove that every submodule of M is a direct summand of M.

6 Let A be any set. Let S be the set of all pairs $(B, \leq)$ such that B is a subset of A and $\leq$ is an ordering of B. Define a partial order on S, and use it to show that there exists an order relation on A.

7 An ordering on a set A is called a ***well ordering*** if every nonempty subset of A has a least element. Prove that every set can be well ordered. (Zorn's Lemma is actually equivalent to the statement that every set can be well ordered.)

8 Let G and H be groups. Prove that there are subgroups A and B of G and H, respectively, and an isomorphism $f: A \to B$ such that f cannot be extended to an isomorphism of subgroups properly containing A and B.

9 Let I be a left ideal in a ring R with identity. Prove that $1 + a$ has a left inverse for all a in I if and only if I is contained in all the maximal left ideals of R.

Bibliography

1. Birkoff, G., and MacLane, S., *A Survey of Modern Algebra,* 3rd ed., New York: Macmillan, 1963.
2. Fraleigh, J. B., *A First Course in Abstract Algebra,* 2nd ed., Reading: Addison-Wesley, 1976.
3. Gallian, J. A., *Contempory Abstract Algebra,* Lexington: D. C. Heath, 1986.
4. Herstein, I. N., *Abstract Algebra,* New York: Macmillan, 1986.
5. Herstein, I. N., *Topics in Algebra,* 2nd ed., New York: Wiley, 1975.
6. Hungerford, T. W., *Algebra,* New York: Holt, Rinehart, and Winston, 1974.
7. Jacobson, N., *Basic Algebra I,* San Francisco: Freeman, 1974.
8. Lang, S., *Algebra,* 2nd ed., Menlo Park: Addison-Wesley, 1984.

Index